Paying for Agricultu

Other Books Published in Cooperation with the International Food Policy Research Institute

Agricultural Change and Rural Policy: Variations on a Theme by Dharm Narain
Edited by John W. Mellor and Gunvant M. Desai

Crop Insurance for Agricultural Development: Issues and Experience
Edited by Peter B. R. Hazell, Carlos Pomareda, and Alberto Valdés

Accelerating Food Production in Sub-Saharan Africa
Edited by John W. Mellor, Christopher L. Delgado, and Malcolm J. Blackie

Agricultural Price Policy for Developing Countries
Edited by John W. Mellor and Raisuddin Ahmed

Food Subsidies in Developing Countries: Costs, Benefits, and Policy Options
Edited by Per Pinstrup-Andersen

Variabilities in Grain Yields: Implications for Agricultural Research and Policy in Developing Countries
Edited by Jock R. Anderson and Peter B. R. Hazell

Seasonal Variability in Third World Agriculture: The Consequences for Food Security
Edited by David E. Sahn

The Green Revolution Reconsidered: The Impact of High-Yielding Rice Varieties in South India
By Peter B. R. Hazell and C. Ramasamy

The Political Economy of Food and Nutrition Policies
Edited by Per Pinstrup-Andersen

Agricultural Commercialization, Economic Development, and Nutrition
Edited by Joachim von Braun and Eileen Kennedy

Agriculture on the Road to Industrialization
Edited by John W. Mellor

Intrahousehold Resource Allocation in Developing Countries: Models, Methods, and Policy
Edited by Lawrence Haddad, John Hoddinott, and Harold Alderman

Sustainability, Growth, and Poverty Alleviation: A Policy and Agroecological Perspective
Edited by Stephen A. Vosti and Thomas Reardon

Famine in Africa: Causes, Responses, and Prevention
By Joachim von Braun, Tesfaye Teklu, and Patrick Webb

Paying for Agricultural Productivity

EDITED BY JULIAN M. ALSTON, PHILIP G. PARDEY, AND VINCENT H. SMITH

Published for the International Food Policy Research Institute

The Johns Hopkins University Press
Baltimore and London

Published 1999
Printed in the United States of America on acid-free paper
9 8 7 6 5 4 3 2 1

The Johns Hopkins University Press
2715 North Charles Street
Baltimore, Maryland 21218-4363
www.press.jhu.edu

International Food Policy Research Institute
2033 K Street, NW
Washington, D.C. 20006
(202) 862-5600
www.ifpri.org

Library of Congress Cataloging-in-Publication Data will be found
at the end of this book.
A catalog record for this book is available from the British Library.

ISBN 0-8018-6185-3
ISBN 0-8018-6278-7 (pbk.)

Contents

Figures

Tables

Foreword

Work conducted at the International Food Policy Research Institute (IFPRI) and elsewhere has long shown that investments in agricultural research and development offer high returns. Improving seed varieties and livestock and designing better farming technologies and practices can boost farmers' productivity enough to jump-start whole economies.

Until recently, government has been the main funder and performer of agricultural R&D in most countries. Past government spending on agricultural research in a number of countries has resulted in tremendous gains in agricultural productivity, greatly increasing the amount of food and other agricultural goods farmers could produce on a given amount of land.

Now, however, mechanisms for paying for and performing agricultural research are undergoing dramatic transformations worldwide. The growth of public spending on agricultural research is slowing, and private companies are assuming greater responsibility for funding and conducting research. Allocation of research funds is becoming increasingly competitive. Also, new areas of study, such as food processing, food safety, and environmental concerns, are shifting resources away from research on production agriculture, at least in the developed countries.

Paying for Agricultural Productivity examines the similarities and differences in the institutional innovations under way in five developed countries with important agricultural research systems. Australia, the Netherlands, New Zealand, the United Kingdom, and the United States together accounted for nearly half of public spending on agricultural research by developed countries in recent years. As the authors document the changes occurring in these countries, they begin to tackle the crucial long-term questions: Will the new ways of doing business make agricultural R&D more effective in increasing agricultural productivity? Have the changes led to efficiencies in the total quantity and quality of research being undertaken?

The changes described in this book have implications for poorer countries as well. The science now being undertaken to produce new agricultural technologies is more proprietary in nature than in the past, meaning that those wishing to use the new technologies will have to pay. Whether this trend will hurt developing countries as they seek to raise agricultural productivity is still unclear. Developing countries may well find some of the changes made by developed countries in funding and organizing agricultural research to be appropriate models for their own research systems. The editors of this volume are actively engaged in a study of the funding and other policy changes concerning agricultural R&D in the poorer countries of the world.

Understanding the changes taking place is an essential first step toward answering the questions raised. This book presents the most comprehensive discussion and analysis yet available on the new directions in financing agricultural R&D. It should be of great interest to policymakers and scholars concerned with ensuring that agriculture is as productive and sustainable as it can be as we move into the twenty-first century.

Per Pinstrup-Andersen
Director General, IFPRI

Preface

Agricultural research spending grew rapidly from the 1950s until the late 1970s and generated extraordinary growth in agricultural productivity around the globe. Since the late 1970s many developed economies have undertaken structural shifts in public funding for agricultural research and the institutional arrangements through which it is carried out.

The rate of growth of public-sector spending has slowed, and the relative importance of the private sector has grown—as a source of funds, as a research performer, and as an influence over public agricultural research and development (R&D). Institutional aspects of national agricultural research policies have changed as well. These institutional innovations include new funding mechanisms, the reorganization and privatization of research institutions, the introduction of managed competition, the broadening of the research agenda, and adjustments in the research mix.

The motivation for the work reported in this book was that these changes seemed to be important and to have common foundations, yet they have been different in different countries and their consequences are not fully understood. The strategy was to combine information from different countries on their recent agricultural R&D policy changes, to compare and contrast the changes and their implications, and then to draw policy lessons. The guiding principle is that knowledge about the design of research policy and its implications in one place will be applicable, with modifications, elsewhere. To take advantage of this knowledge spillover we must understand the similarities and differences among places and the respective roles of the institutions involved in agricultural R&D.

Our ultimate objective is to provide a comprehensive assessment across developed and developing countries. Work is continuing on other countries. Here we review and evaluate changes in financing agricultural R&D in five rich countries—Australia, the Netherlands, New Zealand, the United Kingdom, and the United States. These nations accounted for more than 40 percent of total public investments in agricultural research by member countries of the Organization for Economic Cooperation and Development (OECD) in the mid-1990s. Summary information on 22 OECD countries provides a context for this review.

The primary purpose of the work is to document agricultural R&D funding patterns, institutional arrangements, and their changes. We also begin to assess the changes in terms of whether they may have improved economic efficiency and welfare in terms of the total quantity of research being undertaken; private- and public-sector roles and relationships; the use of less costly sources of funds; the allocation of resources among competing programs, projects, and institutions; the allocation of resources across basic and applied research and extension; the effectiveness with which research funds are used; administrative overhead; and other transaction costs. Much of this assessment work will have to continue as the consequences of policy change unfold.

Acknowledgments

Initial funding for this work was provided by the United States Congress, Office of Technology Assessment. Further specific support was provided by the University of California Pacific Rim Project and the United Kingdom Department for International Development. In addition to these specific project funds, significant general support was provided by the International Food Policy Research Institute (IFPRI), Montana State University, the University of California–Davis, and the University of California Agricultural Issues Center.

The editors gratefully acknowledge the assistance of several IFPRI staff who helped with manuscript preparation, including Beverly Abreu, Nienke Beintema, Mary-Jane Banks, Connie Chan-Kang, and Heidi Fritschel. Last we wish to thank the authors for doing the real work and cheerfully and promptly responding to editorial requests and suggestions.

1 Introduction

JULIAN M. ALSTON, PHILIP G. PARDEY, AND
MICHAEL PHILLIPS

Throughout the twentieth century improvements in agricultural productivity have been closely linked to investments in agricultural research and development (R&D) and to the policies that affect agricultural R&D. Since the late 1970s many countries have implemented major changes in the funding for and the institutional basis of public agricultural R&D and in the incentives affecting private R&D. This book documents and compares these changes in the developed world as a step toward evaluating the social gains and losses stemming from these changes. The stakes are high. Having highly productive agricultural systems is as important now as it ever was for the five countries treated in detail here (Australia, the Netherlands, New Zealand, the United Kingdom, and the United States), and for the rest of the world as well. Throughout much of history, boosting agricultural production to feed more mouths often depended on bringing new land under cultivation. In more recent times, the growth of land in agriculture has stalled or even reversed, so any future increases in production are likely to be driven by technological developments designed to increase agricultural productivity. Thus it is essential to identify whether the changes made in agricultural R&D policies are moving agricultural research systems toward or away from improving the rate of growth of agricultural productivity and increasing the social benefits flowing from agricultural R&D.

Worldwide, in rich and poor countries alike, the public institutions involved in financing and executing agricultural R&D are under pressure to economize. The same forces are at work to different degrees, with different timing, and with different effects in different countries. Research administrators are being asked to do more with less. Moreover, in spite of a decline in direct public funding of agricultural research, the research agenda has broadened beyond conventional productivity improvements and now places a considerable emphasis on issues such as environmental impacts and food safety. Also there are demands for more accountability within the public research system. Other common elements include a renewed requirement to undertake some assessment of R&D policy and to involve economic principles in such assessments, a tendency to shift toward a smaller public-sector role in providing taxpayer funding for research, and a complementary shift toward enhancing incentives for private-sector investments. Responding effectively to these pressures involves changing the policies, procedures, and institutions used to finance R&D, allocate the resources, and carry out the research.

Although the forces at work and the responses to them are similar everywhere, there are some important differences among countries. In this book we first lay out the principles for public-sector roles in financing agricultural R&D. Then, against those broad principles and in the context of the evolving forces for change, we review the recent developments in agricultural R&D institutions, policies, and investments in five more-developed countries: Australia, the Netherlands, New Zealand, the United Kingdom, and the United States. By comparing these countries' responses to often similar, but sometimes different, impetuses for change, we hope to draw some general lessons for countries seeking to improve their agricultural research systems in the face of tighter resources and changing public perceptions and expectations.

Agricultural Research in the Public Domain

New technologies that facilitate economic development of the agricultural sector are not free. In the past, especially during the eighteenth and nineteenth centuries, agricultural productivity increased as a result of the efforts of innovative farmers, who used selection, breeding, and many other activities to improve the yields of plant varieties and the vigor of farm animals. However, the much more rapid agricultural productivity growth rates enjoyed by most developed countries since the 1930s have been driven largely by formal investigations in agricultural research institutions. A wide variety of arrangements have appeared in different countries to provide for the financing of these investigations, typically (although not always) involving substantial government support.

In many countries, formalized agricultural research institutions supported by national governments were first instituted toward the end of the nineteenth century. Developments in these systems have occurred in waves, with common undercurrents in different countries. One such wave was a period of expansion in the early part of the twentieth century, a period of institutional innovation in which the foundations for many of the present agricultural research institutions were laid. Then, especially in the period after World War II, public-sector investment in national agricultural research systems (NARSs) grew enormously. However, since the beginning of the 1980s, growth in public agricultural R&D investment has slowed or stopped in most countries, and for some countries annual investments in research have declined. Also, since the 1960s, domestic research programs have been complemented by a comparatively small but significant international component, largely financed directly by rich-country governments or indirectly via international agencies such as the World Bank. International agricultural R&D funding has also been curtailed in recent years.

The prevailing view is that public agricultural research institutions and the policies that surround them are now at a pivotal point in their history. A long period of sustained growth appears to have ended, and an extended—if not permanent—phase of general fiscal restraint appears to have begun. This, combined with a more skeptical view of the social benefits from investments in science, has led to a tightening of resources available for research and calls for clearer justification for R&D funds and accountability for their use. In addition, the research agenda has broadened. Tighter resources and fundamental shifts in research directions and policies imply changes in institutional cultures as well as operations.

Elements of Change in the Research Policy Environment

Some of the developments that give rise to the perception of a new environment involve a continuation of longer-term trends, and some represent recent and radical departures from past patterns. Examining what has changed and why is a useful first step in formulating appropriate policies. Because many of the influences are common among countries, it can be useful to compare agricultural R&D policies, institutions, and investments among countries. This makes it possible to better understand the events in any particular country, and to draw lessons from the experiences and responses of each for what may be done better in the others. The common elements of change among the countries studied include

- a reassessment of the role of government;
- reduced public funding for R&D or slower growth in such funding;
- changing perceptions of private- versus public-sector roles;
- changing policies, public institutions, and funding mechanisms;
- a broadening of the research agenda to emphasize nontraditional areas; and
- increased requirements for accountability and priority-setting processes.

Some of these elements of the changing environment for public agricultural R&D are driven more by political fashion than economics. However, some reflect changing economic circumstances or an improved understanding of the economics of R&D.

A unifying conceptual framework for the international comparisons is presented in Chapter 2, which examines the economic theory that underpins a public-sector role in agricultural R&D. An economic approach provides a framework and a perspective for disentangling those elements that are properly involved and those that should be ignored or set aside in developing improved R&D policies in a changing economic and political environment.

Purpose of the Book

Our aim in writing this book was to develop and bring together information from a range of countries on each country's policies, institutional structures, and changing investment patterns in agricultural R&D. To do this we critically review and evaluate post–World War II developments in agricultural R&D in Australia, the Netherlands, New Zealand, the United Kingdom, and the United States.

In presenting these country cases we examine several key aspects of agricultural R&D policy. First we give a general account of how public research policies and institutions in these countries are changing in response to changing conditions. Previously, access to this type of information has been sporadic and ad hoc. Here we provide a more complete and comparative record of changing conditions across five important national agricultural research systems. Second, the book describes an economic framework in which to consider these changes. This framework makes it possible to provide an integrated account and assessment of changes in policies affecting research priorities, research funding, and research institutions. We consider the external forces of change to the research system, provide an economic framework to consider alternative approaches to change, identify and assess the resulting choices of each country, and compare and contrast the outcomes among the countries.

Structure of the Book

The book is divided into three parts. The first part (Chapters 2 and 3) is introductory. Chapter 2, "The Economics of Agricultural R&D Policy," restates and reinterprets the economic rationale for government involvement in R&D in general and agricultural R&D in particular. This economic framework provides a unifying theme for assessing alternative policy options and their likely consequences. Chapter 3, "Agricultural R&D Investments and Impacts," begins by briefly summarizing results from published studies of agricultural productivity in the five countries. This provides a quantitative picture of productivity developments throughout the OECD more generally and gives guidance on how to interpret this empirical evidence and understand the links between agricultural productivity and R&D expenditures and policies. The bulk of Chapter 3 describes aggregate R&D expenditures in each of the five countries over the postwar period and identifies commonalties and differences. New data series that track overall developments in private and public agricultural R&D across 22 OECD countries are also presented and discussed.

The second part, which forms the core of the book, presents the five country studies of national agricultural R&D systems. A chapter is devoted to each country: Chapter 4, "Agricultural R&D Investments and Institutions in the United States"; Chapter 5, "Agricultural R&D Policy in Australia"; Chapter 6, "Agricultural R&D Policy in the United Kingdom"; Chapter 7, "Financing Agricultural R&D in the Netherlands"; and Chapter 8, "Agricultural R&D Policy in New Zealand."

The countries chosen for study present a diverse range of economic and political circumstances and institutional changes that enable us to explore the promises and pitfalls involved in revamping agricultural R&D policies and institutions. For example, in New Zealand, as part of that country's recent rounds of general economic reforms, agricultural research has been radically reformed and substantially privatized. In Australia, policies have evolved in the direction of much greater industry support for research conducted in the public sector, which is funded through commodity levies with matching government grants. Questions have been raised about the implications of these types of developments for the total resources available for research, the balance between private- and public-sector roles, and the balance between more-basic and more-applied research—all of which have implications for the effectiveness of and net social benefits from the national R&D effort.

The third part identifies commonalties and differences among the agricultural research systems. Chapter 9, "A Synthesis," comprises a summary of the main themes, a synthesis of the case histories, and a summary of the findings from the book.

The Audience

This book has been written primarily for policymakers and administrators with agricultural research and extension responsibilities in international agencies and national and state governments. It is they who must gauge the adequacy and appropriateness of research activities for which they are responsible and build the new institutions for R&D that may permit sustained growth and development into the next century. The information is also of interest to scholars who seek to know what has happened in agricultural R&D and why and to understand the consequences in ways that may lead to more informed policy choices.

The material in this book is intended to be accessible and of interest to farmers, food processors, wholesalers, retailers, environmentalists, scientists, and all who have a direct stake in or are affected by the research system. Understanding the histories of public agricultural research institutions and the forces of change that are confronting each system and learning from the changes made to address these external forces will provide a better basis for formulating public agricultural R&D policies that are both politically feasible and economically worthwhile.

2 The Economics of Agricultural R&D Policy

JULIAN M. ALSTON AND PHILIP G. PARDEY

The standard economic rationale for government intervention in agricultural R&D is market failure. Where net private benefits are less than net social benefits, from society's viewpoint the private sector will invest too little in research. This is especially likely to be true in industries such as production agriculture, which has large numbers of relatively small firms. Evidence of high social and private rates of return to R&D in agriculture supports the view that society is investing too little in agricultural R&D. On this basis, government intervention is called for to correct the market failure and promote a greater total investment in agricultural R&D, especially in research areas with relatively low private R&D incentives and relatively high social payoff.

Most commentators interpret the call for more government intervention to mean using more taxpayer dollars to finance more public-sector R&D. However, other government policies may also be used to promote an economically more efficient agricultural R&D enterprise by affecting the total resources devoted to R&D, the allocation of those resources among research areas and institutions, and the efficiency with which the resources are managed and used. By going beyond simply "more dollars," the policy options raise a much harder set of questions. Economics provides the unifying theme for addressing these policy problems. The particular virtue of the economic approach is that it provides a coherent, consistent basis for developing, considering, and evaluating alternative policies.

In this chapter we look into the economics of government policies for agricultural R&D, including a review of principles for government involvement, an economic assessment of some practicalities of R&D policy (mainly from a financing perspective), and a brief consideration of some political-economy aspects of R&D policy. First, we discuss why private markets may fail to provide enough research when the resulting knowledge is to some extent a public good whose benefits cannot be appropriated by private firms. Then we discuss environmental externalities and the reasons why, if left to themselves, private economic agents have weak incentives to provide R&D that addresses these problems. We also discuss why redirecting public research may not be sufficient or appropriate. We review the empirical evidence on rates of return to research, which supports the underinvestment hypothesis. To conclude the section on principles, we identify alternative potential forms of government intervention and present a decision tree for optimal policy choice.

The body of this chapter concerns practical aspects of research policies. We discuss the role of economies of size, scale, and scope; diseconomies of distance; and research spillovers and their implications for which research should be performed on a regional, national, or international basis. Then we explore economic efficiency and equity issues with respect to alternative methods of financing agricultural R&D, including general tax funds, R&D tax concessions, commodity levies, and matching grants. Government policies for providing incentives for increased private R&D, including recent initiatives with respect to the establishment of intellectual property rights, are also explored. Alternative procedures for allocating public funds among different research organizations are investigated and the benefits and costs associated with block grants, competitive grants, contracts, and earmarked funds are discussed.

A final section briefly examines the impact of interest groups and rent seeking on the structure of government intervention in agricultural science policy and offers conclusions and summaries.

Principles for Government Involvement in Research

A basic economic principle is that the net benefits to society from production and consumption will be maximized when the social costs are equal to the social benefits from the last (marginal) dollar spent on research (as long as total social benefits exceed total costs). A market failure is said to exist when private incentives provided by market mechanisms lead to a different resource allocation and a different product mix than the socially optimal outcome.

Market failure in agricultural R&D seems to be widely taken for granted: from society's perspective, if left to their own devices, private interests would invest too little in agricultural R&D and would provide the wrong mixture of research investments. To understand why agricultural R&D is and ought to be a policy issue for governments—and what sort of policy approaches should be considered—we need to understand the nature of the market failures that apply to knowledge and innovation.

Divergent Private and Social Costs and Benefits

Market failure in R&D most often arises because private costs and benefits from research do not coincide with social costs and benefits, which leads to an underinvestment in research from society's perspective. That is, R&D opportunities that would be socially profitable are not exploited because they are privately unprofitable.

PUBLIC GOODS. Investment in R&D leads to the development of new knowledge and, as Lindner (1993:208) noted, "Knowledge *per se* is a classic public good in the sense that it is non-rival in consumption and non–price excludable." Nonexcludability occurs when property rights in a good have not been, or cannot be, defined or readily enforced (for example, improved open-pollinated seed varieties that can be saved for planting in subsequent seasons or years). Nonrivalry exists when one person's consumption of a good does not detract (or subtract) from someone else's consumption of it. (One example is the use of new crop management practices such as changed seeding rates or planting times.) These properties should not be thought of in absolute terms; some types of research output may exhibit some degree of nonrivalry or nonexcludability (and thus they will be partially rival or excludable). In practice, the new knowledge arising from R&D is rarely a pure public good. Rather, it exhibits some degree of "public goodness" that reflects the public policy environment in which it is produced and used, along with the characteristics of the technology and the markets for affected environmental goods and commodities.

APPROPRIABILITY. The partial public-good nature of much of the knowledge produced by research means that research benefits are not fully privately appropriable. Indeed, the main reason for private-sector underinvestment in R&D is inappropriability of some research benefits: the firm responsible for developing a technology may not be able to capture (that is, appropriate) all of the benefits accruing to the innovation, often because fully effective patenting or secrecy is not possible or because some research benefits (or costs) accrue to people other than those who use the results. This appropriability problem extends beyond relations among single individuals to relations among collectives such as one producer cooperative or industry group versus another, among regions within a country, or even among countries.

For certain types of research, the rights to the results are fully and effectively protected by patents. Thus, the inventor can capture the benefits by using the results from the research or selling the rights to use them; for instance, the benefits from most mechanical inventions and the development of new hybrid plant varieties, such as hybrid corn, are appropriable. Often, however, those who invest in R&D cannot capture all of the benefits—others can "free-ride" on an investment in research by using the results and sharing in the benefits without sharing in the costs. An agronomist or farmer who developed an improved wheat variety would have difficulty appropriating the benefits because open-pollinated crops like wheat reproduce themselves, unlike hybrid crops, which do not. The inventor could not realize all of the potential social benefits simply by using the new variety himself; even if he sold the (fertile) seed one year, the buyers could keep some of the grain produced from that seed for subsequent use as seed. Hence he is not able to reap the returns to his innovation. In such cases, private benefits to an investor (or group of investors) are less than the social benefits of the investment and some socially profitable investment opportunities remain unexploited. The upshot is that, in the absence of government intervention, investment in agricultural research is likely to be too little.[1]

Appropriability is not a question of absolutes. Complete appropriability of the fruits of invention is never likely to be possible. As long as some private benefits can be obtained, there will be incentives for some private investment. The real issue, then, is the appropriate form and degree of partial correction for problems arising from incomplete appropriability. Because policy intervention is justified only when the benefits exceed the costs, in many cases the best policy is no intervention.

ENVIRONMENTAL EXTERNALITIES. A second type of divergence between private and social returns can be identified: there may be unaccounted-for environmental side effects of the implementation of particular research results, which are called externalities. Externalities arise when one individual's production or consumption activities involve spillover effects on other individuals who are not compensated through markets. Groundwater pollution with agricultural chemicals is an example of a negative environmental externality. Pollination of a neighboring orchard by a beekeeper's bees is a positive externality from honey production (and, if the orchard improves the quality of the honey, the externality may be reciprocated).[2] Generally, negative environmental externalities have attracted more attention.

The existence of externalities means that marginal private costs (or benefits) from economic activities differ from the corresponding marginal social costs (or benefits). As a result, private decisions are not socially optimal; that is, they lead to a market failure. Even in the absence of market failures associated with inappropriability of R&D benefits, there will be distortions in incentives. Thus, the direction of research will be biased against technologies that mitigate environmental externalities and in favor of technologies that exacerbate these problems. However, it is not sufficient to redirect agricultural R&D toward environmental issues. The very nature of negative externalities means that private investors do not have adequate incentives to try to reduce them, either through the choice of production practices with available technology or through the choice of the direction for technology to evolve through research, development, and adoption decisions. If agricultural R&D is to be effective in

1. The appropriability problem may take another form: spillovers through perhaps unforeseen interindustry applications of research results, such as when livestock R&D has applications for human medicine or when fundamental findings related to bioengineering are applicable across a broad range of commodity research. Again, the problem is that these spillovers will not be recognized as benefits by the providers or funders of the R&D, unless there is some way for them to recoup some of those benefits.

2. Free-riding by others on an individual's research results is a type of externality too—a positive externality that has beneficial spillover effects.

reducing environmental externalities, the resulting new technologies must be adopted, and, if they are to be adopted, they must be viewed as profitable by potential adopters. This could happen in one of two ways: a new, environmentally friendly technology may be privately more profitable than the current technology under the current incentives or the government may act to change the adoption incentives as well.

Similar arguments apply to the development and adoption of technologies that consume stocks of unpriced or underpriced natural resources. Private incentives are liable to lead in the direction of the development and adoption of excessively consumptive technologies unless government acts to modify the incentives and "internalize" the externalities (Alston, Anderson, and Pardey 1995). Moreover, if the incentives are not corrected, the redirection of R&D toward resource and environmental issues might not be appropriate (because there will not be any social or private benefits from commercially irrelevant R&D).

Evidence of Market Failure

The payoff to research can be summarized in terms of the private rate of return (comparing private costs and benefits to the investors in the research) and the social rate of return (comparing benefits and costs to society as a whole). Alston, Marra, Pardey, and Wyatt (1998) documented and discussed the results of more than 290 studies of social and private rates of return to agricultural research. The overwhelming conclusion was that estimated real rates of return to agricultural research have been very high, typically well in excess of 20 percent per year. The relevant comparison is with the real rate at which the government borrows money, typically 3–5 percent per year. Because the rate of return to R&D is much greater than the borrowing rate, in general, there appears to have been a gross underinvestment in agricultural research.

It is less clear that there has been an underinvestment in agricultural research relative to other types of sector-specific research. For instance, a number of studies recently documented by the Industry Commission (1995) in Australia showed that rates of return to industrial R&D were comparable to rates of return to agricultural R&D: typically well in excess of 20 percent and often ranging around 100 percent per year.[3] Therefore, the rate-of-return evidence does not support a diversion of resources from industrial R&D to agricultural R&D. Rather, when taken at face value, the evidence on rates of return

3. The commission documented 20 rates of return to industrial R&D (reported in 10 studies of the United States and 4 studies of Japan) to the industry and, where available, to firms in other industries, and to the nation as a whole. The unweighted means of the annual rates of return were 26 percent to the industry (standard deviation of 13 percent), 75 percent to firms in other industries (standard deviation of 27 percent), and 85 percent to the nation (standard deviation of 22 percent). The commission also reported similar evidence on rates of return to industrial R&D in Belgium, Canada, France, Germany, and the United Kingdom.

to industrial R&D and agricultural R&D supports the view that resources should be diverted from other economic activities.[4]

Some reservations can be raised about the evidence on rates of return. Most studies have not adjusted for the effects of price-distorting policies on the measures of research benefits, an omission that might lead to either over- or understatements of the benefits and the rate of return (Alston, Norton, and Pardey 1995). Most studies also have not adjusted for the effects of the excess burden of taxation on the measures of costs, an omission that will lead to a systematic understatement of the social costs and an overstatement of the social rate of return (for example, Fox 1985).[5] Many studies present estimates of average rather than marginal rates of return, and the latter are most relevant for evaluating the social benefits from marginal changes in R&D. Further, Alston, Craig, and Pardey (1998) have shown that typical econometric assumptions about the research lag structure can lead to upward biases in the estimated rates of return to research.[6] However, a number of factors could lead to underestimated rates of return to agricultural R&D, including the omission of spillovers from agricultural R&D into nonagricultural applications (although the mistreatment of R&D spillovers in the opposite direction could have opposite effects) and the consequences of research into areas such as the environment, food safety, and social science that are not reflected in conventional productivity or rate-of-return measures.

The evidence on high rates of return to research has been heavily promoted by those who wish to shore up or expand public funding support for agricultural R&D. However, informal impressions suggest that there has been rising skepticism about whether the estimated rates of return are accurate or meaningful as a justification for more public investment or whether other solutions should be sought for any apparent underinvestment problem. At the same time, in spite of the absence of any supporting evidence in the form of rates of return, there appears to be a tendency to shift public-sector research resources in the direction of environmental and natural resource issues and away from the more traditional agenda. These trends are explored in

4. Estimated rates of return may not be fully comparable between agricultural R&D and industrial R&D (or even within those classes) because different studies make different types of assumptions, use different concepts, and hold different things constant. These differences can have substantial effects on the estimates and are important if relevant or meaningful comparisons are to be made. This also holds for comparisons of the rate of return to agricultural R&D with that for investments in education: Psacharopoulous (1993, 1994) reviewed a large number of studies and reported social rates of return for investments in education of 9–25 percent per year.

5. The "excess burden" is due to the social costs of market distortions arising from taxation as well as the costs of enforcement, compliance, collection, and administration of taxes (for example, Fox 1985; Fullerton 1991).

6. Alston, Craig, and Pardey (1998) suggested a rate of return to public agricultural R&D in the United States of less than 10 percent per year, whereas previous studies using similar data found much greater rates of return.

Chapters 4–8, where case studies of individual countries are presented, and they are assessed in Chapter 9.

Forms of Government Intervention

Government policies to reduce private-sector underinvestment in agricultural R&D include

- improvements in private property rights over agricultural inventions;
- the creation of new public- or private-sector R&D institutions (for example, legal arrangements under which an industry funds research cooperatively);
- enhanced incentives for private R&D (for example, tax breaks, subsidies, or other incentives); and
- the provision of public funds for publicly or privately executed R&D.

The dominant strategy has been to use government revenues to finance mainly public-sector and some private-sector R&D.[7] This strategy includes the provision of tax breaks and other financial incentives for private R&D, which involves a loss of government revenues, as well as the direct use of government funds both to finance private R&D through grants and contracts and to finance the production of knowledge in a variety of publicly administered R&D organizations.

These alternatives may differ in terms of their incentive effects, the net social (deadweight) cost of distortions in the quantity and mix of research, and the total social cost of financing R&D. Further, different interventions will be more or less effective at correcting different types of market failures, and they will have different distributional or equity consequences.

Considerations in Choosing Policies

One key principle in finding solutions to underinvestment in R&D—or, more realistically, arrangements that will reduce its impact at a justifiable cost—is that the solutions or arrangements must deal simultaneously with several sources of market failure. An important and potentially difficult trade-off arises when government attempts to remedy nonexcludability problems directly by strengthening the property rights applying to research results (for example, by patents or copyrights), thus encouraging greater private provision of R&D by enabling research producers to reap the rewards from their efforts.[8]

7. Another way to finance public-sector agricultural R&D is to sell the scientific results. Even public-sector organizations such as universities now often patent their research results where possible and sell the products.

8. Changes in legislation defining property rights may not be sufficient if it is not practicable to enforce the new property rights. For instance, Venner (1997) found no evidence of an increase in private-sector investment in wheat breeding research in response to the introduction of

Even if this approach had no practical difficulties, it would often clash with the most appropriate arrangement for provision of a nonrival good, which involves its provision at zero price because the social marginal cost of providing it is zero.[9] In other words, the price that is high enough to encourage innovation is by definition too high from the point of maximizing the social benefits of the innovation.[10] Public providers attempting to raise revenue often appear to ignore this principle in setting prices for information with public-good characteristics. For these reasons, "privatization," in the sense of making research results excludable by improving the definition of property rights attached to those results, can involve losses in economic efficiency because it does not address the other characteristic of public goods, namely nonrivalry in consumption.

A further distinction between the ideal arrangements for provision (production) of research and the ideal arrangements for financing research must be kept in mind when designing remedies for a private-sector underinvestment in R&D. Certainly, doing the research and paying for it are related and should be considered in conjunction, but it is worth keeping the distinction in mind for the following reason. Research funds should be raised in a least-cost manner; that is, the sum of administrative costs and distortionary losses to the economy should be minimized. The issue of how to spend funds is clearly separate, at least conceptually, from how they should be raised. Raising funds in the least-cost way should not preclude the use of efficient competitive processes for allocating them among research programs. As a practical matter, however, R&D financing arrangements are likely to influence, and be influenced by, the research agenda and the efficiency of research allocation because different financing arrangements will involve different decisionmakers having differing objectives. This tendency is clear from the substantial increases in the amount of money being raised by levies and, apparently, changes in research priorities, when ostensibly small changes were instituted in the membership of the boards of Australia's R&D corporations in 1989 (see Chapter 5).

A third important principle is that, to the extent practicable, the beneficiaries of research should bear the cost. This principle implies that, when most of

---

the U.S. Plant Variety Protection Act. The act served primarily as a marketing tool. More generally, the empirical evidence on the effects of changes in intellectual property protection on the respective private and public roles in R&D is comparatively thin and far from definitive (see, for example, Perrin 1994; Pray, Knudson, and Masse 1994).

9. A standard result in economics is that efficiency in a market is achieved when the price of the good is set equal to the marginal cost of providing it. Thus when a good can be provided to additional people at no extra cost (that is, it is completely nonrival in consumption), efficiency dictates that it be provided at a zero price. In the extreme case, a patent gives its owner a pure monopoly over the patented good and the monopoly price is likely to well exceed the socially optimal price.

10. This is not to mention the problem of potential overinvestment in R&D as firms compete with each other to win the patent first. The idea of "patent races" is discussed in Wright (1983).

the benefits from a research program accrue within an industry, the industry should finance the research; when most of the benefits spill over to the general community, the research should be funded from general government revenues. There are two reasons for employing the beneficiary-pays principle. One is based simply on a notion of fairness or equity. That is, if beneficiaries can be identified, they should foot the bill where that can be facilitated. Further, it may be cheaper, and thus more efficient, to raise money from the relevant industry for R&D rather than using general public funds raised through the tax system (Alston and Mullen 1992; Alston and Pardey 1996:254–255). Thus, in plausible circumstances, there are equity and efficiency arguments in favor of the beneficiary-pays principle.

Finally, an overriding principle is that any remedies for market failures should clearly make matters better in aggregate; that is, the benefits should be greater than the costs associated with the remedies. With this in mind and in recognition of information problems and potential for "government failure" as well as market failure, interventions should be designed to maintain or increase competition in the provision of research services and transparency in the processes through which public R&D resources are allocated. It is important to exercise discretion in judging where the market failures in R&D are important and where they are not because government investment in R&D in a particular area is likely to crowd out or redirect some private-sector R&D. Where private-sector underinvestment in R&D is not otherwise a problem, public-sector R&D can cause private-sector underinvestment.

The R&D Mix

Another way to look at the policy problem is to examine the mix of R&D financed from public sources or undertaken in public institutions in terms of the balance (1) between more-applied and less-applied (or near-market) research, (2) between research areas such as production and environmental issues, or (3) between parts of the food chain such as farming and food processing. Some of these issues can be broached in principle, but the resolution of most rests on the answers to empirical questions.

BASIC AND APPLIED RESEARCH. One common and important argument is that the extent of market failure and the degree of private underinvestment will be greater in more-basic research, whose benefits are by definition less appropriable than those of more-applied or near-market research. It follows that governments should pay greater attention to more-basic research, and there is some evidence that some governments are shifting in that direction. However, this trend raises a number of difficulties. One such difficulty concerns different conceptions of the basic-applied distinction (the relevant one for this purpose is the extent of potential appropriability of results, not the perceived relative importance of the science). Another is the difficulty of monitoring the true nature of the research activity (because scientists may simply reclassify their work without changing its nature if they perceive advantages in doing so). Making these distinctions correctly and meaningfully is important. For instance, Nelson (1997:46) discusses basic-applied aspects of science policy at some length and claims that "there are high social and economic costs of allowing the establishment, or at least the strong enforcement, of intellectual property rights on basic scientific knowledge."

ENVIRONMENTAL ISSUES. As already noted, the existence of environmental externalities and an enhanced public perception of environmental problems do not by themselves justify an increased emphasis of public agricultural R&D on environmental issues at the expense of other issues. The overriding criterion should be the size of anticipated social benefits relative to social costs of the research. The existence of any such benefits may require a change in environmental policies rather than in research emphasis. This distinction has not always been appreciated by those who have led the shift in research priorities toward environmental issues.

R&D BEYOND THE FARM GATE. The importance of appropriability problems in R&D will vary depending upon the nature of the industry, its technology, and the research. Farm input suppliers and other components of the agribusiness industry that transport, process, and market farm products tend to be relatively large firms (at least in developed countries) that are able to appropriate research benefits internally. Such firms can profitably conduct an economically efficient R&D enterprise without having to market their research results and deal with the attendant appropriability problems. Agribusiness technologies tend to be chemical and biochemical or mechanical innovations, which can be protected by patents, or process innovations, which can be protected by secrecy. The technology used by agribusiness is often not specialized to agribusiness and can be adapted from broader industrial technologies (for example, refrigeration or transportation technology). Thus, the potential role for the government (that is, the odds of market failure) is generally greater in R&D pertaining to farming than in R&D pertaining to agribusiness.

However, there are exceptions. Some parts of the farming industry are involved in vertically integrated structures where research benefits can be internalized (for example, the broiler chicken industry); certain types of technology applicable to farming are effectively protected by patents (for example, machinery, hybrid lines of plant varieties). R&D incentive problems are important in some parts of agribusiness (for example, in breeding open-pollinated crops).

A Decision Tree

The decision tree in Figure 2.1 can be used for deciding whether to intervene in agricultural R&D and how. The first step is to establish a prima facie case of market failure. If the market is performing its functions adequately, questions still may arise about whether the market outcomes are satisfactory. If the

FIGURE 2.1 A decision tree for research policy

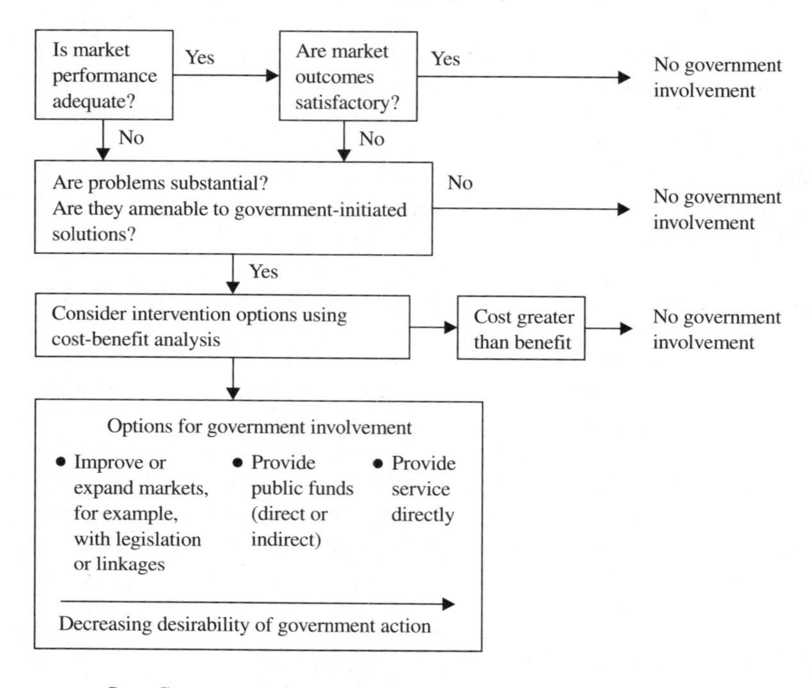

SOURCE: State Government of Victoria (1994:14).

market is working and the outcomes are acceptable, then the government should not be involved. Even if the market is not working ideally or the outcomes are not entirely satisfactory, the case for government involvement depends on whether the problems with the unregulated situation are serious and whether the perceived problems are likely to be amenable to correction by policy. In many cases the answer at this stage would be to accept existing market imperfections rather than make things worse by intervening. In some cases it is worth proceeding to the next stage of decisionmaking and considering explicit policy options. Even at this last stage in the decisionmaking hierarchy, it might be concluded that the costs outweigh the benefits from intervention. However, in some cases intervention is justified by the benefit-cost assessment. The form of the intervention matters too, as summarized in the final stages of Figure 2.1 and discussed in more detail in the following section.

Research Policies in Practice

The appropriate form of government intervention in agricultural R&D varies according to circumstances. In some industries, in some places, and for certain types of R&D, governments may have no role or only a very limited one. In others, the government might have to finance and produce all the R&D using general taxpayer funding. For much agricultural research, a mixture of public and private investments and roles is likely to be the most appropriate.

In addition, as circumstances change, so do the optimal roles for the government and the private sector. Solutions that may have been right at the turn of the century or even 50 years ago may not be right for the twenty-first century. In particular, changes in the nature of agricultural production and farm businesses, the structure of agriculture, and the educational status of farmers and their access to information through modern communications technology have changed the extent and nature of the market failure in farm-level process innovations. At the same time, changes in intellectual property rights, in terms of both the institutional arrangements and their enforceability, have modified the nature of the market failure associated with certain types of R&D. Further, the nature of agricultural science itself has changed—becoming more closely integrated with parent disciplines in biological and physical sciences and, more recently, placing more emphasis on modern biotechnology and information technology than on mechanical and chemical technologies. As the science, the institutions, and the industries change, so do the nature and extent of knowledge spillovers and appropriability problems and so, accordingly, do the roles of government and the private sector. Other relevant changes include the shift in the focus of public opinion and political support away from traditional farming constituencies and toward the environment. These shifts in public perceptions can create opportunities for developing fairer and more efficient ways of financing and managing agricultural R&D, compared with the traditional emphasis on government-financed and government-performed research.

Basic Production Economics of Research

One important and changing element of the equation is the nature of the research enterprise, which can be discussed in terms of the "production economics of research." This topic, which is related to the costs of producing new knowledge, is usefully considered separately from the issues about appropriability and so on. It concerns the implications of changes in the technology of science and the prices of scientific inputs, how the costs of increments to knowledge vary with the size of the agency conducting the research, and its mix of outputs and inputs. Reasonable analogies can be drawn with farm production economics.

ECONOMIES OF SIZE, SCALE, AND SCOPE. In many types of research there are significant economies of size, scale, or scope. Thus, it makes sense to organize relatively large research institutions. Economies of scale refer to reductions in unit costs of production with increases in the size of the organization, which result from things like specialization and appropriate division of labor. Economies of size are exploited when the R&D enterprise is large enough to have appropriate combinations of lumpy fixed factors, such as physical infrastructure and scientific expertise, and other variable inputs, such as equipment and support staff. Economies of scope are cost savings resulting from diversification of the R&D portfolio. Scientists in different fields may be able to make contributions to the research of others; also items of equipment or infrastructure may be used for different projects at different times of day or night or in different seasons of the year. Further, modern biotechnology enables much closer targeting of technologies to specific markets or environments. Research may well move from crop classes (such as the group of varieties classed as the hard-red spring wheats used for breadmaking, the durum wheats used for pasta, or the soft winter wheats used for biscuits and noodles) to crop varieties tailored for specific market purposes across which there will be great potential for economies of scope.

An ideal research institution would be large enough and diverse enough to exploit these various economies so as to minimize unit costs of research output. If these economies become more important over time, so that the optimal size of the institution continues to increase, an ideal institutional arrangement would allow for a corresponding growth in institutional size, perhaps through the consolidation of or improved coordination among preexisting institutions.

DISECONOMIES OF ECONOMIC DISTANCE. The production economics of agricultural R&D has much in common with other industrial R&D but may differ in important ways because of the biological base of the industry, long production cycles, and jointness among agricultural enterprises (shown by, for instance, the raising of sheep for both wool and meat or the growing of grain for feeding to livestock). In particular, much agricultural technology is characterized by site specificity related to agroecological conditions, which defines the size of the relevant market in a way that is much less common in other industrial R&D. Thus the unit costs of making local research results applicable to other locations (for example, by adaptive research) grow with the size of the market and must be added to the local research costs. A close analogy can be drawn with spatial market models of food processing in which processing costs fall with throughput, but input and output transportation costs rise with throughput, so that when the two elements of costs are combined a U-shaped average cost function is derived (for example, Sexton 1990). Economies of size, scale, and scope in research mean that unit costs fall as the size of the R&D enterprise grows, but these economies must be traded off against the diseconomies of distance and adapting site-specific results (the costs of "transporting" the research results to economically "more distant" locations). Thus, as the size of the research enterprise increases, unit costs are likely to decline at first (because economies of size are relatively important) but will eventually rise (as the costs of economic distance become ever more important).

RESEARCH SPILLOVERS. The spatial dimension of agricultural R&D has other implications. Geopolitical boundaries rarely match agroecological zones so closely that research results are confined to the state or nation in which the research costs are incurred (either privately or by the government). For instance, Huffman and Evenson (1993) suggested that about 45 percent of the benefits from research conducted in U.S. state agricultural experiment stations (SAESs) were earned as interstate spillovers. This fact has implications for incentives. The earlier discussion of spillovers among firms within an industry suggested that the government should intervene. But a state or national government can intervene only in relation to the producers within its own boundaries and cannot act on behalf of (or, usually, extract support directly from) producers and consumers in neighboring states or nations for research that spills across the border. Thus the extent to which benefits can be appropriated geopolitically will be a further constraint on the size and location of research institutions and the nature of research undertaken unless institutional arrangements permit research institutions to repatriate benefits into their own state or nation or unless costs can be pooled through cooperative arrangements.

Research Organization

The appropriate institutional structure for organizing research programs should vary according to the nature of the research. Some issues are clearly national issues and are appropriately addressed by national programs. However, the national government can choose whether to address an issue using national funds in national research institutes or universities, in provincial organizations (or, for that matter, in private organizations), or by using incentives to encourage other organizations to take joint action. At the same time, synergies between research and other public-good programs, such as teaching and extension, may call for integrated institutional structures for certain types of work and specialized institutions for others.

INSTITUTIONAL STRUCTURE. In the U.S. land-grant system, for example, SAESs are substantially and physically integrated with colleges of agriculture (and, in many cases, extension agencies). This institutional structure was initially justified on the grounds of "complementarity" between research and teaching and extension (Wright and Zilberman 1993). Although this is still a widely cited rationale for the continued support of the land-grant system, the precise nature and magnitude of these complementary effects is not always clear. In any event, the optimality of the current numbers and structure of land-grant colleges, which have changed little over the years, is an open question, especially with improvements in communications technology and innovations in distance learning and arms-length research collaboration (National Research Council, Board on Agriculture 1995, 1996; Economist 1997). If the land-grant system were to be designed today from the ground up for conditions in the twenty-first century, the results might be very different.

In Australia, the Netherlands, New Zealand, and the United Kingdom, important changes have recently been made in the administration of agricultural R&D, including approaches to its financing, organization, and management. Yet none of those countries have ever had an arrangement like the U.S. land-grant system, which takes advantage of the ostensible synergies provided by combining research, teaching, and extension.

REGIONAL ISSUES. What is the appropriate geopolitical basis for organizing research? Should public-sector research be organized nationally or on a less aggregated basis? What principles are relevant for determining the types of research that are better conducted by a national system versus a subnational organization? Many countries organize research on a subnational (regional) basis, especially those countries where state governments are engaged in the provision of research. In Australia and the United States, for example, state-based organizations spend more on agricultural R&D than federal ones. This creates issues within a country about the balance of types of research undertaken given the potential for geographical spillovers of research results. Similar issues arise between countries as well.

The issues are clearest when research is formally organized and funded by subnational, state, or regional governments. Research spillovers, where results from one state's or region's research are adopted in another state or even overseas, can be important. Also, individual states may not be able to capture economies of size, scale, and scope in research programs that pertain to larger jurisdictions. As a consequence state-level arrangements are often inadequate. At the same time, national funding of national programs is not always the right policy for addressing underinvestment in research issues that involve multiple subnational regions within the country. Because education programs are typically undertaken at a regional level, often by subnational governments such as states or provinces, regional organizations may be better placed to take advantage of the potential synergies between research and education. As with any public good, there is an efficient jurisdiction for the provision of research and education that varies according to the nature of the research and education being provided. Partnerships between national and subnational governments may be necessary for the national government to facilitate the joint production of research and education programs in those countries where research and education are in different jurisdictions.

Financing Options

Agricultural R&D may be a public good in the sense of at least partial nonexcludability and nonrivalness, but this does not mean that everyone in the nation benefits and it does not mean that everyone in the nation should pay. Indeed, for many types of research and common commodity market conditions, commodity prices are not affected, and so the benefits are confined to those producers who are able to adopt the resulting technology. In other cases, the adoption of technology leads to improvements in productivity and therefore to lower commodity prices with benefits distributed between consumers and producers who adopt the new technology, perhaps partially at the expense of producers who do not adopt the new technology and sometimes those who are slower to do so. Sometimes the lower prices are transmitted to producers and consumers in other countries, and sometimes foreign producers can adopt the new technology, which adds further complications to the picture of the distribution of the benefits from the new technology. Citizens who do not consume or produce the commodity in question are not beneficiaries even though they may be taxpayers and asked to support the R&D.[11]

Often, the fairest and most efficient way to fund research is to arrange as much as possible for beneficiaries to bear the costs in proportion to the benefits they receive.[12] This can be accomplished by choosing funding arrangements that reflect the geographic focus and the commodity orientation of the research. Thus, different agricultural R&D programs and projects call for different funding arrangements. However, in most countries the primary source of funding for public-sector agricultural R&D continues to be general tax revenues, which may be an expensive source of revenues.[13] A number of options can be used instead of, or in combination with, the use of general government funds to finance R&D undertaken in the public sector or the private sector. These include tax breaks, commodity levies, and matching grants.

TAX BREAKS. A number of countries have tried tax concessions for private research (for instance, through expensing current R&D costs at rates greater than 100 percent or accelerated depreciation of R&D capital costs). This arrangement is a form of joint venture, with public and private funding of research. This is generally a blunt instrument. It is difficult to minimize the

11. Alston, Norton, and Pardey (1995) elaborate at length on the determinants of the distribution of benefits from research among producers, consumers, middlemen, foreigners, and so on.

12. Incentive problems in agricultural R&D arise from inappropriability of benefits and free-riding and may be serious unless some way can be found to ensure that beneficiaries share appropriately in R&D costs. Hence, as Alston and Pardey (1996) argue, a criterion for efficiency, as well as fairness, is determining to whom the benefits accrue. This issue pertains to the mechanism for allocating research resources among alternatives as well as processes for raising the revenues.

13. Recent studies have shown that it costs society measurably more than a dollar to provide a dollar of general taxpayer revenues to finance public expenditures. The U.S. evidence has been reviewed, summarized, and synthesized by Fullerton (1991), whose results indicate that a dollar of government spending on agricultural R&D may cost society between US$1.07 and US$1.25 when the market distortions induced by taxation are taken into account. Funding from checkoffs (commodity taxes) also involves potential excess burdens for similar reasons, but the required low rates of commodity taxes (less than 1 percent) are likely to involve smaller marginal excess burdens than the prevailing high rates of labor income taxes in most countries, especially when it is considered that such commodity taxes may in fact reduce distortions resulting from commodity support programs in some cases or from the absence of "optimal" trade taxes in others.

transfer effect, wherein forgone taxpayer funds merely substitute for private R&D investments that otherwise would have taken place. More specifically, it is difficult to design tax concessions that discriminate closely among alternative forms of research (that is, additional investments in ongoing lines of research by existing firms versus investments in new research by existing firms versus new, start-up firms or more strategic kinds of R&D with more spillover potential versus applied research) or among providers of research (for example, local versus foreign firms). A blunt tax concession aimed at stimulating new research done locally could simply cause research funds being used elsewhere to be diverted to take advantage of the local tax breaks (Industry Commission 1995). However, while tax breaks involve some transaction costs (in terms of the paperwork involved, auditing costs, and the like), it is a funding approach that is comparatively inexpensive to administer, at least in those places (for example, many developed countries) where the tax system is well equipped for such purposes.

COMMODITY LEVIES. When research benefits are contained entirely within an industry, there is no plausible justification for the use of public funds to support R&D. The important public role is to help the industry raise its own research funds.[14] When the government gives producers the statutory authority to set up an institution such as a U.S. marketing order or an Australian research and development corporation (RDC), with powers to collect a levy or tax from producers to be used to fund research, the problems of nonexcludability and nonrivalry are ameliorated. A greater use of levy funding could enhance economic efficiency in three ways. First, industry funding is a potential complement to other sources of funds that, as a practical matter, are likely to continue to leave total funding inadequate from the viewpoint of both the industry and the nation (in terms of the economically efficient total investment). Second, from the point of view of raising funds in the least-cost way, commodity levies are likely to be a relatively efficient and fair tax base. Third, industry funding arrangements can be organized to provide incentives for efficient use of levy funds and other research resources.

MATCHING GRANTS. Government could encourage a greater use of commodity levies for agricultural R&D by providing matching (or more than matching) support for programs funded using levies. When a combination of industry levy funds and general revenues is used to finance publicly or privately executed R&D, there is a clear case for government involvement in the

14. Nevertheless, dealing with another issue—what research gets done—may still imply some role for the public sector. Where research costs and benefits are industry specific, it may make sense to leave the question of research topics to the relevant industry. However, there may still be problems of intraindustry distribution of benefits and costs and spillovers that lead to distortions in the allocation of industry-based research funds. Once other extraindustry spillovers are present, there are additional reasons for government involvement—possibly both in supplementing the funding and in directing the R&D effort.

administration, management, and allocation of those funds to ensure that the public interest is adequately considered. It is important to understand that industry levy funding is not to be regarded solely as a producer "self-help" arrangement in which producers collectively fund research on their own behalf and to serve their own ends. Consumers and taxpayers are also affected by such enterprises, and they too have a legitimate interest in them.

When spillovers from industry-funded research flow beyond the industry to the general community, the situation is likely to be more complicated. In the case where research results exhibit classic public-good characteristics—that is, both nonrivalry and nonexcludability are severe—then the research should be publicly funded, although it may still be efficient for funding to be provided under contract by the private sector. In this situation it is not possible to devise a way of extracting finance from a section of the community, such as farmers, that is optimal in the sense that problems of nonexcludability and nonrivalry are overcome. However, when a significant proportion of the benefits accrue to an industry, that is, when the research has both public-good and collective private-good characteristics, it is appropriate to fund the research from both public and private sources. Questions arise about whose objectives will determine the allocation of the resources. Because some research has both public- and private-good components, the underinvestment may also be "relative" in the sense that the mix of research may be skewed. The difficulty is to devise a mechanism by which public and private efficiency criteria are simultaneously satisfied. In short, designing a completely fair and efficient commodity levy arrangement for financing research is not simple, and perhaps this may be why most countries have made limited use of these arrangements in the past.

Protecting Intellectual Property

The private and public roles in agricultural R&D hinge largely, but not exclusively, on the degree to which the benefits from R&D are appropriable and the distribution of the benefits. The nature and degree of property rights surrounding agricultural innovations determine these appropriability aspects and, thereby, the incentives to invent and the consequences of those inventions. Thus the pace and focus of biological innovation in agriculture and related industries, who pays for the R&D and how much, and, ultimately, the incidence of the costs and benefits of the research are all affected by the form of the property protection afforded the results of the R&D.

The crux of the access-versus-appropriability dilemma involves balancing access to biological innovations in ways that reveal knowledge that can stimulate further invention, while conferring some degree of monopoly rights that generate revenue streams to reward successful innovation. A longstanding policy response to this dilemma has been to enact and enforce patent protection for certain types of inventions: the first patent act was passed in the United

States in 1790, and patent systems were instituted even earlier elsewhere, especially in Europe (Huffman and Evenson 1993). Government-sanctioned property protection over living things is a much more recent phenomenon.

National efforts to protect the intellectual property of biological innovations are increasingly being shaped and circumscribed by international laws and conventions. Some of these international initiatives (for example, aspects of the 1993 Convention on Biological Diversity) seem to be driven more by concerns about the equitable distribution of the benefits from biological inventions (both in space and time—within the current population and across generations) than by concerns about concepts of economic efficiency implicit in much of the earlier policy responses to this problem. There are widespread perceptions that "northern" firms (farmers or agribusiness concerns in richer countries) are benefiting at the expense of "southern" farmers (poor farmers in less-developed countries) from the unregulated use of "southern" germ plasm in breeding new varieties that are sold commercially under the protection of national systems of property rights. Other changes in property-rights regimes are related to broader efforts to strengthen the property-rights regulations, which form part of the package of internationally agreed-on policies that underpin the trading arrangements enforced by the World Trade Organization (WTO). Indeed, the Marrakesh agreement signed by 132 countries to date, which was part of the Uruguay Round General Agreement on Tariffs and Trade/WTO negotiations that came into force in January 1995, essentially commits all developed countries to have a functioning system of property protection for biological inventions by 1999; less-developed countries have until 2005 to enact such legislation.

Many of the details regarding the property-rights policies and laws covering biological innovations are far from settled. If past history is any guide, these details will continue to evolve as political, economic, and scientific circumstances dictate.[15] Precisely what life forms are to be protected, the scope of the protection (for example, in terms of the progeny from new varieties, the genetic parts of a new plant variety such as specific DNA sequences, or the R&D processes that underpin this area of research), and the specific aspects of novelty and utility that are to be covered (for example, the phenotypic or genotypic characteristics that are to be covered and precisely what constitutes an inventive step) are all subject to increasingly intense legal, public, and policy scrutiny. All of these details may vary markedly in their economic effects. Specifically, the form of the property protection may have significant efficiency as well as equity effects, with important consequences for the structure of the R&D market in terms of the research that gets done and who does it. Although there is much speculation and discussion about the nature and magni-

15. For a good description of some of these details relating to the WTO agreement, see Leskien and Flitner (1997).

tude of these effects, there is little hard evidence from which to make any definitive statements. Moreover, the rapidly changing science and legislative environment surrounding these issues, especially in recent years, compounds the problem of drawing meaningful policy insights from what little evidence is available.

Allocating Research Resources

The institutional arrangements used to apportion research funds among different research-executing agencies often result in resource allocations that are not based on strong economic foundations. High measured rates of return notwithstanding, a sizable share of the potential benefits from agricultural research enterprises may be wasted in inefficient resource allocation.

ROLES FOR ECONOMIZING. Some would say that in most countries the system has worked very well (citing high reported rates of return) and, by implication, that it would be unwise to spoil a good thing. There is some truth to that view. The public-sector agricultural R&D system has achieved a great deal, and it would be undesirable to change it in ways that would diminish its capacity to contribute to the economy in the future. By the same token, because the public-sector agricultural R&D system has done well in the past does not mean that it could not have done better. Moreover, past success does not guarantee continued future success. The economic environments in which national research systems find themselves, including research technology and research opportunities, are changing rapidly. Approaches that worked in the past may not work in the future.

Allocating scarce research resources is an economic problem. In practice, too little use is made of economic analysis, economic incentives, and the economic way of thinking about the allocation problems. Rather, systems typically emphasize processes and politics, the input side, and pay scant attention to actual performance, the output side. In most countries, there is a notable lack of any systematic attempt to undertake meaningful economic evaluations of agricultural research investments as an integral part of the resource-allocation process. Resources are mostly allocated according to ad hoc approaches that may simply serve to ratify prior prejudices.

FUNDING FORMS. Another issue is how the funding should be provided. The possibilities take the following general forms: gifts, which are funds provided with no particular strings attached; grants, which specify some general commitments by the researchers; and contracts, which entail specific obligations. There is a perception that recent years have seen moves toward proportionately greater use of contracts and grants and a reduction of gifts (that is, formula funding). Has the desirable balance been overshot?

Competitive grants have a great deal to recommend them as a way of allocating public-sector research resources. However, competing for grants is hard and expensive work. If competitive grants are to deliver the promised benefits of greater allocative efficiency, they must be allocated according to efficiency criteria. A poorly administered and corrupt system of competitive grants could easily be worse than an inflexible system of block grants or funding according to some formula, which is unrelated to past or prospective performance.

MANAGED COMPETITION. Managed competition has been proposed as a way of making science and scientists more responsive to changing public research priorities that may, in turn, enable an expansion of (or stave off a contraction of) available funds. Some (for example, Just and Huffman 1992; Huffman and Just 1994) have argued that the transaction costs involved in competitive-grants programs—in terms of the costs to individual scientists of preparing proposals and reporting to granting bodies and the costs of evaluating the proposals and deciding which ones to support—are so high that the programs cannot be economic.[16] That charge could be correct, but relevant alternatives must be evaluated on a comparable footing.

Every method of allocating research resources involves four types of costs: (1) information costs (the costs of obtaining relevant information on the benefits from different types of R&D projects, on which to base decisions); (2) other transaction costs (the costs of applying for, managing, and administering grants); (3) opportunity costs of inefficient resource allocation, which arise when research resources are not used in the projects and programs with the highest social payoff; and (4) rent-seeking costs (costs of resources being spent wastefully in an attempt to redistribute grant resources).

Different research resource allocation processes will involve different amounts of particular types of costs. For instance, through the proposal process competitive grants generate information about research alternatives for decisionmakers. Although competitive grants may lower the cost of certain types of information, they also involve relatively high transaction costs. They might also involve relatively high rent-seeking costs (for instance, scientists do lobby for support). However, these additional costs may be justified if competitive grants lead to a lower overall social cost because they reduce the (opportunity) cost of resource misallocation. On the other hand, formula funds involve relatively high resource misallocation costs, which tend to get higher the longer a formula stays fixed (because circumstances change), and relatively low transaction costs. This is not to say the transaction costs are zero or that rent-seeking costs are zero with formula funds. There is a fair bit of bureaucracy associated with the administration of the funds; the formulas do, or at least may, change from time to time; and some resources are spent simply to preserve the status quo. Earmarked funds may involve the greatest rent-seeking and resource distortion costs, but they may also involve relatively small transaction costs. In

16. There can be substantial value added in the careful, thorough preparation of proposals reinforced by—and then evaluated through—critical (usually peer) review before funding is awarded.

short, the full costs should be considered when comparing research resource allocation procedures.

A middle ground is likely to be best for many situations: enough competition to ensure a vigorous and adaptable research program that optimally exploits the available information on scientific opportunity and economic implications; enough security and confidence in future funding so the scientists will take appropriate risks and pursue long-term opportunities; not too much cost in terms of the time scientists spend in drafting proposals, justifying expenditures, and reporting results; and not so narrow-minded so that curiosity and flair are stifled.

Such a Goldilockian optimum, with every element just right, may be hard to achieve. Part of the solution is likely to involve relatively long-term funding of particular people or research teams, rather than particular projects, based more on their past performance than their promised future research, perhaps especially for the more basic types of scientific work. Competition can be effective as a resource allocation and incentive mechanism without requiring a morass of planning processes and committees, which to some represent the antithesis of competition.

Positive Economics and Government Intervention

The discussion so far has been normative. It generally prescribes policies based on economic efficiency (interpreted broadly) as the criterion for government intervention in agricultural R&D. More detailed arguments justifying this approach have been spelled out elsewhere (for example, Alston, Norton, and Pardey 1995; Alston and Pardey 1996). But not even all economists accept the arguments for basing research resource allocation decisions solely on economic efficiency. Also, many research administrators and policymakers attempt to use agricultural research as an instrument to pursue a host of other, nonefficiency objectives.

The list of objectives that have been used for research is long. It often includes income distribution objectives, including an emphasis on benefits to the poor and malnourished, especially in less-developed countries (both in their own national agricultural research organizations and in international development agencies); urban versus rural people; farmers versus others; or people in particular regions. In some countries one objective for agricultural research has been to improve the balance of trade. In others preservation of the natural environment has been an objective for research in its own right, regardless of whether research will even achieve that outcome, or, in less extreme cases, regardless of the costs involved in using research policy rather than some other policy to pursue environmental objectives.

Thus, despite what may or may not be justified by economic arguments, accounting for and understanding changes in agricultural R&D policy must extend beyond a consideration of changes in the nature of market failures in research. In particular, a political-economy perspective suggests that to understand policy change it is necessary to consider not just net social benefits, but also who benefits from any change and who is powerful.[17] Also, it is important to account for the changing political influence of different interest groups over time.

In richer countries, such as those covered in this book, agriculture has been a protected sector—although to a lesser extent in Australia and New Zealand, which have relatively small and open economies and enjoy a relatively strong comparative advantage in agriculture. In all rich countries, the share of agriculture in the economy has been declining over time and the political power base of agriculture has been shrinking. This suggests a shrinking of support for R&D to the extent that it is seen as a transfer to agriculture rather than as a correction for market failure (see Alston and Pardey 1994, 1996). At the same time there have been some general shifts in the direction of smaller government roles, or at least reduced government spending and enhanced private-sector roles. These types of forces have been at least as important in dictating changes in public policy for agricultural R&D as changes in the nature and extent of market failures in research. In this chapter we have set aside these political-economy issues and concentrated on the implications of economic efficiency arguments for the role of government. In later chapters we return to some of the insights from political-economy models to help understand the causes of changes in agricultural R&D policies over time.

Conclusion

Market failure leads to private-sector underinvestment in agricultural R&D. This phenomenon can account for the major result from the empirical literature across different commodities and different countries that agricultural R&D has been, on average, a profitable investment from society's point of view. In turn, this suggests that research may have been underfunded and that current government intervention may be inadequate. However, this is not to say that the amount of government spending necessarily should increase.

Changes in government intervention can take many forms. Some commentators focus on increased funding of R&D from general government revenues, but this is only a part of the picture. Government can also act to change the incentives for others to increase their investments in private or public R&D (as well as what research is done, by whom, and how effectively). A premise that government intervention is inadequate implies simply that the nature of the

---

17. See Busch et al. (1991) for a sociological assessment of modern biotechnologies from some of these perspectives.

intervention should change to stimulate either more private investment or more public investment.

Policy options available to the government for stimulating private R&D activity include improving intellectual property protection; changing institutional arrangements to facilitate collective action by producers, such as establishing levy arrangements; and encouraging individual or collective action through the provision of subsidies (or tax concessions) or grants in conjunction with levies.

Intellectual property rights are applicable or enforceable only for certain types of inventions and come at the cost that privately optimal prices may exceed socially optimal prices. Commodity-specific levy arrangements are most applicable for commodity-specific R&D, which is of a relatively applied nature, although more general agricultural R&D could be funded by a more general agricultural levy, as in the Netherlands. In those cases where the fruits of invention are only partially appropriable, a case can be made for partial support from general government revenues through subsidies or matching grants in conjunction with commodity levies, as used in the Australian R&D corporations. To some extent the arrangements for financing agricultural R&D can be separated from who conducts the research and what research is undertaken and how the R&D process is managed. It is useful to consider these elements as separate issues, but inevitably they become intertwined.

In addition to efficiency gains from increasing the total R&D investment, the government can also intervene with a view to improving the efficiency with which resources are used within the R&D system. Changes over time in economic circumstances imply changes in R&D institutions. Some research activities that were once perceived clearly as the province of the government have become part of the private domain. Examples include much applied work on the production and evaluation of agricultural chemicals and new plant varieties. Governments can respond by privatizing certain public functions or institutions or, more gradually, by phasing itself out of the roles where the private sector is assuming an ever greater role. Some restructuring or consolidation of agricultural R&D institutions, in some instances on a geographic basis, is implied by the changing nature of the research being undertaken; its focus relative to agriculture, agribusiness, and the environment; and the spatial and economic applicability of the results, as well as by the changing nature of economies of size, scale, and scope in research. In addition to changes in the organization of research institutions, there is also scope for, and evidence of a growing tendency toward, more economic rationalism in the processes for managing research and allocating research resources and in the structure of incentives for scientists.

The optimal intervention by a government whose aim is to reduce distortions arising from inadequate private-sector incentives for agricultural R&D would

- optimize the total investment in public-sector agricultural R&D and the mix of R&D, while minimizing the distortions from crowding out private R&D;
- minimize the costs of raising revenues to finance public-sector R&D by using least-cost sources of funds;
- organize public-sector R&D institutions so that they can conduct R&D in the least-cost way, according to their comparative advantages, with a minimum of wasteful replication of facilities and programs, but taking account of synergism between public-sector R&D and other public-sector activities such as higher education; and
- allocate and use research resources efficiently among programs and projects (that is, according to economic criteria, not political criteria), minimize transaction costs and administrative and bureaucratic overhead, and allow decentralized decisionmaking where effective incentive mechanisms are possible.

These four points relate to economic efficiency of R&D in terms of the total funding, the sources of funds, institutional organization, and resource allocation and management.

An assessment of the evolution of agricultural R&D policy can evaluate qualitatively whether changes in policy appear to be steps in directions consistent with economically more efficient agricultural R&D, but quantitative assessments are unlikely to be feasible. A useful first step would be to compare policy choices against the decision tree in Figure 2.1. In prescribing agricultural R&D policies, an overriding criterion perhaps should be first to do no harm. Scientists and scientific institutions are complicated and in some senses fragile. Because it is not well understood how successful research happens, policymakers should exercise care and patience in modifying institutions and policies that have been successful in the past. However, inaction is not an option. As the external environment changes, policies and institutions must change too. The discussion of principles and practical implications provided in this chapter is meant to be useful in its own right for those who choose policies, but also valuable for assessing the recent policy changes discussed in Chapters 4–8.

3 Agricultural R&D Investments and Impacts

PHILIP G. PARDEY, JOHANNES ROSEBOOM, AND
BARBARA J. CRAIG

This chapter serves as background to the country case studies included in the following section. We present and discuss new data regarding agricultural R&D and productivity developments in the five countries examined in this book and place these developments in a broader perspective. The links between public investments in R&D and measured rates of productivity growth are briefly described, and productivity developments in developed countries over the past three decades (including evidence on the rates of return to R&D) are presented in summary form. To place agricultural research in a broader, science-policy perspective, we present summary data on overall trends in developed-country R&D before providing the stylized facts regarding funding trends for agricultural research.

Links Between R&D and Productivity

Economists define productivity as the quantity of output per unit of input. In this book, we approximate total factor productivity with total measured output divided by total measured inputs (that is, multifactor productivity or MFP). Growth in MFP is linked to such factors as education, R&D, investments in infrastructure, and changes in the quality of inputs (some of which is attributable to R&D).

To understand the consequences of domestic, public R&D for productivity, it is important to control for other sources of productivity growth (for example, foreign or private research or extension and investments in education and infrastructure) and to account for the effects of R&D-induced improvements in quality. This is not easy. Growth in measured MFP is not entirely attributable to public R&D and is not always a good thing.[1] However, many studies have identified public agricultural R&D as a primary source of measured growth in agricultural productivity with large social benefits.

1. Contemporary, measured growth in productivity may be at the expense of longer-run gains if it involves the consumption or permanent degradation of unmeasured biological, natural, and other inputs into agriculture, such as soil and water (see Alston, Anderson, and Pardey 1995).

Empirical Evidence on Productivity Growth

Many discussions of agricultural productivity use partial measures (output per unit of a particular input, rather than all inputs). For most discussions of productivity in relation to technical change, broader, multifactor indexes are more relevant. The partial productivity indexes provide information about changing factor mixes over time that are only partially attributable to new technology.

Partial Factor Productivity Measures

The most familiar partial productivity indexes are yields, which are given in terms of output per hectare or per worker. Empirical evidence on productivity growth in rich countries indicates that output has grown even though inputs of labor and land have not increased. In countries of the Organization for Economic Cooperation and Development (OECD), measured output per unit of labor grew, on average, by about 5.0 percent per year from 1961 to 1993 (Table 3.1).[2] The corresponding growth rate of output per unit of land was much lower—only 1.5 percent per year. The total agricultural labor force declined, while land use did not change much. Thus, the growth in output was accompanied by increases in land-to-labor ratios: hectares per unit of labor grew from 23.7 to 70.2 over the period 1961–1993.

These annual average growth rates mask a great deal of variation over time and across countries. Data for the five countries covered in this volume are included in Table 3.1. To illustrate the trends in land and labor productivity, we have also adopted the graphical technique used by Hayami and Ruttan (1985); in Figure 3.1 the log-transformed ratios of output per hectare and output per worker are graphed for the five countries and others.

Both of these partial productivity measures are constructed using measures of gross agricultural output based on the agricultural production indexes reported in FAO (1995). To put them in comparable units, the output indexes were scaled using agricultural purchasing power parity indexes (defined in international agricultural 1980 dollars) taken from FAO (1986). The land productivity measure relates gross output to the total hectares of land in agriculture, whether irrigated or nonirrigated cropland, pastureland, or rangeland. Labor productivity measures gross output relative to the economically active agricultural population, both male and female. Land- and labor-quality differences are ignored.

Starting in 1961 and ending in 1993, all of the productivity paths move in a northeasterly direction, a trend that indicates growing productivity. A longer productivity locus means a greater percentage change in productivity in that

2. In this chapter the OECD designation refers to the 22 (of the total of 29) countries now belonging to the Organization for Economic Cooperation and Development listed in Appendix Tables A3.1 and A3.2.

TABLE 3.1 Average annual output, input, and partial productivity growth in OECD agriculture, 1961–1993

	1961–1971	1971–1981	1981–1993	1961–1993
		(percentages)		
Agricultural output				
Australia	3.01	1.98	1.77	1.94
Netherlands	3.12	2.85	1.04	2.48
New Zealand	3.02	1.60	–0.05	1.46
United Kingdom	1.88	2.15	0.42	1.78
United States	1.74	2.52	0.47	1.51
Developed countries (21)[a]	*1.79*	*1.97*	*0.60*	*1.45*
Agricultural land				
Australia	0.45	–0.11	–1.49	–0.04
Netherlands	–0.69	–0.56	–0.10	–0.50
New Zealand	–0.10	1.04	–0.56	0.2g
United Kingdom	–0.48	–0.24	–0.40	–0.37
United States	–0.28	–0.12	–0.10	–0.10
Developed countries (21)[a]	*0.03*	*–0.04*	*–0.24*	*–0.07*
Agricultural labor				
Australia	–0.75	0.56	–1.49	–0.38
Netherlands	–3.26	–0.51	–3.04	–1.97
New Zealand	–0.27	1.23	–0.93	0.19
United Kingdom	–3.26	–0.12	–2.33	–1.59
United States	–2.70	0.24	–2.98	–1.48
Developed countries (21)[a]	*–3.90*	*–2.84*	*–3.62*	*–3.34*
Labor productivity				
Australia	3.80	1.41	3.31	2.33
Netherlands	6.59	3.38	4.21	4.54
New Zealand	3.30	0.36	0.90	1.27
United Kingdom	5.32	2.28	2.82	3.42
United States	4.57	2.27	3.56	3.04
Developed countries (21)[a]	*5.92*	*4.96*	*4.38*	*4.96*
Land productivity				
Australia	2.55	2.09	2.09	1.98
Netherlands	3.83	3.43	1.14	2.99
New Zealand	3.12	0.55	0.52	1.16
United Kingdom	2.37	2.40	0.82	2.16
United States	2.02	2.64	0.56	1.62
Developed countries (21)[a]	*1.76*	*2.01*	*0.84*	*1.52*

SOURCE: Authors' calculations based on data in FAO (1995).

[a]Weighted average. Includes all the countries listed in Appendix Table A3.1, except Iceland.

FIGURE 3.1 Land and labor productivity developments in OECD agriculture, 1961–1993

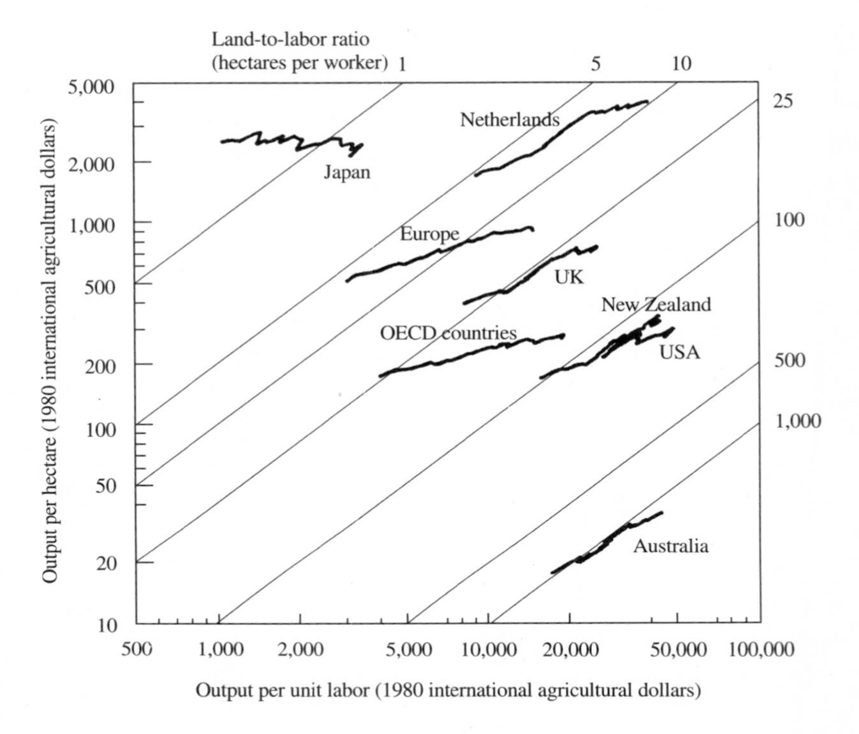

SOURCE: FAO data (1986, 1995).

country. Of the five countries, the Netherlands experienced the fastest rate of growth for both land (3.0 percent) and labor (4.5 percent) productivity. New Zealand had the lowest rates of growth for both measures (both are less than 1.3 percent per year), which are well below the corresponding OECD average growth rates.

The diagonal lines in Figure 3.1 indicate constant factor ratios. A productivity locus flatter than these diagonal lines indicates an increase in the number of hectares per worker in that country moving up the locus from left to right. In Australia and New Zealand the land-to-labor ratios have changed little, whereas in the Netherlands, United Kingdom, and United States, agriculture has become much more land intensive and less labor intensive, and thus land-to-labor ratios have increased.

Although measured levels of output per worker are not markedly different across countries at the end of the sample period (the high in the United States is only 1.92 times the low in the United Kingdom), output per hectare in the

Netherlands is 112 times higher than that in Australia. These productivity differences might be related to technological differences, but they are more clearly associated with substantial differences in the mix and quality of inputs—especially the natural fertility of the soil and climate. For example, in 1993 in the Netherlands there were 9.7 hectares per worker in agriculture, while in Australia there were more than 1,000 hectares per worker. Agriculture in the Netherlands also uses inputs other than labor more intensively per unit of land. For instance, fertilizer inputs per hectare on Dutch farms are among the highest in the world, whereas on Australian farms fertilizer intensity is among the lowest in rich-country agriculture. The fundamental reason is the difference in land quality, that is, considering both the soil and its associated climate and environmental conditions. Given the much higher average quantity of rain falling on the much better soil on Dutch farms, one would expect to get more output from a hectare in the Netherlands, and this expectation is reinforced by an induced greater intensity of labor and fertilizer inputs.

These partial factor productivity indexes provide insight into some aspects of agriculture in these countries but the indexes cannot adequately account for differences in levels or trends in productivity. To get a better picture of productivity developments that are more closely related to public R&D efforts, we turn to published MFP indexes.

Multifactor Productivity Measures

The MFP measures presented in Table 3.2 were constructed using a variety of techniques and data sources, an approach that introduces problems of comparability and interpretation. The baskets of goods included in output and the sets of measured inputs are not completely consistent across studies of different countries. The researchers who developed the MFP measures also did not use the same aggregation techniques. The quantitative significance of these differences is difficult to judge, but some likely biases should be kept in mind when comparing MFP growth rates across countries and methodologies.

The sets of goods included in output baskets differ across countries because, first and foremost, these countries have different product mixes. The value shares of crops (including horticulture) and livestock (including livestock products) have been fairly stable over time in these countries but differ markedly across countries. Livestock accounts for 71 percent of the value of U.K. agricultural output but only 52 percent of U.S. agricultural output. The livestock share in the Netherlands is closer to that of the United Kingdom, whereas Australia and New Zealand look more like the United States in output composition. Given these differences in output composition, the scope for technological change over time may be quite different in the various countries. The limited evidence currently available on U.S. R&D (Huffman and Evenson 1993) suggests that productivity change may be easier to promote in crops than

TABLE 3.2 Annual average growth rates in multifactor productivity

	1949–1954	1955–1959	1960–1964	1965–1969	1970–1974	1975–1979	1980–1984	1985–1990
				(percentages)				
Australia[a]	na	na	na	na	na	–2.09	7.69	2.89
Netherlands[b]	2.07	3.30	2.03	4.56	4.36	1.92	3.09	3.07
New Zealand	–0.64	3.49	0.80	1.22	2.56	–0.66	2.10	na
United Kingdom[c]	na	na	na	0.34	0.66	3.79	1.83	1.57
United States	1.91	1.67	1.10	2.27	3.52	1.31	4.02	0.47

SOURCE: See Table 3.3 for details.

NOTE: na indicates not available. Averages based on arithmetic averages of annual growth rates over period.

[a]Australian data begin in 1977.
[b]Netherlands data end in 1989.
[c]United Kingdom data begin in 1967.

in livestock. Knopke, Strappazzon, and Mullen (1995) also find evidence that productivity increases were larger for crop production than for livestock production in Australian broadacre agriculture, at least from 1977 to 1994.

Differences in input measurement as part of the MFP calculations are also substantial. The studies included here differed with respect to both the time period covered and the degree of detail on inputs and outputs. An overview of the differences is given in Table 3.3. The quantitative consequences of these differences are hard to predict, but some qualitative differences can be anticipated based on studies of U.S. agriculture (Craig and Pardey 1996a,b). Craig and Pardey's U.S. MFP indexes were constructed with far more detail than the other MFP indexes, especially within the broad categories of land, labor, and capital. Craig and Pardey (1996b) found that controlling for quality changes in agricultural land, labor, and capital inputs resulted in lower estimated productivity growth rates. These adjustments for changes in input quality are not mirrored in the other country studies.

The labor force in U.S. agriculture has moved toward fewer but better-educated farmers. The aggregation procedure in Craig and Pardey (1996a,b) treats farmers with different age and education profiles as different workers. Therefore, when an increase in output is paralleled by a shift from unskilled to more skilled workers, part of the increase in output is attributed to the use of a mix of labor of higher average quality. Total acres in U.S. agriculture changed very little over the period, but the use of irrigated cropland increased at the expense of lower-quality land.

The effects of quality adjustment on capital inputs are harder to anticipate because investment patterns cause capital service measures to be more changeable than either land or labor input measures. In U.S. agriculture adjusting for

TABLE 3.3 Comparison of methods and coverage in estimating multifactor productivities

Country	Study	Years	Outputs	Inputs	Aggregation Methodology
Australia	Knopke, Strappazzon, and Mullen (1995)	1977–1993	12 commodities • crops • livestock • other • broadacre agriculture is 50% total value	24 subgroups • labor, 4 • land, 1 • livestock, 3 • intermediate, 12 • capital, 4	Törnqvist-Theil divisia price indices Implicit quantity indices
The Netherlands	Rutten (1992)	1949–1989	2 commodities • crops (including horticulture) • livestock (including livestock products)	10 subgroups • intermediate, 6 • depreciation, 1 • land, 1 • labor (hours), 1 • capital, 1	Törnqvist-Theil divisia price indices Fisher fixed-base quantity indices
New Zealand	Scobie and Eveleens (1987)	1926–1984	Total value	Total value for 6 components • capital, 4 • labor, 1 • nonfactor, 1	Laspeyres fixed-base quantity indices
United Kingdom	Thirtle and Bottomley (1992)	1967–1990	4 commodities • crops • horticulture • livestock • livestock produce	10 components • intermediate, 5 • capital from outside, 1 • capital from inside, 2 • land, 1 • labor, 1	Törnqvist-Theil divisia quantity indices
United States	Craig and Pardey (1996b)	1949–1991	54 commodities • field crops, 14 • fruit and vegetables, 26 • livestock, 9 • nursery and greenhouse products, 1 • tree nuts, 3 • other, 1	58 subgroups • land, 3 • labor, 32 • capital, 12 • purchased inputs, 11	Törnqvist-Theil divisia quantity indices

quality of capital does not have the same effect on MFP growth rates in every time period. For example, from 1949 to 1959 the number of tractors on U.S. farms increased by nearly 4 percent per year. Taking the average horsepower of the tractors into account reduces the measured effective increase in tractor services by a full percentage point per year during that decade. In contrast with this earlier period, quality-adjusted tractor services increased at a much higher rate than tractor numbers did in the 1960s and 1970s because of a substantial increase in the average size of tractors purchased. Consequently, controlling for tractor quality results in MFP growth rates that are faster in the 1950s but slower in the 1960s and 1970s.

It is difficult to generalize about the likely direction and magnitude of bias in the MFP indexes reported for other countries that do not control for quality change in inputs, but it is possible to anticipate what would happen if more detailed data on input quality were available. Similar changes in the agricultural labor force have occurred in all these countries; consequently, the measured productivity growth rates probably overstate those that would be found if the studies had accounted for changes in the mix of labor. The quantity of land in agriculture has declined in all of these countries in the postwar period, but we suspect the use of a quality-adjusted land series would reduce the reported rate of decline and thus would reduce measured productivity growth rates.

Compared with land and labor, quality change in agricultural capital services is far less uniform over time and space in the United States; similar patterns are likely to be found in Australia, the Netherlands, New Zealand, and the United Kingdom. In all these countries, the long-term trend has been to substitute capital for labor. However, the irregularity of investment patterns means that there may be no uniform trend in capital quality. In years with low rates of investment, the average quality of the capital stock in use may decline when deterioration of aging machines is not compensated for by the purchase of new and improved machines. In the absence of information concerning the trend in quality-adjusted capital services, there is no simple way to predict the consequences for productivity measurement of controlling for quality change in capital inputs.

Returns to Research

Whether public-sector R&D expenditures generate returns that justify their claims on government budgets depends on the links between R&D spending and agricultural productivity and output growth. A large number of studies have used data on agricultural output and inputs and the effects of research on the input-output relationship to estimate the aggregate returns to research (see Alston, Marra, Pardey, and Wyatt 1998).

Studies have either based estimates of research benefits on MFP indexes or used the same data in a related dual or primal multivariate regression model of industry technology. Such estimates of returns to research are at best only as reliable as the underlying data. In either approach, omitting inputs or failing to adjust for changes in input quality results in a distorted picture of the consequences of R&D for the input-output relationship and returns to past R&D investments. If labor quality in agriculture has improved over time in ways not captured by the measure of the agricultural workforce, some of the increase in output will be attributed inappropriately to other variables included in a regression that are correlated with the unmeasured labor-quality information. Likewise, an MFP index will have a larger unexplained productivity increase when labor force measures do not reflect labor-quality improvements. In both cases, the research-spending variable may be credited with an impact on output that should more properly be attributed to other inputs.[3]

The empirical evidence suggests relatively high returns to public spending on agricultural R&D. Rates of return for OECD countries in excess of 40 percent are not uncommon. Alston and Pardey (1996) describe the likely bias in these rates of return but go on to argue that, although improved data and empirical techniques would reduce these estimated rates of return, any adjusted estimates would likely indicate a social rate of return to public agricultural research in rich-country agriculture that more than justifies the investment made in it.

Trends in Total R&D

Funding for agricultural R&D is increasingly intertwined with funding for research more generally, with concerns about the appropriate role of government in R&D and economies more broadly, and with significant shifts in the role of the public and private sectors in science. Thus, measuring R&D investments is more challenging as the institutions involved in funding, managing, and performing research evolve and become more complex. Distinguishing between those who fund R&D (and, perhaps, who manage the funds) and those who perform the research provides a clearer picture of changing institutional roles within programs of national research. It also facilitates international comparisons and allows for important policy insights to be drawn from the data.

The research expenditure series reported below are separated into public and private categories on the basis of research performers and, where possible, research funders. Care was taken to maintain a consistent institutional coverage over time and to make the data comparable across countries.[4] Although

3. Alston, Norton, and Pardey (1995) discuss approaches to measuring returns to research and the pitfalls associated with input quality adjustments.

4. Agricultural research is taken here to include crop, livestock, forestry, and fisheries research. See OECD (1994) for additional, definitional details related to the compilation of science indicators.

some unavoidable discrepancies remain, these data provide a reasonable basis for making broad international comparisons. The more detailed data reported in Chapters 4–8 provide for more complete insights into the complex and changing institutional arrangements surrounding the funding and conduct of agricultural R&D.

Public-Private Performance of R&D

Overall investment in R&D has been on the rise in most OECD countries over the past several decades, with the pace of privately performed R&D (increasing by 4.3 percent annually from 1981 to 1993) being greater than the growth in public R&D (3.5 percent; Table 3.4). The OECD countries spent a total of $286.6 billion (1985 international dollars) on all public and private R&D in 1993, which is 1.6 times the corresponding 1981 figure. Australia, the Netherlands, New Zealand, the United Kingdom, and the United States accounted for $151.9 billion (53 percent) of the 1993 OECD total. Notably, the rate of growth in both publicly and privately performed research has slowed in recent years. Nonetheless, by 1993 the private sector undertook $190.8 billion of research—about two-thirds of all the public and private R&D performed throughout the OECD that year.

Country data reveal a good deal of variation about the OECD average in the relative roles of the public and private sectors.[5] Most countries have a lower-than-average share of private-sector research (16 of 22), given the preponderance of private R&D in Japan and the United States, two countries that account for a significant share of total research in the OECD. Portugal, Greece, and New Zealand rank lowest in terms of their private shares (22, 27, and 30 percent, respectively, in 1993), whereas the United States, Sweden, and Switzerland rank highest (71, 70, and 70 percent, respectively).

Flow of Funds

Distinguishing between the sources of funds for R&D and those who spend those funds provides an even greater contrast in the public and private roles in research and gives some perspective on the imputed net flow of public funds to private R&D providers. In 1993 private agencies undertook $190.8 billion of R&D, whereas public agencies performed $95.8 billion of research (Table 3.4). Table 3.4 also reports a separate series on the amount of private and public funding for R&D. Public spending is considerably higher than the public provision of R&D services ($120.0 billion public dollars spent in 1993 versus $95.8 billion of research), whereas the private sector undertakes much more

5. The country case studies in Nelson (1993) provide comparative insights into developments regarding innovation systems in a number of important developed and some developing countries.

TABLE 3.4 Overall R&D by performer and source of funds in OECD countries

	1981	1986	1991	1993	Annual Growth 1981–1993
	(billions 1985 international dollars)				(percentages)
R&D performers					
Private	120.6	167.7	199.2	190.8	4.3
Public	64.4	78.1	91.3	95.8	3.5
Total	185.1	245.8	290.5	286.6	4.0
Source of funds					
Private	93.6	131.6	169.7	166.6	5.4
Public	91.6	114.3	120.8	120.0	2.4
Total	185.1	245.8	290.5	286.6	4.0
Imputed net flow of funds from public to private performers	27.0	36.2	29.5	24.2	na
Net flow of public funds to private performers as a share of private R&D			(percentages)		
	22.4	21.6	14.8	12.7	na

SOURCE: OECD (1991, 1995c, 1996).

NOTE: na indicates not available. Derived from gross expenditure on R&D (GERD) data included in OECD sources. Data include the 22 developed countries listed in Appendix Table A3.1.

R&D ($190.8 billion) than it funds ($166.6 billion). The imputed net flow of public funds to private research providers shows that a considerable amount of privately performed R&D is underwritten with public dollars.[6] However, the amount of public funds flowing to private R&D has declined in recent years—from a peak of $37.4 billion (1985 international dollars) in 1987 to $24.2 billion in 1993. The publicly funded share of privately performed R&D has declined steadily over the years, down to 12.7 percent in 1993 from a high of 22.4 percent in 1981.

The decline in public support for private research, especially over the recent past, is widespread throughout the OECD, and the magnitude of the

6. Unfortunately the data on public funding of R&D represent "direct" government expenditures on research and appear to exclude implicit transfers to the private sector through targeted tax concessions for R&D (OECD 1994:98).

decline corresponds to cutbacks in defense-related R&D. For example, France, the United States, and the United Kingdom—countries where the share of total public research dollars spent on defense research in 1985 was 67.5, 51.0, and 32.5 percent, respectively—account for the greatest share of the drop in public funds flowing to private R&D providers (for example, $31.9 billion of the $36.2 billion 1986 total in Table 3.4 came from these three countries, and they accounted for $21.0 billion of the $24.2 billion of public funds flowing to private R&D providers in 1993). Correspondingly, the public funds committed to defense R&D declined from a three-country total of $42.9 billion (1985 international dollars) in 1986 to $37.8 billion in 1993.

Overall Research Intensities

Figure 3.2 shows the intensity with which countries invest in R&D. In 1993 OECD countries collectively invested 2.59 percent of their gross domestic products (GDPs) in R&D, an increase of 0.35 percentage points over the corresponding 1981 figure but little changed since the mid-1980s. There is a substantial spread of research intensities around this OECD average; the over-time pattern varies among countries and so the rank order of countries has correspondingly changed over the years. In 1993 Sweden had the highest R&D intensity ratio (3.66 percent) and Greece the lowest (0.69 percent).

Among the five countries highlighted in this book, the United States committed 2.86 percent of its GDP to research while Australia (1.72 percent) and New Zealand (1.18 percent) had the lowest research intensities in 1993. Australia has shown a steady increase in its R&D intensity since 1981. New Zealand's intensity stagnated throughout much of the 1980s but began to gain ground in the 1990s. In contrast, the Netherlands, United Kingdom, and United States have all experienced declines in their respective R&D intensities since the mid-1980s. Indeed, according to these data, the intensity with which funds are invested in R&D in the United Kingdom is now less than it was in 1981.

Socioeconomic Objectives of Public R&D

There have been substantial shifts in public R&D spending priorities over the past several decades, which in some important instances reflect shifts in public funds earmarked for defense. Defense research has shrunk considerably as a share of public R&D spending in the United States and the United Kingdom during the 1990s but still accounts for more than 60 percent and 50 percent, respectively, of the central government's research budget in those two countries (Table 3.5). For most other OECD countries (aside from France and Sweden) defense research accounts for less than one-fifth of all public R&D spending; thus, other factors have played a greater role in accounting for changes in R&D priorities in those countries. Nevertheless, the share devoted to defense research has shrunk in Australia and hardly changed in New Zealand

FIGURE 3.2 Overall R&D intensities of OECD countries

SOURCE: Data from OECD (1991, 1995c, 1996) and the World Bank (1995).

and the Netherlands. Health research garnered a greater share of the public R&D purse in Australia, the United Kingdom, and especially the United States; obtained a shrinking share in the Netherlands; and changed little in New Zealand.[7]

Using a more inclusive measure of the socioeconomic orientation of public R&D, the lower section of Table 3.5 shows that 7.5 percent of total publicly performed R&D in OECD countries focused on agriculture in 1993. However, there are marked differences over time and across the five countries studied here in the agricultural orientation of their public research. In 1993, just over one-third of all publicly performed R&D in New Zealand was directed

7. These trends are based on the OECD's (1994) assessment of the socioeconomic objectives of the R&D spending reported in central government budgets. To the extent that other branches of government spend a considerable amount on R&D (for example, state governments in the United States), these trends will misrepresent overall public-sector spending patterns. Also, the reported figures do not fully differentiate the socioeconomic objectives of the public dollars spent in the "general university sector."

TABLE 3.5 Perspectives on publicly performed R&D

	1981	1986	1991	1993
		(percentages)		
Share of defense R&D in central government R&D expenditures[a]				
Australia	17.1	16.1	19.8	17.6
Netherlands	6.7	6.0	6.5	6.8
New Zealand	na	na	2.1	2.1
United Kingdom	58.9	58.3	56.1	53.6
United States	56.7	71.9	62.2	61.4
Share of health R&D in central government R&D expenditures[a]				
Australia	7.6	10.5	13.0	13.9
Netherlands	10.8	5.5	6.3	5.5
New Zealand	na	na	6.8	6.7
United Kingdom	5.0	5.4	7.6	9.1
United States	12.4	11.2	14.9	15.7
Share of agricultural R&D in total public R&D expenditures[b]				
Australia	19.9	18.0	14.1	14.8
Netherlands	15.4	13.8	13.7	14.1
New Zealand	49.1	49.4	46.7	35.5
United Kingdom	7.1	7.2	6.6	6.6
United States	6.2	5.5	5.7	5.6
Subtotal (5)[c]	*7.6*	*6.8*	*6.8*	*6.6*
Other OECD (17)[c]	10.3	9.6	8.5	8.3
Total OECD (22)[c]	*8.9*	*8.1*	*7.6*	*7.5*

SOURCE: See Table 3.4 and Appendix Table A3.1.

NOTE: na indicates not available.

[a]General university funds (GUF) and advancement of knowledge research can be a significant component of central government R&D spending but are not classified by socioeconomic objective. These estimates express central government R&D spending on the relevant socioeconomic objective (minus GUF and advancement-of-knowledge research) relative to total R&D spending by the central government. Using central government R&D spending minus corresponding GUF and advancement of knowledge research to form the spending shares changes the cross-country relativities a bit but has little effect on the pattern of change over time.

[b]These estimates are not directly comparable with the shares presented in the upper part of the table. They express all public agricultural R&D spending as a share of all government (central or otherwise) spending on R&D.

[c]Number of countries in respective totals.

toward agriculture. In the United States, agriculture accounted for less than 6 percent of all public research. Agriculture's share of total public R&D has shrunk most markedly in Australia and New Zealand, whereas in the Netherlands, United Kingdom, and United States the agricultural share of public research declined at a slower rate. This pattern of a steady decline in the share of public R&D funds directed toward agriculture is common throughout the OECD countries.

Agricultural R&D Investments

Agricultural research had its institutional beginnings around the middle of the nineteenth century when various agricultural experiment stations were established in France, Germany, and the United Kingdom—partly in response to the emergence of the formal agricultural sciences through the work and writings of Boussingault, von Liebig, and Lawes.[8] The practice of providing public funds to support national agricultural research agencies staffed with professional scientists (introduced for the first time in Möckern, Saxony, Germany, in 1852) spread rapidly throughout the Western and, eventually, the developing world. By 1875 more than 90 agricultural experiment stations had been established throughout Europe and soon spread to countries like Canada, the United States, and Japan (True and Crosby 1902; Grantham 1984).

By the turn of the century, the scientific foundations of agricultural R&D were becoming well developed. Darwin's theory of evolution, the pure-line theory of Johannson, the mutation theory of de Vries, and the rediscovery of Mendel's genetic laws all contributed to the rise of plant breeding. Pasteur's germ theory of disease and the development of vaccines opened up lines of research in the veterinary sciences. This provided the scientific basis for a steady expansion in the public commitment to agricultural R&D during the first half of the twentieth century.[9] The evolution of the science of genetics gathered pace around the middle of this century, with Hershey and Chase, Watson and Crick, and others uncovering the role and structure of DNA, findings that led directly to the modern biotechnologies based on recombinant DNA techniques, monoclonal antibodies, and new cell and tissue culture technologies. These changes paralleled efforts by governments to enact legislation that strengthened the intellectual-property protection applied to living organisms, such as new plant varieties and related genetic material, and to implement

8. One aspect that confounds international comparisons of agricultural R&D concerns differences in the meaning given to the term "agriculture." Agricultural production has become much more reliant on purchased chemicals, seeds, and machinery inputs, and a good deal of post-harvest processing and handling occurs beyond the farm gate. Many statistics fail to distinguish between farm-focused R&D and R&D directed toward the input supply, food, and fiber-processing sectors. In addition, as the environmental emphasis given to agricultural R&D has increased, nontraditional agencies have begun to carry out research of relevance to agriculture, while some agricultural research spills over to sectors beyond commercial agriculture (for example, pest- and weed-control methods from agriculture are used in urban gardens and golf courses).

9. These are but a few of the examples that underpinned technical progress in agriculture during the earlier part of this century. See Salmon and Hanson (1964) for the details of these discoveries.

a range of public science policies that stretched well beyond agriculture.[10] Together these developments fundamentally changed the nature of the agricultural sciences, the public and private roles, and the balance between locally provided and internationally traded R&D goods and services and continue to affect all these aspects of the agricultural sciences.

Public Agricultural Research in Developed Countries

GENERAL TRENDS. Publicly performed agricultural R&D in developed countries grew by 2.2 percent per year from $4.3 billion (1985 international dollars) in 1971 to $7.2 billion in 1993 (Table 3.6). During the 1970s real spending grew annually by 2.7 percent but slowed substantially during the period 1981–1993 to average only 1.8 percent per year. The five countries of focus here expanded over the whole period at a slower rate than the other OECD countries (1.8 percent per year compared with 2.5 percent).

Figure 3.3 gives shares of developed-country spending on public agricultural R&D in 1993. Country shares for many OECD countries were comparatively stable over the previous two decades, although Belgium, Germany, Ireland, New Zealand, and the United Kingdom had significantly smaller shares in 1993 than in 1971. Finland, Iceland, Italy, Norway, Portugal, and Spain increased their respective shares by more than 25 percent over this same period.

The public commitment to agricultural research varies markedly among countries. The United States invests more than any other country in the world in public agricultural R&D—over 430 times more than Iceland, the country with the smallest spending on agricultural research in our group of 22 developed countries.[11] Only 10 developed countries spend more than $200 million per year on agricultural research. With the exception of New Zealand, this group includes the countries examined in this book, as well as Canada, France, Germany, Italy, Japan, and Spain. Together these 10 countries account for nearly 90 percent of the developed world's public agricultural R&D spending and for 42 percent of the global total.

INSTITUTIONAL STRUCTURES. Public agricultural research systems vary in terms of who funds, manages, and performs the research. R&D is carried out by national, regional, and state government agencies (some with

10. For example, the United States has had plant patents for asexually reproduced cultivars since the 1930s and Plant Variety Protection Certificates since the 1970s, and utility patents for living organisms have been available since the landmark ex parte Hibberd decision of 1985. The agreement on Trade-Related Intellectual Property Protection (TRIPS)—specifically annex 1c of the 1995 Marrakesh Agreement establishing the World Trade Organization—mandates the extension of intellectual property protection to plants.

11. In terms of the number of researchers, China is the largest public agricultural research system in the world. Fan and Pardey (1992) report more than 55,000 scientists and engineers working in Chinese agricultural R&D in 1988, compared with nearly 10,800 full-time-equivalent scientists working for the USDA and SAESs in 1993.

TABLE 3.6 Public agricultural research expenditures in developed countries, 1971–1993

	1971	1981	1991	1993	1971–1981	1981–1993	1971–1993
					Annual Growth Rates		
	(millions 1985 international dollars)				(percentages)		
Australia	237.5	281.8	308.3	311.8	2.1	0.3	1.3
Netherlands	140.9	211.6	248.5	258.5	4.0	1.7	2.0
New Zealand	115.2	134.7	111.0	108.8	2.2	−2.2	0.2
United Kingdom	274.4	372.3	364.4	371.1	2.7	−0.2	1.2
United States	1,229.5	1,621.7	2,028.2	2,073.7	2.5	2.3	2.2
Subtotal (5)[a]	*1,997.5*	*2,622.1*	*3,060.4*	*3,124.0*	*2.6*	*1.5*	*1.8*
Other OECD (17)[a]	2,322.3	3,121.5	3,896.1	4,044.5	2.8	2.0	2.5
Total OECD (22)[a]	*4,319.8*	*5,743.6*	*6,956.5*	*7,168.5*	*2.7*	*1.8*	*2.2*

SOURCE: See Appendix Table A3.1.

[a]Number of countries in respective totals.

FIGURE 3.3 Shares of developed-country spending on public agricultural R&D, 1993

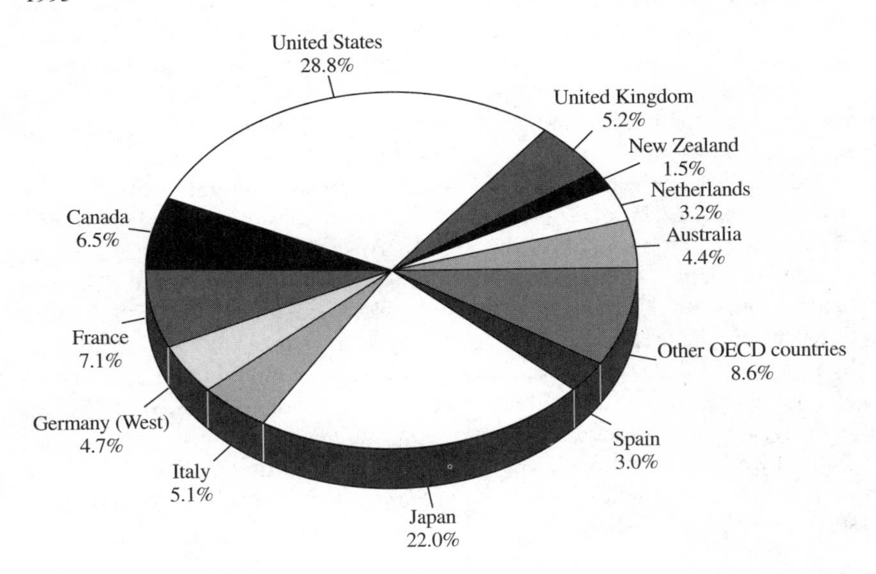

SOURCE: See Appendix Table A3.1.

roles extending well beyond research and well beyond agriculture), some semipublic entities, and various university faculties and departments.

The five countries reviewed in detail in this volume have quite diverse institutional setups. The government agencies in the Netherlands, New Zealand, and the United Kingdom are mostly administered by national departments, whereas in Australia and the United States public agricultural R&D is supported by a mix of state and national institutions. In Australia the tendency has been to rationalize state government agencies in recent years, which, combined with cutbacks in rural R&D by the principal national research organization (Commonwealth Scientific and Industrial Research Organization [CSIRO]), has shifted the balance of effort to the universities. The United States has national government facilities (the U.S. Department of Agriculture, USDA) with the state-level R&D being carried out in experiment stations located in land-grant colleges. The U.S. trend has involved relatively less spending in national laboratories and relatively more by state agricultural experiment stations (SAESs).[12]

Public agricultural R&D has undergone various degrees of restructuring in recent years. In the Netherlands, New Zealand, and the United Kingdom, there has been a substantial revamping of the public agencies involved in carrying out research. In England and Wales, as well as the Netherlands, the agencies that perform public R&D have been consolidated and, in some important instances, commercialized. In New Zealand a number of new Crown Research Institutes were established in an effort to develop a public market for R&D services that distinguishes between buyers and sellers of such services. There has been comparatively little structural change in the public agencies that undertake agricultural R&D in Australia and the United States, although the funding and management structures within Australia have been radically revised in recent years. Considerable details regarding these institutional changes are provided for each country in Chapters 4–8 and are summarized in Chapter 9.

The role of universities as providers of public agricultural research varies widely across OECD countries. In 1993, universities accounted for less than 15 percent of public agricultural research in Australia, New Zealand, and the United Kingdom; 74 percent in the United States; and 28 percent in the Netherlands. Sweden and Germany were the only OECD countries besides the United States where the preponderance of public agricultural research was performed by universities, although university research also accounted for sizable shares of publicly performed agricultural R&D in Denmark, Greece, Italy, Japan, Norway, and Switzerland. On average the university share of public agricultural R&D expenditures crept up only slowly over time (Table 3.7). However, for the Netherlands, United Kingdom, and United States the university share grew quite substantially.

---

12. In 1971 US$2.06 was spent in the SAESs for every USDA dollar; in 1993 the ratio was US$2.86 for every USDA dollar.

TABLE 3.7 University share of public agricultural R&D spending

	1971	1981	1991	1993
		(percentages)		
Australia	10.7	8.2	11.5	11.9
Netherlands	14.1	21.0	27.8	28.0
New Zealand	13.0[a]	13.0[a]	17.8	13.6
United Kingdom	2.3	2.9	9.7	14.7
United States	67.3	67.5	74.0	74.1
Subtotal (5)[b]	*44.8*	*45.4*	*54.3*	*54.9*
Other OECD (15)[b]	32.4	30.3	32.5	32.6
Total OECD (20)[b]	*38.6*	*37.8*	*42.8*	*43.1*

SOURCE: See Appendix Table A3.1.

[a]Estimated.
[b]Number of countries in respective totals.

Private Agricultural Research in Developed Countries

Table 3.8 provides our current best estimates of privately provided agricultural R&D and the respective private share of total agricultural R&D spending from 1981 to 1993. A common perception is that agricultural research is primarily the domain of the public sector, whereas research in other sectors of the economy is the province of the private sector. However, these new data reveal that privately performed R&D has been a prominent feature of contemporary agricultural R&D in rich countries. Indeed, the private share has trended up significantly since 1981, and now almost half the OECD's agricultural R&D is performed by the business sector. Privately performed agricultural R&D totaled $7 billion in 1993 compared with $4 billion in 1981; this reflects an annual rate of growth of 5.0 percent compared with 1.8 percent for publicly performed agricultural R&D and 4.3 percent for all private research in the OECD.

The relative importance of private R&D in total agricultural R&D varies across the OECD countries. In the United Kingdom, the business sector performs more than 60 percent of the agricultural research, and in the Netherlands and the United States the private share is now in excess of 50 percent. Australia and New Zealand still have significantly smaller private involvement in agricultural research, but privately performed R&D is rapidly becoming a major component of the national agricultural research effort in both countries.

Private and public laboratories do different types of R&D (Table 3.9). Approximately 12 percent of private research focuses on farm-oriented technologies, whereas more than 80 percent of public research has that orientation. Food and other post-harvest research accounts for 30–90 percent of private

TABLE 3.8 Privately performed agricultural R&D

	1981	1986	1991	1993	Annual Growth 1981–1993
	(millions 1985 international dollars)				(percentages)
Privately performed agricultural R&D					
Australia	25.2	68.4[a]	112.6[a]	137.3	15.1
Netherlands	183.7	210.7	241.9	281.3	3.7
New Zealand	9.9	12.2	26.9	40.0	13.8
United Kingdom	414.6	471.8	593.5	614.8	4.8
United States	1,417.9	1,966.7	2,261.4	2,391.5	4.3
Subtotal	*2,051.2*	*2,729.9*	*3,236.2*	*3,464.9*	*4.6*
Other OECD (17)[a]	1,963.9	2,691.5	3,406.2	3,543.8	5.4
Total OECD (21)[a]	*4,015.2*	*5,421.3*	*6,642.4*	*7,008.8*	*5.0*
Privately performed agricultural R&D as a percentage of total agricultural R&D					
	(percentages)				
Australia	8.2	17.9	26.8	30.3	
Netherlands	46.5	49.1	49.3	52.1	
New Zealand	6.8	9.1	19.5	26.9	
United Kingdom	52.7	55.7	62.0	62.4	
United States	46.6	52.1	52.7	53.6	
Subtotal	*43.9*	*49.1*	*51.4*	*52.6*	
Other OECD (17)[b]	38.6	43.5	46.6	46.7	
Total OECD (22)[b]	*41.1*	*46.1*	*48.8*	*49.4*	

SOURCE: Alston and Pardey (1996); Mullen, Lee, and Wrigley (1996a); OECD (1991, 1996).

NOTE: Data derived mainly from OECD's intramural business sector series. Calculated as the sum of R&D performed by industries classified as "agriculture, forestry, and fisheries," "food and beverages," and 10 percent of total research performed by the "chemical and pharmaceutical industries." The agricultural chemical series was constructed using an approximate rule-of-thumb procedure that was chosen after consulting various other relevant sources. For comparability purposes, agricultural mechanization R&D was excluded from the U.S. private sector series.

[a]Estimated.

[b]Number of countries in respective totals.

agricultural R&D, and in countries like Australia, Japan, the Netherlands, and New Zealand it is the dominant focus of privately performed research related to agriculture. Chemical research is of comparatively minor importance in Australia and New Zealand but accounts for more than 40 percent of private research in the United Kingdom and the United States and nearly 75 percent of private agricultural research in Germany.

There is a clear concentration of certain lines of private R&D in particular countries. Japan, the United States, and France account for 33, 27, and 8

TABLE 3.9 Focus of public and private intramural agricultural R&D, 1993

	Publicly Performed		Privately Performed		
	Agriculture	Food and Kindred Products	Agriculture	Food and Kindred Products	Animal Health and Agricultural Chemicals
	(millions 1985 international dollars)				
Expenditures					
Australia	na	na	36.6	87.3	13.4
The Netherlands	225.3	33.2	52.7	163.8	64.8
New Zealand	87.7	19.7	4.8	34.6	0.6
United Kingdom	325.6	45.1	106.3	211.4	297.1
United States	na	na	316.2	821.2	1,254.0
Subtotal (5)[a]	*na*	*na*	*516.7*	*1,318.4*	*1,629.9*
Other OECD (16)[a]	na	na	300.4	1,736.1	1,448.5
Total OECD (21)[a]	*na*	*na*	*817.1*	*3,054.4*	*3,078.4*
	(percentages)				
Shares of respective subtotals					
Australia	na	na	26.7	63.6	9.8
The Netherlands	87.2	12.8	20.2	58.2	23.0
New Zealand	81.7	18.3	12.1	86.5	1.4
United Kingdom	87.8	12.2	17.3	34.4	48.3
United States	na	na	13.2	34.3	52.4
Subtotal (5)[a]	*na*	*na*	*15.1*	*38.0*	*47.0*
Other OECD (16)[a]	na	na	8.6	49.8	41.6
Total OECD (21)[a]	*na*	*na*	*11.8*	*43.9*	*44.3*

SOURCE: See Table 3.8 and OECD (1996).

NOTE: na indicates not available. OECD 21-country total excludes Switzerland owing to lack of data.

[a]Number of countries in respective totals.

percent, respectively, of all food-processing research carried out by the private sector in the OECD. Chemical research related to agriculture is even more concentrated: the United States, Japan, and Germany represent 41, 20, and 10 percent, respectively, of all reported private-sector research.[13] This pattern of concentration of private agricultural research would be unlikely to change significantly if counterpart research in developing countries was also considered.

13. These data exclude Switzerland, whose share of agricultural chemical R&D is likely to be substantial but unlikely to place it in the top three performers.

FIGURE 3.4 Focus of private agricultural R&D in the United States

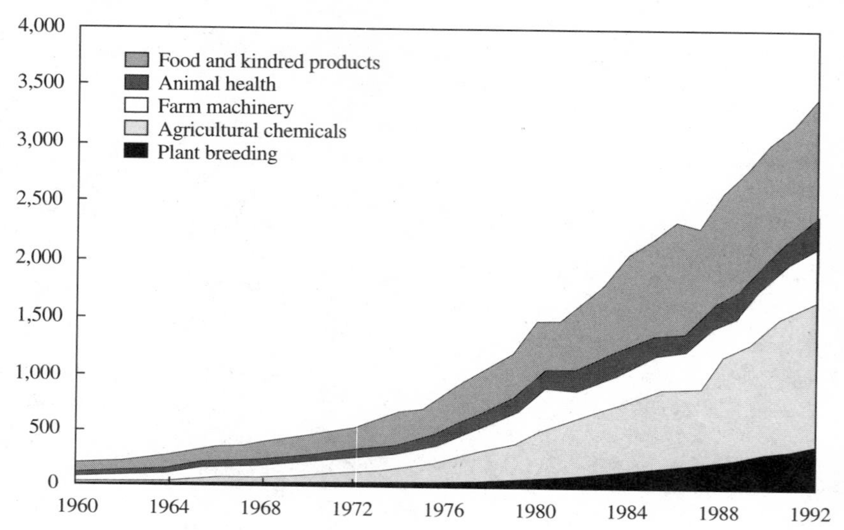

SOURCE: USDA, ERS (1995) revised data.

Contemporary, cross-country evidence on private R&D is revealing, but for the United States more detailed time-series data are available. The focus of private research in the United States has changed considerably in the past several decades (Figure 3.4). In 1960 agricultural machinery and post-harvest food-processing research accounted for more than 80 percent of total private agricultural R&D. By 1992 these areas of research collectively accounted for only 42 percent of the total, with the share of total private research directed toward agricultural machinery having declined from 36 percent in 1960 to less than 12 percent. Two of the more significant growth areas in private R&D have been plant breeding and veterinary and pharmaceutical research. Spending on agricultural chemicals research grew most quickly and now accounts for more than one-third of total private agricultural R&D. These data reveal a dramatic shift in the pattern of publicly and privately performed agricultural R&D in the United States over the past three decades. These patterns of change seem likely to persist; if they do, they will have major implications for the structure, conduct, and content of public research at both state and federal levels.

Public Agricultural Research Intensities

An alternative perspective on agricultural R&D spending is provided by the agricultural research intensities presented in Table 3.10. The most commonly

TABLE 3.10 Total public R&D versus public agricultural R&D intensities

	1971	1981	1991	1992	1993
			(percentages)		
Total publicly performed R&D relative to total GDP					
Australia	na	0.84	1.00	0.98	0.91
The Netherlands	na	0.95	0.99	0.99	0.99
New Zealand	na	0.86	0.71	0.83	0.82
United Kingdom	na	1.07	0.86	0.87	0.86
United States	na	0.78	0.84	0.84	0.83
Subtotal (5)[a]	*na*	*0.82*	*0.85*	*0.85*	*0.84*
Other OECD (17)[a]	na	0.74	0.84	0.86	0.89
Total OECD (22)[a]	*na*	*0.78*	*0.85*	*0.86*	*0.87*
Publicly performed agricultural R&D relative to agricultural GDP					
Australia	2.88	3.02	4.09	3.82	3.66
The Netherlands	2.26	3.32	3.11	3.42	3.92
New Zealand	3.46	4.03	3.25	3.03	3.09
United Kingdom	2.26	3.34	3.16	3.19	2.90
United States	1.61	1.64	2.30	2.22	2.45
Subtotal (5)[a]	*1.88*	*2.03*	*2.58*	*2.51*	*2.69*
Other OECD (16)[a]	1.25	1.71	2.37	2.53	2.81
Total developed countries (21)[a,b]	*1.48*	*1.84*	*2.46*	*2.52*	*2.75*
Total developing countries	*0.37*	*0.51*	*0.50*	*na*	*na*

SOURCE: See Appendix Table A3.1, OECD (1991, 1995c, 1996), World Bank (1995), and data used by Alston, Pardey, and Roseboom (1998).

NOTE: na indicates not available.

[a]Number of countries in respective totals.
[b]Excludes Switzerland owing to lack of agricultural GDP data.

constructed intensity ratios express public agricultural research expenditures as percentages of agricultural output measured in terms of agricultural GDP.[14] In 1993, developed countries as a group spent $2.75 on public agricultural R&D for every $100 of agricultural output, a sizable increase over the $1.48 they spent per $100 of output two decades earlier.

14. It would also be desirable to calculate intensities involving private research, but there are serious problems in obtaining commensurable measures of agricultural output. Private agricultural R&D spans economic activity in the input supply and post-harvest sectors that are excluded from value-added measures of agricultural output like the agricultural GDP figures used.

Agricultural research intensities vary substantially among the developed countries included in Table 3.10. For example, the United States had the lowest public agricultural research intensity ratio (2.45 percent) in 1993 and the Netherlands (3.92 percent) had the highest. Nonetheless, among developed countries, agricultural research intensities vary much less than the amount spent on agricultural R&D, which varies by a factor of 430. When differences in the sizes of agricultural sectors are taken into account, cross-country differences in spending are substantially reduced but by no means eliminated. The remaining differences are due to a host of factors, including the policy stance each country takes in public support of its agricultural sector and economies of scale and scope in agricultural R&D programs.[15]

Table 3.10 also provides comparisons of agricultural research intensities (publicly performed agricultural R&D relative to agricultural GDP [AgGDP]) and *total* R&D intensities (total publicly performed R&D relative to GDP). In developed countries agricultural R&D intensities consistently exceeded total R&D intensities; in 1993, the intensity with which OECD governments invested in agricultural research was about three times the corresponding public intensity for all R&D. The data also show that this difference in intensity widened considerably over the past two decades. Therefore, the relative decline in the economic importance of the rural sector has not been matched by a corresponding decline in public support for rural research. This may reflect differences in the political clout of the rural sector, but it may also reflect differences in the appropriate public research roles in the rural and nonrural sectors.

The other striking feature of Table 3.10 is the marked difference in public agriculture research intensities among developed and developing countries. Developed-country governments invest much more intensively in their agricultural sectors than do developing countries, and this gap has widened considerably over the past two decades.

Two alternative research intensity or spending ratios are reported in Table 3.11. One measures agricultural R&D spending relative to the size of the economically active agricultural population and the other relative to total population. In 1993 developed countries spent $466 (1985 international dollars) per agricultural worker, an increase of more than three times the corresponding 1971 figure. Research spending per capita has risen too, from $6.13 per capita in 1971 to $8.83 in 1993 (a 38 percent increase).[16]

Global Agricultural R&D Patterns

The international context for agricultural R&D has changed substantially during the past few decades. Global investments in public agricultural research

15. See Alston and Pardey (1994) for a discussion of these issues in this context.

16. Roe and Pardey (1991) report comparable measures for developing countries and use these indicators to discuss the political-economy aspects of public investment in agricultural R&D.

TABLE 3.11 Intensity ratios for publicly performed agricultural research, 1971–1993

	1971	1981	1991	1993
		(1985 international dollars)		
Agricultural R&D spending per economically active agricultural worker				
Australia	539	609	771	824
Netherlands	446	712	1,134	1,261
New Zealand	880	916	822	824
United Kingdom	384	535	660	710
United States	331	434	729	798
Subtotal (5)[a]	*376*	*491*	*748*	*814*
Other (17)[a]	86	167	311	350
Total (22)[a]	*133*	*239*	*418*	*466*
Agricultural R&D spending per capita				
Australia	18.2	18.9	17.8	17.9
Netherlands	10.7	14.9	16.5	16.9
New Zealand	40.4	43.1	32.8	31.5
United Kingdom	4.9	6.6	6.3	6.4
United States	5.9	7.1	8.0	8.1
Subtotal (5)	*6.8*	*8.2*	*8.8*	*8.9*
Other (17)[a]	5.6	7.1	8.5	8.8
Total (22)[a,b]	*6.1*	*7.6*	*8.7*	*8.8*

SOURCE: Appendix Table A3.1, FAO (1995), and World Bank (1995).
[a]Number of countries in respective totals.
[b]Includes the countries listed in Appendix Table A3.1.

performed by national agencies have more than doubled since the early 1970s—from $7.3 billion (1985 international dollars) in 1971 to about $15.0 billion in 1991 (Table 3.12).[17] For all regions of the world, however, real R&D spending grew at a much slower pace during the 1980s than in the 1970s. As a group, in 1971 developed countries accounted for 59 percent of the spending (Figure 3.5). By 1991 the situation had changed markedly. Developed-country

17. These "global" totals are preliminary estimates that exclude Eastern European and former Soviet Union countries. The principal data source for the 1961–1985 period is Pardey, Roseboom, and Anderson (1991). These data were revised and updated for African countries using various ISNAR *Statistical Briefs;* for most of the principal Asian countries (including China and India) with data from Pardey, Roseboom, and Fan (1997); for Latin America with data from Cremers and Roseboom (1997); and for the developed countries with data from Appendix Table 3.1. Semiprocessed data from numerous other sources were obtained for most of the mid-sized to larger NARS and a number of smaller systems. The less-developed countries for which we have direct estimates are thought to account for about 85 percent of the less-developed country total.

TABLE 3.12 Global trends in public agricultural research expenditures

	1971	1981	1991
	(millions 1985 international dollars)		
Expenditures			
Developing countries (131)[a]	2,984	5,503	8,009
Sub-Saharan African (44)[a]	699	927	968
China	457	939	1,494
Asia and Pacific, excluding China (28)[a]	861	1,922	3,502
Latin America and Caribbean (38)[a]	507	981	944
West Asia and North Africa (20)[a]	459	733	1,100
Developed countries (22)[a]	4,320	5,744	6,956
Total (153)[a]	*7,304*	*11,247*	*14,966*
	1971–1981	1981–1991	1971–1991
		(percentages)	
Average annual growth rates			
Developing countries	6.4	3.9	5.1
Sub-Saharan African	2.5	0.8	1.6
China	7.7	4.7	6.3
Asia and Pacific (excluding China)	8.7	6.2	7.3
Latin America and Caribbean	7.0	–0.5	2.7
West Asia and North Africa	4.3	4.1	4.8
Developed countries	2.7	1.7	2.3
Total	*4.3*	*2.9*	*3.6*

SOURCE: For developing countries, IFPRI-ISNAR Agricultural Science and Technology Indicators (ASTI) database; for developed countries, Appendix Table A3.1.

NOTE: Regional groupings of countries are given in the appendix table in Pardey, Roseboom, and Anderson (1991); the Sub-Saharan Africa group reported here includes South Africa. The data from the series published in 1991 form the basis for the revised and updated ASTI data reported in the current table.

[a]Number of countries in the respective totals.

R&D spending had slipped to less than half (46 percent) of public-sector R&D spending worldwide. In 1991, Asian countries accounted for 62 percent of developing-country expenditures (19 percent for China alone), West Asian and North African countries for 14 percent, and Latin America and Caribbean countries and Sub-Saharan African countries (including South Africa) each accounted for about 12 percent.

International Agricultural Research

There is some internationally funded and conceived agricultural research that complements the work of national research agencies. This international R&D represents a relatively recent institutional innovation. The first such venture

FIGURE 3.5 Regional shares of global agricultural R&D spending, 1971 and 1991

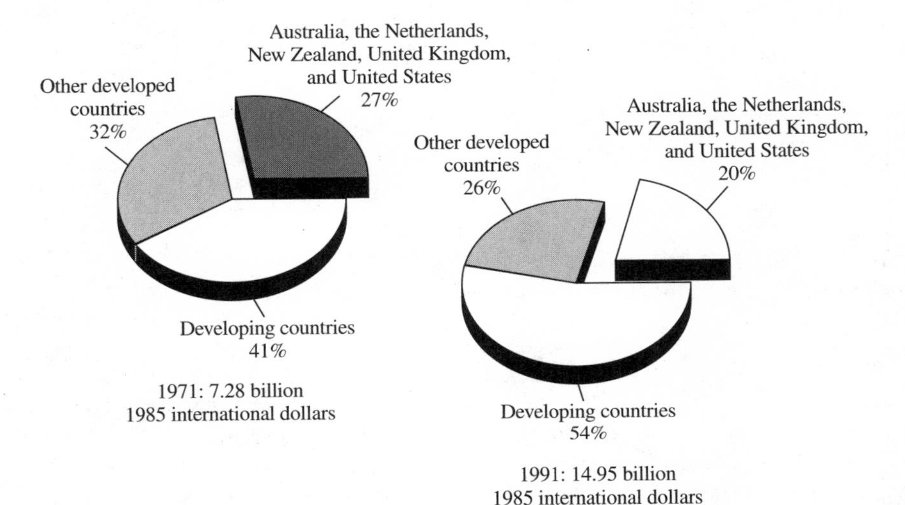

SOURCE: See Table 3.12.

was the cooperative Mexican government–Rockefeller program established in 1943 to conduct wheat research. This effort became the model for many of the subsequent programs in international agricultural research, and it later evolved into the Centro Internacional de Mejoramiento de Matz y Trigo (CIMMYT). Another notable example was the initiative of the Rockefeller and Ford Foundations to establish the International Rice Research Institute (IRRI) at Los Baños, the Philippines, in 1960. Other international centers were established at Ibadan, Nigeria (International Institute of Tropical Agriculture [IITA]), in 1967 and at Cali, Colombia (Centro Internacional de Agricultura Tropical [CIAT]), in 1968.

The further development of international agricultural research centers took place largely under the auspices of the Consultative Group on International Agricultural Research (CGIAR), which was established in 1971.[18]

18. For more details on institutional developments related to the CGIAR see Baum (1986) and Anderson and Gryseels (1991). Anderson and Gryseels (1991) identify an additional 17 multilateral agricultural research agencies that were operating outside the CGIAR system in the mid-1980s. In addition, L'Institut Français de Recherche Scientifique pour le Développement en Coopération (ORSTOM) and Centre de Coopération Internationale en Recherche Agronomique pour le Développement (CIRAD) are two French-funded institutions that undertake research of direct relevance to tropical and subtropical agriculture in developing countries. In 1995 these two agencies collectively spent $158 million (1985 international dollars) on agricultural R&D.

FIGURE 3.6 Consultative Group on International Agricultural Research (CGIAR) expenditures, 1972–1997

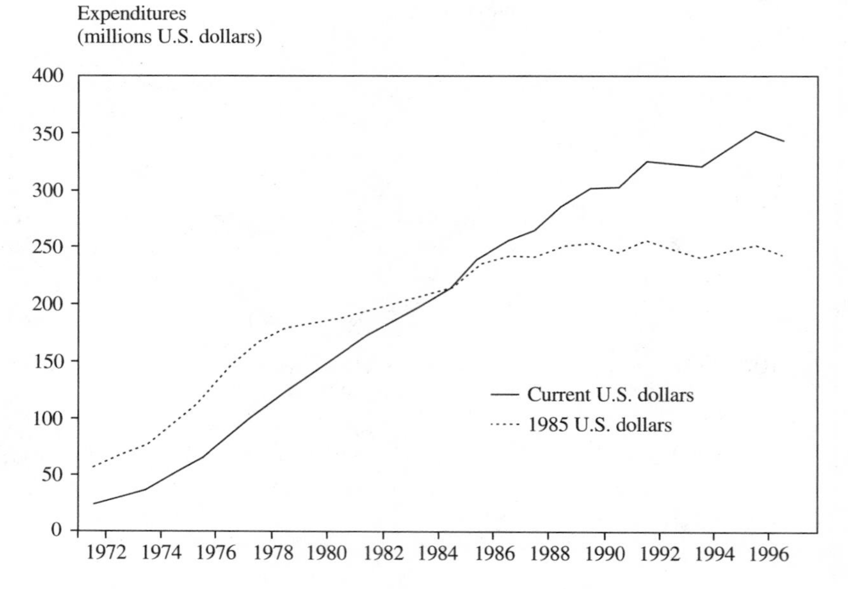

SOURCE: Appendix Table A3.3.

Figure 3.6 shows overall funding trends for the CGIAR (data for each institution are included in Appendix Table A3.3). Funds for the CGIAR system come mainly from the foreign aid (not agricultural R&D) budgets of developed countries given directly to the international centers or via contributions to agencies such as the World Bank, the Inter-American Development Bank, the Asian Development Bank, and the European Union.

The chronology of the CGIAR system involved three phases. The first was the pre-CGIAR period when the four founding centers (CIMMYT, IRRI, IITA, and CIAT) developed independently. In the second phase, which lasted from 1971 to the early 1980s, the geographical, commodity, and research focus of the CGIAR broadened; the number of centers increased from 4 to 13; and real spending grew rapidly (14.5 percent per year from 1972 to 1981). In the third phase, which began in 1991, five more centers were added (consolidated to 16 centers by 1996) and the research mandate continued to broaden to include additional commodities and a greater emphasis on environmental R&D. However, real spending stagnated after 1991 and decreased by 0.66 percent per year from 1990 to 1997. This recent trend was an exacerbation of a slowdown that had begun in the mid-1980s. Between 1985 and 1991, real expenditures grew by just 2.2 percent per year, so the addition of new centers involved cutbacks in contributions to the existing centers. In 1991, the latest year for which comparable figures are available, the CGIAR spent a total of US$246.3 million (1985 prices), which represents about 1.6 percent of the global spending and 3.1 percent of the developing-country spending on public agricultural research.

Conclusion

Agricultural productivity growth has been substantial, but the measures of productivity presented in this chapter indicate that it is far more erratic than R&D spending patterns. Because productivity measures are not direct proxies for technological change, linking R&D expenditures and growth in agricultural output is not straightforward. However, the available empirical evidence does suggest that there is a robust, positive relationship between R&D and productivity growth. Translating this statistical relationship into an economically meaningful measure of the returns to research is also difficult, but the existing evidence suggests that the returns to public R&D in agriculture have been high—certainly high enough to justify the investments that have been made.

In the following chapters, how this empirical evidence could inform policy decisions and institutional designs will be discussed for each of the five countries surveyed. Separate chapters are presented for the United States (Chapter 4), Australia (Chapter 5), the United Kingdom (Chapter 6), the Netherlands (Chapter 7), and New Zealand (Chapter 8). These chapters explore in more detail the patterns of R&D investments and the associated institutions.

In this chapter, a number of stylized facts regarding public and private R&D investments, in general, and agricultural R&D, in particular, have been developed and presented. We found the following global trends:

- For all regions and many countries in the world, growth in overall public investments in agricultural research slowed substantially during the 1980s compared with the 1970s (from 6.4 percent per year during 1971–1981 to 3.8 percent during 1981–1991 for developing countries and from 2.7 percent to 1.7 percent for developed countries).
- As a group, developing countries spent more on public agricultural R&D in 1991 ($8.0 billion in 1985 international dollars) than developed countries ($6.9 billion), which is a reversal of the relative shares that prevailed only a decade earlier.
- The growth of investments in international agricultural research has slowed too, declining by 0.7 percent per year since 1990 compared with 3.6 percent growth throughout the 1980s.
- The gap between agricultural research intensity ratios (that is, public agricultural R&D spending relative to agricultural GDP) in developed

and developing countries widened considerably since 1971; by 1991 developed-country intensities averaged 2.46 percent compared with 0.50 percent for developing countries.

The following are trends in developed countries:

- The OECD countries as a group spent a total of $286.6 billion (1985 international dollars) on all public and private R&D in 1993; about two-thirds ($190.8 billion) of this research was performed by the business sector. There was a substantial net flow of public funds to private research providers, although the rate of flow has diminished in recent years in line with cutbacks in public funding for defense research.
- Public R&D spending priorities vary widely across countries. The share of central government R&D spending earmarked for defense research is down in most countries, and the share of all public R&D funds going to agriculture shrank steadily from 8.9 percent in 1981 to 7.5 percent in 1993; Australia and New Zealand experienced a marked decline, while the rate of decline in the Netherlands, United Kingdom, and United States was much more modest.
- Public plus private agricultural R&D spending in the OECD totaled $14.2 billion (1985 international dollars) in 1993, up from $9.8 billion in 1981.
- Privately performed R&D has been a prominent feature of contemporary agricultural research in rich countries. Private R&D totaled $7 billion (1985 international dollars) in the OECD, accounting for about half of all agricultural research spending in 1993, and it grew at an annual average rate of 5.0 percent from 1981 to 1993.
- Private and public laboratories focus on different research. Less than 15 percent of public research is on food and kindred products, but 30–90 percent of private research has that orientation. More than 40 percent of private agricultural R&D in the United States and the United Kingdom focuses on chemical products.
- Across the developed world, public spending on agricultural R&D grew by 2.7 percent per year from 1971 to 1981 (slowing substantially to 1.8 percent per year from 1981 to 1993) to total $7.2 billion in 1993; just 10 developed countries (Australia, Canada, France, Germany, Italy, Japan, the Netherlands, Spain, the United Kingdom, and the United States) now account for 90 percent of the developed world's public spending on agricultural research and 44 percent of global spending.

APPENDIX TABLE A3.1 Publicly performed agricultural R&D, 1971–1993

	1971	1972	1973	1974	1975	1976	1977	1978	1979	1980
	(millions 1985 international dollars)[a]									
Australia	237.5	239.8	227.0	230.7	255.2	257.1	256.6	320.5	255.3	268.3
Austria	*19.7*	*18.9*	*18.1*	*17.3*	16.4	20.8	20.3	20.6	21.4	21.7
Belgium	31.4	33.3	*35.3*	*37.2*	39.1	*41.0*	*42.9*	*44.7*	*46.6*	48.4
Canada	354.7	315.7	265.8	322.4	346.9	331.1	331.2	411.8	436.0	436.4
Denmark	35.2	34.2	32.6	31.2	29.8	29.3	28.9	28.2	27.9	34.0
Finland	23.4	25.5	26.3	24.8	28.3	30.9	37.5	32.9	37.5	37.7
France	295.7	275.7	315.1	326.4	340.1	355.2	359.6	357.8	383.2	394.7
Germany	308.6	335.8	337.3	353.2	341.4	326.9	309.7	300.7	291.6	301.5
Greece	*24.7*	*27.8*	*30.9*	*34.1*	*37.2*	*40.2*	43.1	*43.2*	44.1	34.0
Iceland	3.3	*3.3*	*3.4*	*3.4*	3.5	*4.1*	4.8	*5.2*	5.7	*5.4*
Ireland	29.7	31.8	32.6	29.4	33.8	35.6	38.9	35.6	35.9	28.6
Italy	67.7	85.0	109.2	73.1	81.1	83.6	90.7	111.7	101.2	124.3
Japan	926.2	991.3	1,028.1	1,029.6	1,034.1	1,033.3	1,078.5	1,119.3	1,204.0	1,203.2
Netherlands	140.9	157.0	165.7	178.9	190.9	197.9	203.8	218.5	206.3	210.6
New Zealand	115.2	89.9	101.5	117.3	119.0	108.4	107.7	123.8	114.4	123.1
Norway	34.0	39.8	41.4	42.5	47.7	53.7	54.6	60.0	59.8	56.2
Portugal	28.9	22.5	*24.4*	*26.3*	*28.2*	30.3	*30.5*	31.9	*31.1*	*30.2*
Spain	*47.4*	*50.8*	59.1	60.6	*58.4*	*64.7*	*67.4*	88.4	90.3	92.7
Sweden	57.3	66.4	80.9	96.6	110.3	*99.5*	88.8	87.5	90.5	85.7
Switzerland	34.4	38.7	42.9	46.8	32.7	36.3	29.0	*30.0*	31.1	33.9
United Kingdom	274.4	286.4	288.5	312.9	324.4	332.1	323.9	330.9	332.7	353.8
United States	1,229.5	1,362.1	1,386.6	1,387.6	1,396.2	1,583.8	1,545.5	1,570.7	1,578.9	1,583.8
Total	*4,319.8*	*4,531.9*	*4,652.4*	*4,782.0*	*4,894.9*	*5,095.8*	*5,093.9*	*5,373.7*	*5,425.6*	*5,508.5*

(continued)

APPENDIX TABLE A3.1 *Continued*

	1981	1982	1983	1984	1985	1986	1987	1988	1989	1990
					(millions 1985 international dollars)[a]					
Australia	281.8	294.2	316.7	290.3	299.6	313.5	292.1	282.5	269.1	289.4
Austria	*23.0*	*23.8*	*24.6*	*25.4*	26.2	*26.0*	25.7	*25.4*	*25.1*	*25.1*
Belgium	*47.2*	*45.3*	45.9	46.7	48.1	45.9	46.7	41.8	41.3	40.3
Canada	451.5	467.0	519.0	537.2	544.2	506.3	420.6	423.0	432.8	443.7
Denmark	38.4	41.5	41.1	41.3	41.9	45.3	48.2	54.4	59.5	58.8
Finland	39.1	42.1	39.5	40.7	43.1	48.0	50.9	50.8	52.8	55.1
France	409.5	418.5	434.8	433.0	443.7	433.0	442.8	432.1	431.4	449.2
Germany	299.8	301.7	307.0	307.0	312.5	304.7	296.9	302.9	308.8	311.6
Greece	49.8	37.1	37.1	51.9	57.3	52.3	47.7	47.4	48.6	43.2
Iceland	5.2	*5.0*	4.7	*5.2*	5.6	*6.1*	6.6	*7.4*	8.2	8.1
Ireland	33.9	33.8	28.2	28.5	26.0	23.4	25.1	26.5	23.8	25.3
Italy	187.9	198.3	220.4	212.9	266.2	316.0	305.3	352.7	322.4	346.5
Japan	1,232.1	1,230.5	1,249.4	1,273.7	1,265.1	1,268.1	1,281.6	1,290.0	1,338.8	1,411.9
Netherlands	211.6	217.4	214.9	217.7	213.8	218.4	224.4	223.0	234.2	240.5
New Zealand	134.7	135.2	130.9	130.2	*125.9*	*122.3*	*118.3*	*114.6*	*111.2*	107.3
Norway	58.3	63.1	66.0	65.4	65.3	71.6	74.8	74.2	83.0	86.2
Portugal	*29.4*	28.5	*28.8*	29.1	*32.1*	35.0	38.0	41.5	50.6	49.3
Spain	98.6	103.5	103.2	120.9	127.3	138.9	158.7	175.9	189.4	200.6
Sweden	81.2	92.0	101.6	112.0	118.8	132.9	139.9	142.9	135.4	134.5
Switzerland	36.6	*40.1*	43.6	*44.7*	*45.9*	47.0	*47.4*	*47.8*	*48.1*	*48.5*
United Kingdom	372.3	382.1	372.0	369.6	370.3	374.6	356.6	357.7	360.1	356.5
United States	1,621.7	1,657.5	1,667.1	1,678.7	1,760.1	1,804.8	1,860.2	1,907.2	1,956.8	1,995.6
Total	*5,743.6*	*5,858.1*	*5,996.6*	*6,062.3*	*6,239.2*	*6,334.1*	*6,308.5*	*6,421.7*	*6,531.3*	*6,727.2*

	1991	1992	1993	Annual Growth 1971–1991	Annual Growth 1981–1993
	(millions 1985 international dollars)[a]			(percentages)	
Australia	308.3	312.1	311.8	2.1	0.3
Austria	*25.1*	*25.1*	*25.1*	2.2	1.1
Belgium	40.9	40.3	36.2	4.5	–1.9
Canada	471.7	450.2	466.6	4.1	–1.0
Denmark	58.3	*58.3*	*59.2*	–0.3	4.3
Finland	65.2	67.3	64.2	5.7	4.9
France	462.0	493.6	509.7	3.6	1.2
Germany	*314.4*	*317.2*	*320.1*	–1.3	0.7
Greece	39.8	29.5	30.7	5.9	–2.3
Iceland	8.6	9.4	*9.8*	6.2	7.8
Ireland	*25.3*	*25.3*	*25.3*	1.0	–1.5
Italy	368.2	360.9	*360.9*	6.8	6.1
Japan	1,472.6	1,471.9	1,548.4	2.8	2.1
Netherlands	248.5	255.8	258.5	4.0	1.7
New Zealand	*111.0*	106.5	108.8	2.2	–2.2
Norway	90.5	103.4	101.2	6.0	5.0
Portugal	58.4	*53.5*	*63.6*	2.4	7.6
Spain	*212.6*	*225.2*	*238.6*	7.2	7.7
Sweden	133.5	135.8	*135.2*	2.4	4.0
Switzerland	*48.9*	49.3	*49.7*	–2.0	2.1
United Kingdom	364.4	382.3	371.1	2.7	–0.2
United States	2,028.2	2,060.9	2,073.7	2.5	2.3
Total	*6,956.5*	*7,034.0*	*7,168.5*	*2.7*	*1.8*

(*continued*)

APPENDIX TABLE A3.1 *Continued*

SOURCE: Australia: Mullen, Lee, and Wrigley (1996a), Chapter 5, this volume; Austria: OECD (1985, 1995c); Belgium: OECD (1974, 1979, 1995c), Grauls (1987); Canada: OECD (1987a, 1997); Denmark: Forskningssekretariatet (1986), Forsknings-og Teknologiministeriet (1993), OECD (1995c); Finland: OECD (1987a, 1991, 1995c), Central Statistical Office of Finland (various years); France: INRA (1986a,b, 1988, 1991, 1994); Germany (West): Bundesminister für Forschung und Technologie (1979), Rost (1988, 1993), OECD (1991); Greece: OECD (1981, 1987a, 1995c); Iceland: OECD (1972, 1974, 1979, 1983, 1987b, 1991, 1995c); Ireland: OECD (1974, 1979, 1987a, 1995c); Italy: Galante and Sala (1988), Sala (1993); Japan: Boyce and Evenson (1975), OECD (1979, 1987b, 1995c), Ohashi (1987), Business Bureau of Agriculture, Forestry, and Fishery Technical Conference (1994); the Netherlands: Roseboom and Rutten (1995); New Zealand: National Research Advisory Council (various years), MRST (1993, 1995, 1996); Norway: Sundsbo and Villa (1987), Vaage and Bjorgum (1988), OECD (1995c); Portugal: Junta Nacional de Investigacao e Tecnologica (1981, 1986, various undated issues), OECD (1991, 1995c); Spain: INE (1978, 1986, various undated issues), Herruzo, Fernandez, and Echeverría (1993); Sweden: Croon (1986, 1988), OECD (1987a,b, 1995c); Switzerland: Bundesamt für Statistik (1982, 1983), OECD (1987a), Buri, Suarez, and Walder (1988), Kurath (1994); United Kingdom: Chapter 6, this volume; United States: Alston and Pardey (1996).

NOTE: Italicized figures are interpolated or constructed.

[a]Research expenditures in current, local currency units were first deflated to base year 1985 using local, implicit GDP deflators from the World Bank (1995) and then converted to 1985 international dollars using the corresponding purchasing power parity indices reported in Summers and Heston (1995, Mark 5.6 diskette version). See Pardey, Roseboom, and Craig (1992) for more discussion on this method of currency conversion.

APPENDIX TABLE A3.2 Privately performed agricultural R&D, 1981–1993

	1981	1982	1983	1984	1985	1986	1987	1988	1989	1990	1991	1992	1993	Annual Growth 1981–1993
	(millions 1985 international dollars)[a]													(percentages)
Australia	25.2	31.0	36.7	42.5	55.5	68.4	79.3	90.2	90.6	91.0	112.6	136.9	137.3	15.1
Austria	19.8	20.5	21.2	22.0	21.7	21.4	21.2	20.9	20.6	20.6	20.6	20.6	20.6	–0.1
Belgium	65.1	69.5	73.9	73.8	73.7	78.9	83.8	89.2	91.6	95.8	100.0	100.0	100.0	3.9
Canada	75.5	86.4	81.6	83.5	91.6	100.6	97.3	97.9	93.6	98.9	103.2	106.0	118.7	2.9
Denmark	25.7	28.1	30.5	33.6	36.7	39.0	41.3	43.7	45.6	46.7	47.9	48.3	49.7	5.7
Finland	22.9	27.4	31.9	32.1	32.4	31.1	29.8	40.1	50.4	61.3	72.1	61.4	50.7	8.7
France	255.6	267.9	271.4	326.8	352.8	390.2	412.1	444.6	479.6	507.3	510.1	540.3	572.4	7.2
Germany (West)	426.1	433.2	440.2	459.5	478.8	492.3	505.7	515.4	525.0	522.6	520.2	489.5	458.9	1.3
Greece	2.8	3.5	4.2	4.9	5.6	6.2	5.7	5.2	4.3	5.4	6.4	8.6	10.8	7.8
Iceland	0.1	0.3	0.5	0.5	0.4	0.5	0.6	1.5	2.3	2.3	2.3	2.4	4.5	30.6
Ireland	15.1	15.4	14.0	13.5	17.5	20.9	21.6	21.5	21.9	23.5	32.6	40.4	48.3	10.5
Italy	90.6	96.2	95.1	93.8	110.1	111.0	134.0	141.1	144.2	148.8	151.3	159.3	157.2	5.5
Japan	801.8	887.6	942.6	1,010.5	1,082.2	1,145.5	1,344.1	1,396.5	1,506.3	1,610.0	1,559.4	1,587.1	1,639.7	6.7
Netherlands	183.7	177.5	181.9	183.4	206.9	210.7	228.1	230.8	260.1	247.5	241.9	237.8	281.3	3.7

(*continued*)

APPENDIX TABLE A3.2 *Continued*

	1981	1982	1983	1984	1985	1986	1987	1988	1989	1990	1991	1992	1993	Annual Growth 1981–1993
	(millions 1985 international dollars)[a]													(percentages)
New Zealand	9.9	9.8	9.8	10.6	11.4	12.2	13.0	19.6	26.1	26.0	26.9	33.8	40.0	13.8
Norway	6.2	6.3	7.2	7.7	8.9	12.6	16.2	18.1	19.9	19.0	18.1	20.5	22.9	13.3
Portugal	1.7	1.9	3.1	4.2	4.1	4.0	4.2	4.4	4.8	5.2	5.3	5.4	5.4	8.2
Spain	30.2	37.0	34.1	41.5	49.3	62.8	75.7	90.3	82.3	84.2	86.5	94.5	94.5	10.9
Sweden	73.5	81.2	88.9	96.3	103.7	102.9	102.1	105.6	109.1	103.9	98.6	106.5	114.3	2.7
Switzerland	51.2	51.2	50.0	57.4	62.8	71.6	73.3	76.3	78.8	75.8	71.7	72.4	75.3	3.8
United Kingdom	414.6	389.9	365.1	415.3	465.1	471.8	489.7	495.3	577.7	623.0	593.5	641.3	614.8	4.8
United States	1,417.9	1,536.9	1,620.1	1,795.5	1,863.3	1,966.7	1,879.9	2,084.6	2,141.4	2,223.0	2,261.4	2,391.5	2,391.5	4.3
Total (22)	*4,015.2*	*4,258.6*	*4,403.8*	*4,808.8*	*5,134.4*	*5,421.3*	*5,658.5*	*6,032.7*	*6,376.2*	*6,641.8*	*6,642.4*	*6,904.7*	*7,008.8*	*5.0*

SOURCE: See Table 3.4 and, for the United States, USDA, ERS (revised 1995) and Alston and Pardey (1996). All other OECD countries (except Switzerland, for which there are no data available), OECD (1991, 1996), unpublished OECD data files (February 1997). Calculated as total privately performed R&D as a percentage of gross expenditure on R&D.

NOTE: Data derived from OECD's intramural business sector series. Calculated as the sum of R&D performed by industries classified as "agriculture, forestry, and fisheries," "food and beverages," and 10 percent of total research performed by the "chemical and pharmaceutical industries." The agricultural chemical series was clearly constructed using an approximate rule-of-thumb procedure that was chosen after consulting various other relevant sources. For comparability purposes, agricultural mechanization R&D was excluded from the U.S. private sector series.

[a]See notes to Appendix Table A3.1 for details of currency conversion method.

APPENDIX TABLE A3.3 Consultative Group on International Agricultural Research (CGIAR) funding by center

	IRRI[a]	IITA[a]	CIMMYT[a]	CIAT[a]	CIP[a]	ICRISAT[a]	ILRI[a]	WARDA[a]	IPGRI[a]	ICARDA[a]	IFPRI[a]	ISNAR[a]	ICLARM[a]	ICRAF[a]	IWMI[a]	CIFOR[a]	Total
								(millions current U.S. dollars)									
1960	7.405	—	—	—	—	—	—	—	—	—	—	—	—	—	—	—	7.405
1961	0.188	—	—	—	—	—	—	—	—	—	—	—	—	—	—	—	0.188
1962	0.434	—	—	—	—	—	—	—	—	—	—	—	—	—	—	—	0.434
1963	0.872	—	—	—	—	—	—	—	—	—	—	—	—	—	—	—	0.872
1964	0.566	—	—	—	—	—	—	—	—	—	—	—	—	—	—	—	0.566
1965	1.014	0.250	—	—	—	—	—	—	—	—	—	—	—	—	—	—	1.264
1966	1.131	0.350	0.506	—	—	—	—	—	—	—	—	—	—	—	—	—	1.987
1967	1.202	1.000	1.179	—	—	—	—	—	—	—	—	—	—	—	—	—	3.381
1968	2.414	2.624	1.768	0.207	—	—	—	—	—	—	—	—	—	—	—	—	7.013
1969	2.524	4.722	2.964	1.429	—	—	—	—	—	—	—	—	—	—	—	—	11.639
1970	2.913	4.440	5.107	2.312	—	—	—	—	—	—	—	—	—	—	—	—	14.772
1971	3.689	6.818	6.113	3.563	—	—	—	—	—	—	—	—	—	—	—	—	20.183
1972	4.482	6.410	6.606	4.466	0.500	0.400	—	—	—	—	—	—	—	—	—	—	22.864
1973	4.636	6.448	7.662	6.410	1.300	2.700	—	—	—	—	—	—	—	—	—	—	29.156
1974	7.884	7.253	7.527	6.102	2.217	3.800	1.029	0.475	—	—	—	—	—	—	—	—	36.287
1975	10.546	9.832	9.155	6.713	2.886	6.203	3.880	0.555	0.487	—	—	—	—	—	—	—	50.257
1976	12.302	11.131	10.946	7.010	4.740	7.372	9.008	0.900	0.915	1.377	—	—	—	—	—	—	65.701
1977	15.419	12.776	11.404	10.230	5.931	11.240	12.116	1.300	1.258	4.708	—	—	—	—	—	—	86.420
1978	15.772	17.412	13.925	13.006	5.842	14.131	15.396	1.900	1.715	7.609	—	—	—	—	—	—	106.708
1979	18.672	19.535	16.667	15.182	7.450	13.534	16.372	2.848	2.369	10.578	1.925	—	—	—	—	—	125.132
1980	21.053	19.231	18.345	17.333	8.123	14.419	19.127	3.456	3.047	13.116	2.456	1.200	—	—	—	—	140.906
1981	22.645	22.784	20.292	18.933	9.618	15.720	20.515	3.488	3.568	15.784	3.174	1.610	—	—	—	—	158.131

(continued)

APPENDIX TABLE A3.3 *Continued*

	IRRI[a]	IITA[a]	CIMMYT[a]	CIAT[a]	CIP[a]	ICRISAT[a]	ILRI[a]	WARDA[a]	IPGRI[a]	ICARDA[a]	IFPRI[a]	ISNAR[a]	ICLARM[a]	ICRAF[a]	IWMI[a]	CIFOR[a]	Total
								(millions current U.S. dollars)									
1982	25.170	27.300	20.440	21.610	9.970	19.590	19.480	5.190	3.090	15.700	3.920	2.910	—	—	—	—	174.370
1983	24.620	26.010	20.570	23.130	10.960	20.940	21.320	6.360	4.520	20.500	4.960	4.110	—	—	—	—	188.000
1984	26.920	27.850	24.860	24.030	11.750	21.110	23.770	5.880	4.200	20.880	5.880	4.300	—	—	—	—	201.420
1985	31.330	33.520	27.210	23.280	11.050	24.410	25.240	4.850	4.510	21.770	6.300	4.650	—	—	—	—	215.770
1986	29.810	36.410	27.210	24.330	16.431	31.630	29.180	6.940	5.030	21.830	7.150	6.130	—	—	—	—	242.081
1987	32.750	35.840	28.344	29.520	15.250	41.010	28.430	5.950	5.120	23.630	6.000	5.600	—	—	—	—	257.444
1988	30.990	37.820	33.170	28.310	17.760	35.430	32.260	5.450	6.250	23.340	8.400	6.300	—	—	—	—	266.480
1989	36.540	32.500	33.650	31.290	22.010	38.780	33.900	6.800	7.510	24.430	11.080	9.260	—	—	—	—	288.750
1990	41.730	35.210	32.990	32.740	21.930	39.110	37.210	8.340	7.450	23.230	12.450	11.260	—	—	—	—	304.650
1991	38.693	34.331	34.403	34.117	23.569	36.523	35.049	13.696	8.064	22.000	13.451	10.793	—	—	—	—	304.689
1992	41.690	35.662	31.702	32.176	21.744	32.933	32.861	10.130	12.306	20.586	13.381	10.678	6.102	13.052	9.126	—	326.129
1993	44.774	34.330	32.759	33.336	21.527	31.799	26.041	9.090	13.620	21.150	12.463	10.359	7.169	13.777	8.875	2.400	323.469
1994	40.000	33.800	29.000	35.100	22.400	29.600	23.900	8.700	16.300	22.700	13.100	10.500	6.500	16.700	8.800	4.800	321.900
1995	38.120	33.200	27.100	34.600	24.000	33.400	25.700	9.200	19.600	23.400	13.800	11.500	7.100	16.800	9.400	8.900	338.000
1996	40.030	37.180	30.240	36.800	26.060	31.070	25.900	9.800	19.970	23.200	16.200	11.200	8.600	17.400	10.180	9.680	353.510
1997	34.970	31.860	30.400	33.300	25.530	27.520	26.700	9.200	19.580	27.600	18.100	10.400	8.500	22.200	10.080	10.600	346.540

SOURCE: Authors' calculations based on financial data files obtained from the CGIAR Secretariat, June 1998.

NOTES: — indicates not applicable. Funding includes all unrestricted (that is, core) and earmarked (that is, complementary, restricted core, and special project) support. ILRI includes funding for ILRAD and ILCA, which were formed in 1973 and 1974, respectively, and merged operations in 1995. IPGRI includes funding for INIBAP beginning 1992.

[a]IRRI, International Rice Research Institute; IITA, International Institute of Tropical Agriculture; CIMMYT, Centro Internacional de Majoramiento de Maíz y Trigo; CIAT, Centro Internacional de Agricultura Tropical; CIP, Centro Internacional de la Papa; ICRISAT, International Crops Research Institute for the Semi-Arid Tropics; ILRI, International Livestock Research Institute; WARDA, West Africa Rice Development Association; IPGRI, International Plant Genetic Resources Institute; ICARDA, International Center for Agricultural Research in the Dry Areas; IFPRI, International Food Policy Research Institute; ISNAR, International Service for National Agricultural Research; ICLARM, International Center for Living Aquatic Resources Management; ICRAF, International Centre for Research in Agroforestry; IWMI, International Water Management Institute; CIFOR, Center for International Forestry Research.

4 Agricultural R&D Investments and Institutions in the United States

JULIAN M. ALSTON, JASON E. CHRISTIAN, AND
PHILIP G. PARDEY

In this chapter we review the history, institutions, and recent trends in the funding and performance of agricultural R&D in the United States. Changes in the sources of research funds and allocation mechanisms for these funds are considered. Important issues include the balance between state funds and federal funds and among federal appropriations for intramural research, formula-fund support of specified institutions, contracts, competitive grants, and special appropriations in support of specific institutions and projects. On the expenditure side, our analysis covers the distribution of funding between more-basic and applied research and between research within the U.S. Department of Agriculture (USDA) agencies and in the state agricultural experiment stations (SAESs) and at other institutions. We also consider the distribution of funding across disciplines and by commodity orientation and changing patterns of direct support for scientific personnel as opposed to infrastructure investments in buildings and equipment. Agricultural research funded and performed by private companies is also identified and discussed.

The chapter is organized as follows. We first review the development of U.S. agricultural R&D policies and institutions over the past century, paying particular attention to developments of the past 50 years. We next describe the main R&D performing institutions in the current federal and state agricultural research systems and the various federal funding mechanisms. Then we examine funding trends, including a review of aggregate trends and a detailed examination of the changing funding mechanisms for research in the SAESs. Finally we analyze spending patterns by region, type of expenditure (spending per scientist), scientific nature (basic, applied, or development), and disciplinary orientation.

U.S. Agricultural Research Policy: A Historical Overview

Since the USDA and the land-grant colleges were founded over a century ago, U.S. public-sector agricultural science has evolved into a major enterprise. An important element has been the longstanding close association and integration of agricultural research with extension and higher education.[1] The 1862 Act of Establishment created the USDA and emphasized the discovery and diffusion of knowledge as its primary function. Policymakers thus created a substantial scientific institution, ultimately organized as the Agricultural Research Service (ARS) within the USDA. In 1862 Congress also passed the first Morrill Land Grant College Act. Thus, the formalization of intramural research within the USDA was accompanied by the introduction of the land-grant colleges, which would eventually become the main providers of extramural research funded by the USDA.

Extramural Research and the SAESs

Federal funding of extramural USDA research was initiated through the provisions of the 1887 Hatch Experiment Station Act, which authorized annual appropriations to SAESs "established under the direction" of the land-grant colleges. The two streams of agricultural research—intramural USDA laboratories and the SAESs—have since developed in parallel, with some tension in the USDA's dual role as research funder and research performer. The history of USDA support for extramural research is almost entirely the history of USDA support for research carried out at the SAESs, at least until the recent introduction of the National Research Initiative (NRI), which provides funds for agricultural research irrespective of the institutional affiliation of supported researchers.

The Smith-Lever Act of 1914 established a federal role in agricultural extension, which previously had been provided only by state and local governments. The act introduced formula funding and required that federal funds be matched by state appropriations (innovations that would later be adopted for research as well as extension). The formula established the amount of federal support for extension according to each state's share of the total U.S. rural population. Formula funding for research has also evolved over time, so that now a part of the Hatch funds are divided equally among states, part is distributed on the basis of rural population, and part is based on farm population.[2]

The overall structure of federal policy toward agricultural science and technology changed little between the Smith-Lever Act of 1914 and the New Deal legislation of the 1930s, although the Purnell Act of 1925 authorized a substantial increase in federal funding of the SAESs. The political environment

1. These events have been documented in much greater detail by Moore (1967), Rasmussen and Baker (1972), and Kerr (1987), among others. More recent studies include the National Research Council, Board on Agriculture (1994, 1995, and 1996) and Congress of the United States, OTA (1995). Huffman and Evenson (1993) and Alston and Pardey (1996) provide economists' perspectives.

2. "Hatch funds" refers generally to federal formula funds for SAES agricultural research, reflecting the origins of the system in the Hatch Act of 1887, as modified and extended by subsequent legislation. Text of much of this legislation is reprinted in Kerr (1987).

of the period 1916–1919 was dominated by World War I and its immediate aftermath. The 1920s was a decade of low commodity prices and, consequently, reduced payoffs to increased productivity in agriculture. During the first part of the period, the American farm sector benefited from strong demand for agricultural products; World War I decimated agricultural production in Europe, and the North American farm sector was called upon to make up the balance. The value of U.S. farm sales increased from about US$87 billion (in 1993 dollars) in 1914 to US$129 billion in 1918. However, once the war ended, markets for U.S. agricultural commodities contracted radically. Farm sales declined to US$65 billion in 1921, about 50 percent less than in 1918. Although there was some recovery in commodity prices in the latter part of the 1920s, farm sales remained well below US$100 billion and during the early 1930s declined again to just over US$50 billion in 1932.

The 1935 Bankhead-Jones Act was the major component of the New Deal that directly addressed research in the SAESs. The act's provisions included a number of important innovations. First, it created a "special research fund" controlled by the secretary of agriculture to support several regional research centers. These centers, which received 40 percent of the Bankhead-Jones funding, were to be (and still are) managed by the USDA and are properly considered intramural research. However, they are located close to SAES facilities, are administered by the USDA Office of Experiment Stations, and are managed to "serve as focal points for cooperative research projects between neighboring states" (Kerr 1987:75). The remaining 60 percent of the Bankhead-Jones funds were allocated to experiment stations. Two further important innovations were associated with these funds. First, the funds were distributed to the states according to each state's share of total rural population, an arrangement similar to the formula introduced for federal support of extension under the Smith-Lever Act. Second, a matching requirement was introduced: to receive Bankhead-Jones funds, each state was required to appropriate an equal amount from its own budget. The use of formulae and matching requirements persists in USDA support of the SAESs, although the matching requirement is almost never binding.

The post–World War II period generally has been characterized by an increase in the total federal budget for agricultural R&D, revisions and additions to the legislative basis for that support, a sequence of changes in the institutional arrangements for administering those funds (including attempts to make the experiment stations more accountable to the USDA), and an expanding congressional role in decisions about how the funds are allocated. First, research was directed to finding new uses for agricultural commodities, with 20 percent of the new formula funding authorized in the Research and Marketing Act of 1946 set aside for "marketing research projects." This modification and the earlier innovations in formula funding were incorporated in the revised Hatch Act of 1955, which combined and replaced the various authorizing legislation for federal support of the SAESs. Forestry research in the SAESs was supported by the McIntire-Stennis Forestry Research Act of 1962, with funding allocated among states according to a formula based on forest land, value of timber cut, and contribution of nonfederal forestry research. Formal congressional involvement in research management began with the passage in 1965 of the Special Grants Law; this involvement has included the designation of specific research projects at identified facilities.

The trend toward stronger congressional direction of research funding, including both intramural activities and research in the SAES, and the continuance of a substantial federal management superstructure were extended in the early 1980s with the reauthorizations of the 1981 Amendments to Title XIV and the Farm Bills of 1985 and 1990.

The major innovation of the 1990 Farm Bill in the area of research funding was the NRI Competitive Grants Program, which authorized a substantial increase in the area of competitive grants (first introduced in 1977). The NRI enabling legislation identifies six high-priority research areas and directs the secretary of agriculture to allocate available funds among these areas, following input from outside experts. The legislation also includes three major set-asides: 30 percent for research by multidisciplinary teams; 40 percent for "persons conducting mission-linked systems research"; and 10 percent for "research and education strengthening and research opportunity," including predoctoral and postdoctoral fellowships, equipment grants, and grants to the faculty of small and midsized institutions. However, Congress has funded the NRI at less than the authorized amounts.

Reflecting increased demands for accountability, the 1990 Farm Bill also established the Agricultural Science and Technology Review Board (ASTRB) to assess agricultural research and technology-transfer activities and to recommend how research could best be directed to achieve the purposes outlined in the Farm Bill. The ASTRB was abolished by the 1996 Farm Bill, with its functions superseded by the National Agricultural Research, Extension, Education, and Economics Advisory Board, made up of 30 representatives of the national food research, production, and marketing system.[3]

Intramural Research

Research carried out directly under USDA managers by USDA employees has been a part of the USDA's operations since the creation of the agency in 1862.[4] The first commissioner of agriculture—a noncabinet post that originally directed the USDA—was Isaac Newton. Newton previously had been the super-

3. For details of the National Advisory Board's membership, mission, and activities, see http://www.reeusda.gov.ree/advisory/charter1.htm

4. This section is drawn from Alston and Pardey (1996:10–23). For a more detailed treatment, see Moore (1967) and Rasmussen and Baker (1972).

intendent of agriculture at the Patent Office, with responsibility for the collection and distribution of seeds for new plant varieties. One of his first acts as commissioner was to appoint a superintendent of the propagating garden, the USDA's first research facility, which was located on what is now part of the mall in Washington, D.C. Divisions of Botany, Microscopy, Veterinary Science, and Pomology and a Section of Vegetable Pathology were formed during the second half of the nineteenth century. Although these organizations largely performed "service" rather than R&D (Moore 1967:9), they ultimately evolved into the ARS.

Between the 1880s and the end of World War II, the USDA's intramural research grew rapidly, with employment of scientists within the USDA surpassing total SAES employment in 1904. This growth took place not only at the USDA's central research facility (which eventually moved to its present site at Beltsville, Maryland, outside Washington, D.C.), but also at regional sites. Nine regional research laboratories were created under the 1935 Bankhead-Jones Act, and in 1938 four regional "utilization-research" laboratories were established. Through these initiatives the USDA became the primary provider of applied research that was expected to benefit producers in several states—a role it continues to play.

U.S. Agricultural R&D Performing Institutions

A sizable share of agricultural research in the United States is carried out in the SAESs and other organizations affiliated with the land-grant colleges and universities or in the research institutions of the various agencies of the USDA. A somewhat larger amount of agricultural R&D is carried out by private research institutes, for-profit businesses, and other nongovernmental entities.[5]

Universities and SAESs

The SAES system comprises institutions in all 50 states and the District of Columbia, including land-grant colleges originally endowed by the Morrill Act of 1862, historically black colleges endowed by the Second Morrill Act of 1890, and agricultural experiment stations affiliated with the land-grant colleges established by the Hatch Act of 1887. These organizations are constituted and directed at the state level and receive significant funds from state sources. Indeed, state support for research in the SAESs is the major channel for state and local funding of R&D at colleges and universities. In 1993, when academic

5. USDA, ERS (1995) as well as Klotz, Fuglie, and Pray (1995) include in their definition of private agricultural research expenditures research on food and kindred products, which in 1992 was about US$1 billion, or 30 percent of total private agricultural R&D expenditures. Although there is undoubtedly some food-oriented research performed in the public sector, it likely represents a much lower share. It is quite likely that a majority of production-oriented research continues to be carried out by either the SAESs or the USDA intramural facilities.

institutions were ranked by amounts of state and local research funding, 20 of the top 25 institutions were all members of the SAES system (National Science Board 1996:169–171).[6]

In part, state funding is encouraged by the matching requirements for federal support under the formulae of the Hatch Act and subsequent acts.[7] These arrangements are consistent with the view that local direction of research into local agricultural problems is desirable and the fact that benefits from research are distributed between local producers and domestic and foreign consumers, with spillovers to nonlocal producers.

The choice between intramural performance of R&D and funding of more independent efforts in the university sector is fundamentally important. The United States is more oriented toward university performance of publicly funded R&D than most other major industrialized countries. Among the seven largest members of the OECD, only in Canada, Germany, and the United States do colleges and universities account for more government-funded research (including agricultural and nonagricultural R&D) than intramural government-operated facilities, and in 1993 only in Canada, Japan, and the United States did total expenditures in the higher-education sector substantially exceed intramural government expenditures (by 40 percent in Japan, 47 percent in Canada, and 52 percent in the United States; see National Science Board 1996:Appendix Tables 4.35, 4.36).

From some perspectives, a decentralized system of R&D management may be a logical response to the great variety of environmental conditions facing U.S. agriculture. The problem of sufficient incentives for agricultural R&D seems national; its solution may be quite local, as the technical problems and opportunities facing farmers in one state or region may have little to do with situations in other parts of the continent. However, this may be less true today than it was during the formative years of the U.S. agricultural R&D system early in this century.

USDA Research Facilities

A significant portion of the U.S. agricultural R&D effort takes place within the USDA itself, primarily in the various facilities of the ARS, and the Economic Research Service (ERS). In 1998 the ARS operated 291 research units and laboratories, many of which have close ties to various SAESs. Typically, these

6. The National Science Board tabulation reports R&D expenditures broken down by source of funds (federal, state, and local government, industry, academic institutions, and other). In some cases the reports are systemwide (such as all campuses of Texas A&M University or of Cornell); however, in the few cases where systems include both campuses affiliated with the SAES and nonagricultural sites (such as the Universities of California, Illinois, and Maryland), individual campuses are reported separately.

7. The matching requirement has usually not been binding: most states have spent more than required to receive the federal funds.

facilities focus on crops and problems specific to the region in which they are located. Unlike the SAES system, however, management of the ARS is centralized, with the research units and research centers managed by the ARS administrator in Washington, D.C.[8]

In 1995, 67.6 percent of USDA obligations for R&D were reported to support intramural research, which in the case of the USDA means primarily ARS facilities (National Science Foundation 1996:Appendix Table 4.18). This is a high percentage compared with other U.S. government agencies. Among agencies with larger total R&D budgets in 1995—the National Science Foundation (NSF); the National Aeronautics and Space Administration (NASA); and the Departments of Defense (DOD), Energy (DOE), and Health and Human Services (HHS)—only NASA (at 28.9 percent) spent more than one-quarter of its R&D budget on intramural research. NSF and HHS spent the bulk of their budgets on university research, whereas the DOD, DOE, and to a lesser extent NASA used most of their budgets to support research by private firms.

A curious feature of the USDA intramural effort is how it is classified. As noted, ARS facilities are spread around the country and carry names associated with specific commodities and problems. This feature would seem to indicate an orientation toward more-applied research, which is generally defined as research with a particular end and application in sight. Yet in 1995 almost half of all USDA appropriations for intramural research reported to NSF were identified as "basic research" (National Science Foundation 1996:Appendix Table 4.18). Only NASA and HHS reported more appropriations in support of intramural basic research appropriations than did the USDA.[9] Both NASA and HHS support much larger basic-research portfolios in universities and colleges. In contrast, USDA support for basic research in the university sector is less than half the USDA intramural basic-research activity. It is not clear that the dispersed and commodity-oriented ARS facilities are really carrying out basic research. If they are, it is not clear that they are appropriate institutions to do that kind of work.

Private Research Performers

In the United States, most publicly funded agricultural research is performed by public institutions. However, a small part of the USDA research budget funds research at private institutions (including the USDA component of the Small Business Innovation and Research [SBIR] program and various USDA grants and contracts, including competitive grants). In addition, substantial research programs funded by the private sector are carried out by private firms.

---

8. A detailed list of the various ARS facilities is available at http://www.ars-grin.gov/ars/loc.html

9. It should be noted that HHS is the parent agency of the National Institutes of Health, which operates a large and highly regarded research facility at its Bethesda headquarters.

This research is oriented not directly toward on-farm agricultural activity, but toward either agricultural inputs (for example, farm machinery, seeds) or toward industries using farm products. USDA, ERS (1995) reported private-sector research expenditures in 1992 of US$3.4 billion. Forty-nine percent of these outlays were in the farm machinery and agricultural chemicals industries, 30 percent were in the food and kindred products industry, with only 21 percent in the farm-level animal health and plant breeding product areas. The concentration of private R&D spending in industries concerned with the off-farm aspects of agriculture is not surprising. To appropriate sufficient benefits from research, profit-seeking R&D performers typically need to embody innovations in products; in most cases this involves products that are either sold to farmers or processed from farm products.

A variety of intellectual property structures may have encouraged the growth of private-sector agricultural R&D performance. The intellectual property system in the United States, as it affects agricultural research, has several key elements. First, traditional utility-patent protection has covered machinery, chemical, pharmaceutical, and other inventions since the authorization of a patent system in Article 1, Section 8 of the U.S. Constitution. The Plant Patent Act of 1930 protected asexually reproduced plants (that is, those propagated through cuttings, grafting, and so on). Intellectual property rights concerning sexually reproduced plants such as seed crops were first introduced in the United States by the Plant Variety Protection Act of 1970 (amended in 1980 and again in 1994). More recently, the Board of Patent Appeals ruled in ex parte Hibberd (1985) that both sexually and asexually reproduced plants could be protected under the stronger utility patent, rather than by the plant patent or Plant Variety Protection Certificates issued under the Plant Variety Protection Act. In addition, hybrid seed carries natural secrecy protection provided that the parents of the hybrid innovation are secret; hybrid seed technology is widespread in the corn, sorghum, and sunflower industries.

Between 1970 and 1992, private agricultural R&D expenditures increased by nearly US$1.8 billion (1993 dollars). Over half of this increase went to research on agricultural chemicals, with substantial real increases occurring as well in plant breeding and animal health. All three industries have effective intellectual property regimes. In the case of agricultural chemicals, patent protection is highly effective; Cockburn and Griliches (1987:Table C2) found that across all manufacturing industries, only in the pharmaceuticals industry was patent protection more effective than in the agricultural chemicals industry. Substantial private research in the animal health industry is probably embodied in the products of the pharmaceutical industry.

Cockburn and Griliches (1988) found that the effectiveness of patent protection contributes significantly to stock markets' valuation of firms' past research efforts. The contributions of the Plant Variety Protection Act and other forms of intellectual property to research and innovation in the plant breeding area also have been studied. Butler and Marion (1985) found that both private R&D spending and new variety releases increased in the wheat and soybean industries, and Pray and Knudson (1994) found an increase in private wheat-breeding expenditures after the act was introduced. Venner (1997) found no effect on wheat research investments, wheat yields, or wheat seed prices from the act.

U.S. Agricultural R&D Funding Institutions

Federal Funds

Federal funds for agricultural R&D are divided between direct federal appropriations in support of intramural federal research facilities (mostly ARS) and federal payments to extramural performers (mostly in the SAES system). Federal support of extramural research is mostly part of the budget of what is now called the Cooperative State Research, Education, and Extension Service (CSREES).[10] Some funds are also spent by other USDA branches, such as the U.S. Forest Service (USFS) and the ERS, in the form of grants and contracts to support the specific missions of those agencies.

The mission of the CSREES includes the support of the nation's higher-education, research, and extension programs in agriculture and natural resources.[11] Three types of funding instruments are used to carry out this mission: formula funds to the SAESs; investigator-initiated grants (via the NRI and so on); and the direct funding of SAESs and related institutions through the Hatch Act and related appropriations.

The federal legislation authorizing funding generally includes a requirement for matching funds from the receiving states. This requirement is consistent with the observation that the benefits from agricultural research are shared between (local) producers and (nationwide) consumers and with an equity-based argument that those who benefit should pay. Note that in many cases the matching requirement is not a binding constraint. When states spend more than is required to receive federal funds, a reduction in federal funding may not necessarily induce a further reduction in state funding of agricultural R&D.

One consequence of a system that assigns management responsibility for research to local institutions is a tendency to favor projects with local impact. This may well lead to a bias against basic research, the specific benefits of which cannot in general be identified beforehand and hence must be expected

10. In the Department of Agriculture Reorganization Act of 1994, the USDA Cooperative State Research Service and the Extension Service were merged to form the Cooperative State Research, Education, and Extension Service. In this chapter the current designation is used to describe the activities of both CSREES and its predecessor agencies.

11. http://www.reeusda.gov/new/about/csreesa2.htm contains the mission statement.

to generate substantial "spillovers." In the United States, two primary policy responses have been made to mitigate a bias in favor of local applied research and against more generally applicable basic research.

First, substantial national research laboratories have been established which are either wholly intramural (such as the intramural biomedical research of the National Institutes of Health [NIH]) or operate under contract by universities or private operators (for example, the Lawrence Laboratories at Berkeley and Livermore and the Oak Ridge National Laboratory). However, as noted, although the USDA operates a substantial intramural research program, that program does not differ substantially from the locally managed research programs, which may well be biased in favor of locally appropriable research.

Second, to give extra support to poorly appropriable research, the investigator-initiated grant mechanism is utilized. Such grants are generally awarded following merit-review competitions that involve substantial input from extragovernmental peer review procedures. The NSF uses this approach almost exclusively in its support of basic research. The NIH also has a very large competitive-grants program. The USDA has a much smaller competitive-grants program, which currently operates as the NRI Competitive Grants Program. The NRI, like the competitive-grants programs of other agencies, is open to all researchers, including researchers not affiliated with the SAES system.[12] Alston and Pardey (1996) and others have argued for an expansion of this form of support for agricultural R&D.

Third, USDA supports state-performed agricultural R&D through grants, contracts, and cooperative agreements between other agencies of the USDA and state-level research organizations (primarily SAESs). These are essentially procurements by USDA agencies in support of their missions rather than vehicles for direct support of agricultural R&D. The activities they fund are usually intended to solve agencies' mission-related problems; other contributions to the agricultural (or other) sciences are incidental.

Fourth, USDA also supports state-level agricultural R&D through congressionally mandated special grants and set-asides under which, explicitly or implicitly, CSREES or other USDA funding agencies must fund specific projects, typically specific research facilities. This form of research support results from the intricacies of congressional appropriations policy rather than concerns about economic efficiency. In contrast to what some have argued, the

12. Additional details regarding the operations of the NRI are reported in National Research Council, Board on Agriculture (1989, 1994). In fiscal year 1997, the NRI peer reviewed 2,840 proposals requesting a total of US$591.5 million in funding (USDA, CSREES 1998). Awards were granted to 712 proposals totaling US$87.3 million. The average grant awarded in 1997 (excluding conferences and continuations) was for US$141,834 for 2.6 years. Partial support for 26 conferences was provided for a total outlay of US$175,500—an average of US$6,750 per conference—as well as US$15.9 million for postdoctoral fellowships, new investigator awards, and so-called strengthening awards.

special grants channel does not appear to be a mechanism for diversifying research strengths. Rather, it is a means of expressing congressional district-level interests in both state and national science policy.

State Funds

The fundamental characteristics of the SAES system are state-level management of research and shared state and federal funding of research. This arrangement has been a part of federal formula-funding legislation since the Research and Marketing Act of 1946 (Kerr 1987:220–231), which required that federal funds provided to the SAESs be matched on a one-for-one basis by state governments. However, for several states the matching requirement has not been close to binding. In addition, on an aggregate basis, since 1920 total state support for SAES research has generally been more than 1.5 times total federal support.

Other Funds

The SAESs also fund research from a variety of other sources. First, historically the SAESs have had some internal funds, such as gifts and endowments, income from seed testing and certification programs, and sales of the output of experimental farms. Second, funding from various industrial and private sources has become increasingly important. In many states marketing orders and commissions fund production research (Lee, Alston, Carman, and Sutton 1996).

U.S. Public Funding of Agricultural R&D

The primary instruments of science policy are funding mechanisms. By choosing total funding and the mechanisms to support agricultural R&D, policymakers influence both the amount of R&D undertaken and the ways in which R&D programs are carried out. Here we examine the amounts of funds from different sources that have been used to support agricultural R&D. In addition we consider how funding patterns vary across different regions. It is worth noting that these geographic patterns of support may themselves have been influenced by choices of R&D funding mechanisms (for example, competitive grants versus formula funds versus industry financing), different intellectual property regimes, or other variations in local and regional science-policy environments.

Public-Sector R&D Context

By almost any measure public-sector R&D in the United States is big business.[13] In 1995 the federal government obligated about US$65.0 billion (in

13. Several recent studies have documented trends in agricultural R&D expenditure, including National Research Council, Board on Agriculture (1989, 1994), Huffman and Evenson (1993, 1994), Schweikhardt and Whims (1993), Weaver (1993), Zulauf and Tweeten (1993), USDA, CSRS (1993), Miller and Harris (1994), Alston and Pardey (1996), and Fuglie et al. (1996).

FIGURE 4.1 U.S public-sector R&D expenditures by socioeconomic objective

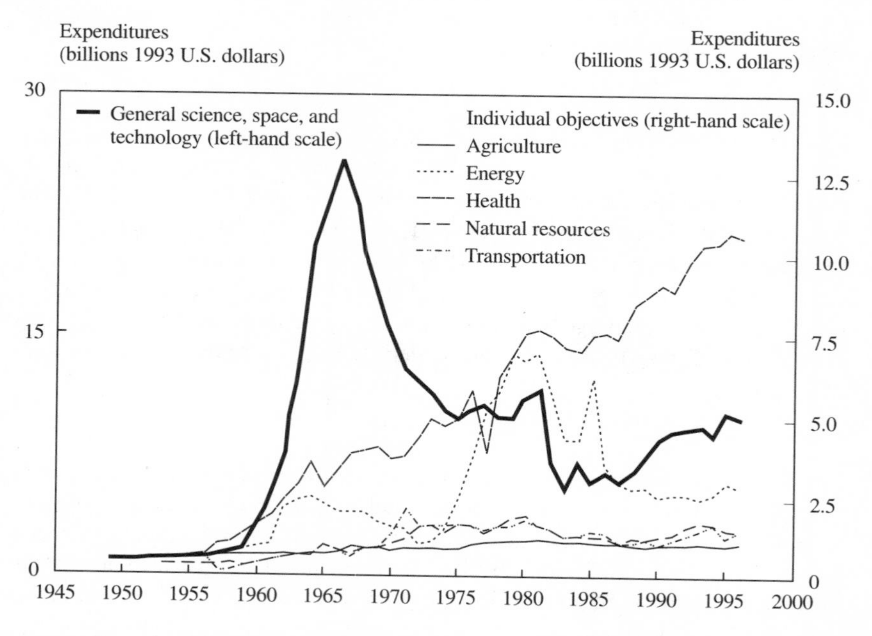

SOURCE: Executive Office of the President (1994); National Science Foundation (1995).

1993 prices) to fund R&D, whereas in 1949 this figure was only US$10.0 billion. The largest share of the public R&D budget supports defense-oriented work. In 1995 US$35.9 billion, or 55.1 percent of total public R&D, was spent on defense-related research, although this is less than the peak share of defense in total federal R&D of 69.3 percent in 1986. In 1995 about US$29.3 billion was budgeted for nondefense research, whereas federally funded agricultural research accounted for just US$1.1 billion, or 1.7 percent of the total budget.[14] In 1995 research on natural resource issues accounted for a larger share (2.4 percent) of total federal R&D funding than agricultural R&D.[15]

Figure 4.1 illustrates that agricultural R&D has been a relatively small share of the total national public-sector R&D effort since 1945. During this period, public R&D expenditure has been dominated by national defense and

14. In 1994 total federal spending on research of US$68.3 billion consisted of US$38.0 billion on defense, US$10.9 billion on health, US$2.9 billion on energy, US$2.0 billion on natural resources, US$1.2 billion on agriculture, and US$13.3 billion on other functions (National Science Foundation 1995).

15. These shares are only indicative. Clearly a portion of agricultural R&D expenditures are also directed toward natural resource issues.

space and technology R&D. Public outlays in several other research areas have also grown relative to agricultural R&D, in particular health, energy, and natural resources R&D.

Agricultural Research in a Long-Term Perspective

OVERALL TRENDS. Table 4.1 provides a detailed, comprehensive, long-term perspective on agricultural R&D spending in the United States.[16] In 1889, shortly after the Hatch Act was passed, federal and state spending appropriations totaled US$860,000. A little over a century later, public-sector agricultural R&D had increased to more than US$2.9 billion at an annual rate of growth of 7.8 percent in nominal terms and 4.2 percent in real terms. Real growth in the components of public-sector R&D spending is illustrated in Figure 4.2, which shows the recent slowdown in the real growth rate, especially regarding intramural USDA research.

One important feature of the U.S. public-sector agricultural R&D system over the past 25 years has been a reduced role of federal agencies as opposed to state agencies in the performance of agricultural R&D. Over the first half century intramural research by the USDA was dominant. In contrast in 1997 the SAESs accounted for 77 percent of total public spending on agricultural R&D, with federal laboratories operated by the USDA accounting for almost all of the remaining 23 percent.

SAES FUNDING. Between 1972 and 1997 total support for SAES research grew by 7.6 percent per year in nominal terms, but only by 2.1 percent in real terms (Appendix Table A4.1). Sources of funds for SAES research also changed (Figure 4.3). Current shares are approximately as follows: 47.4 percent comes from state government, 31.5 percent from the federal government, and 21.1 percent from other sources, including direct industry funding and miscellaneous fees and sales.[17] During their early years the SAESs received a small but growing share of their funds from state sources. State funding has been the slowest-growing source of revenue since 1972; the proportion of funds received from state sources peaked at 69.0 percent in 1970 but has fallen steadily since then.

Funding from other sources (including funds from grants and industry checkoffs, as well as from fees and sales) has grown steadily as a share of total outlays since the early 1970s. Revenues from the sale of services and products (including royalties earned from patented technologies) account for

16. Agricultural R&D is defined as total R&D spending either by the SAESs or by the research agencies of the USDA. Some of the included SAES research is funded by agencies other than the USDA, in particular the National Institutes of Health and the National Science Foundation. Privately performed R&D, such as by seed and fertilizer companies, is not included.

17. The SAESs have some internal funding sources, including fee income from seed certification programs and other testing and inspection services, in particular, and crop and animal sales from experimental farms.

TABLE 4.1 Public-sector agricultural R&D spending

Year or Decade Average	State Agricultural Experiment Stations (SAESs)[a]				Intramural U.S. Department of Agriculture[c] (USDA)		U.S. Total[d]	
	State	Federal	Miscellaneous Fees and Sales	Total	Forest Service[b]	Other USDA	Nominal	Real[e]
	(millions U.S. dollars per annum)							(millions 1993 U.S. dollars per annum)
1889	0.08	0.59	0.06	0.72	na	0.14	0.86	
1890–1899	0.22	0.70	0.11	1.04	na	0.21	1.24	32.79
1900–1909	0.65	0.87	0.31	1.84	na	1.04	2.88	75.32
1910–1919	2.24	1.43	1.09	4.76	na	4.48	9.24	199.90
1920–1929	6.01	2.11	2.09	10.21	na	18.44	28.65	444.25
1930–1939	8.25	4.88	2.60	15.72	na	30.68	46.40	753.15
1940–1949	15.81	7.42	5.44	28.67	na	40.97	69.64	907.95
1950–1959	56.17	19.10	14.27	89.55	na	46.08	135.63	1,026.95
1960–1969	132.10	42.87	25.20	200.18	33.66	109.32	309.50	1,522.44
1970–1979	289.13	131.14	63.41	483.68	82.99	258.58	742.26	2,043.68
1980–1989	646.44	359.41	207.04	1,212.89	127.09	500.37	1,713.25	2,485.20
1990	927.15	500.86	338.07	1,766.07	157.41	614.08	2,380.15	2,614.85
1991	961.73	532.15	358.72	1,852.59	193.45	650.62	2,503.22	2,619.52
1992	956.29	582.06	376.52	1,914.87	217.12	695.87	2,610.74	2,658.67
1993	960.41	632.39	387.54	1,980.33	198.27	642.82	2,623.16	2,623.16

1994	987.45	683.12	415.17	2,085.74	207.01	652.44	2,738.18	2,663.55
1995	1,024.25	708.01	415.45	2,147.71	196.97	658.48	2,806.20	2,636.93
1996	1,025.19	698.46	445.51	2,169.16	186.96	649.96	2,819.12	2,576.87
1997	1,059.07	703.46	472.02	2,234.55	183.23	653.68	2,888.23	2,561.80
Annual growth rates								
					(percentages)			
1889–1997	9.25	6.79	8.73	7.73	na	8.12	7.81	4.16
1980–1990	7.65	6.98	9.76	7.82	2.97	4.85	6.96	0.72
1990–1997	1.92	4.97	4.88	3.42	2.19	0.90	2.80	–0.29

SOURCE: SAESs: 1889–1993, Alston and Pardey (1996, Table 2-1); 1994–1997, USDA, Current Research Information System (various years, Table IV-E); USDA Forest Service: 1969–1997, USDA, Current Research Information System (various years, Table III-E); Other USDA: 1889–1990, Huffman and Evenson (1993, Appendix Table 4A.1); 1991–1997, USDA, Current Research Information System (various years, Table III-E).

[a]Data include experiment stations and cooperating institutions for the 48 contiguous U.S. states. That is, they exclude Alaska and Hawaii (which in 1997 spent a total of US$24.7 million on public agricultural R&D) plus Washington, D.C., and various U.S. protectorates (US$14.9 million in 1997).

[b]Missing data for years prior to 1968 as well as 1971–1973, 1975, and 1976.

[c]Series approximates intramural research by the USDA and consists of total appropriations to the Agricultural Research Service, the Economic Research Service, and the Agricultural Cooperative Service less appropriations to contracts, grants, and cooperative agreements with the SAESs made by these USDA agencies. This treatment ensures the post-1990 data are consistent with data for prior years taken from Huffman and Evenson (1993) but overstates intramural R&D to the extent the USDA provides grants and contracts to agencies other than the SAESs.

[d]Includes SAESs and Other USDA.

[e]Deflated with a revised and updated version of the U.S. agricultural R&D deflator reported by Pardey, Craig, and Hallaway (1989).

FIGURE 4.2 U.S. public-sector agricultural R&D spending by performing sector

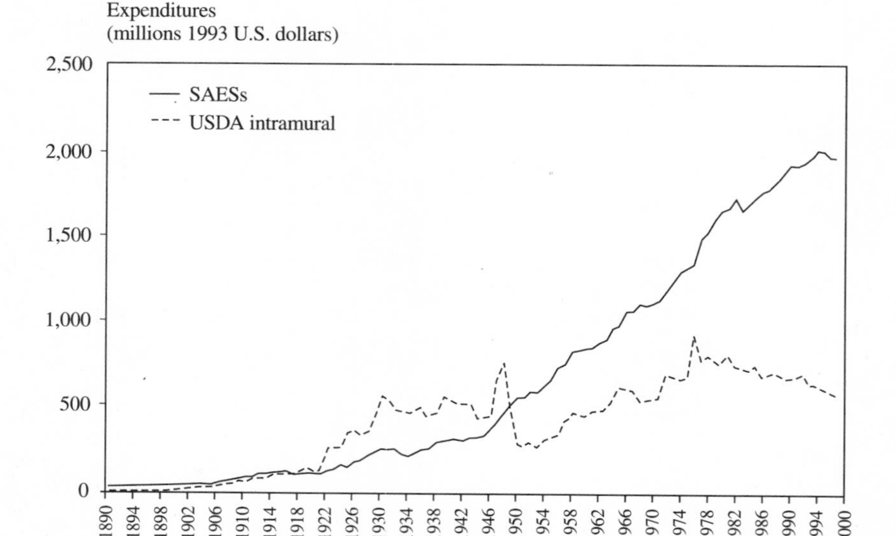

SOURCE: See Table 4.1.

NOTE: Deflated with a revised and updated version of the U.S. agricultural R&D deflator from Pardey, Craig, and Hallaway (1989).

only 5.6 percent of total funds. Industry funds from grants, checkoffs, and similar sources provided about US$178 million in 1997, 7.8 percent of the total, up from 4.6 percent in 1972.

The federal government is no longer the primary source of funds for SAES research. However, over the past 20 years the federal share of public agricultural R&D funding has remained fairly stable at about 30 percent. About 47 percent of federal funds are administered by CSREES. These CSREES resources include funds dispersed on a formula basis, some earmarked funds, and funds made available to the states as part of the competitive-grants program. The remainder of the federal funds received by the states are earmarked; derived from USDA grants, contracts, and cooperative agreements; or received from agencies such as the NSF, NIH, and DOD. The last, which are mostly awarded through competitive procedures, increased from 8.4 percent to 12.8 percent of total SAES funding between 1980 and 1997. Overall, funds from sources other than CSREES have accounted for an increasingly large share of total SAES budgets, rising from one-third of federal support in 1972 to about one-half in 1993.

FIGURE 4.3 Sources of funds for state agricultural experiment station research

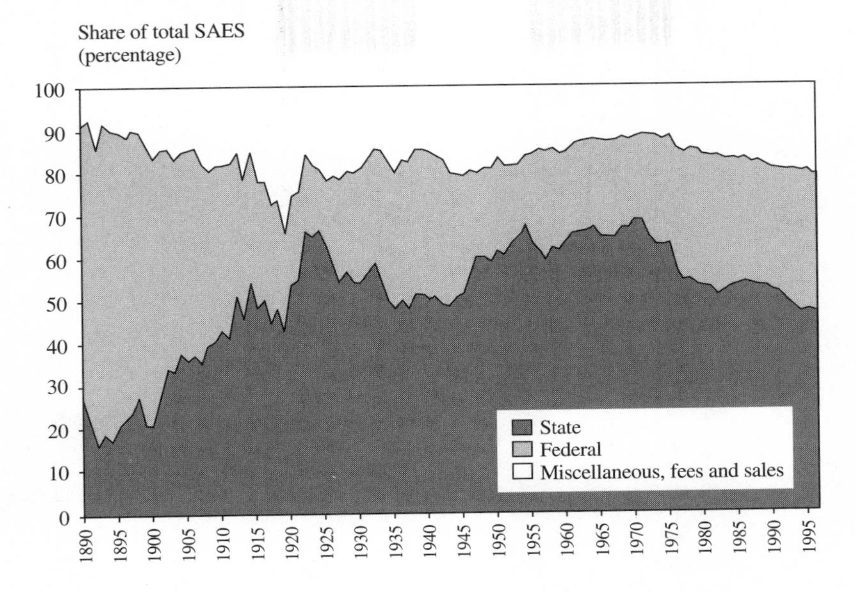

SOURCE: See Table 4.1.

USDA APPROPRIATIONS. The USDA disperses and also utilizes federal R&D funds. Appendix Table A4.2 details the deployment of federal appropriations to the USDA. Since 1970 an increasing share of USDA's total resources allocated to both research and education has been directed to research. Correspondingly, the share allocated to education and extension services has contracted and now accounts for just over one-fifth of total research and education funds, compared with one-third of these resources in 1970. The ARS now accounts for about one-third of total USDA research and extension expenditures, a share that has remained fairly constant in recent years. Most of the USDA's funds are spent on intramural research by agencies such as ARS, ERS, and USFS. Almost one-fifth of the USDA's research resources are administered by CSREES. Most of these CSREES-administered funds go to SAESs and other cooperating institutions, although some of the competitive-grant funds that CSREES oversees are spent by agencies within the USDA. Figure 4.4 provides a recent perspective of total SAES and USDA expenditures on agricultural R&D broken down according to the sources of funds.

Agricultural Extension in a Long-Term Perspective

State and local extension activities predate the passage of the Smith-Lever Act in 1914. However, this was the first year in which federal funds were made

FIGURE 4.4 Funding channels for U.S public-sector agricultural R&D, 1997

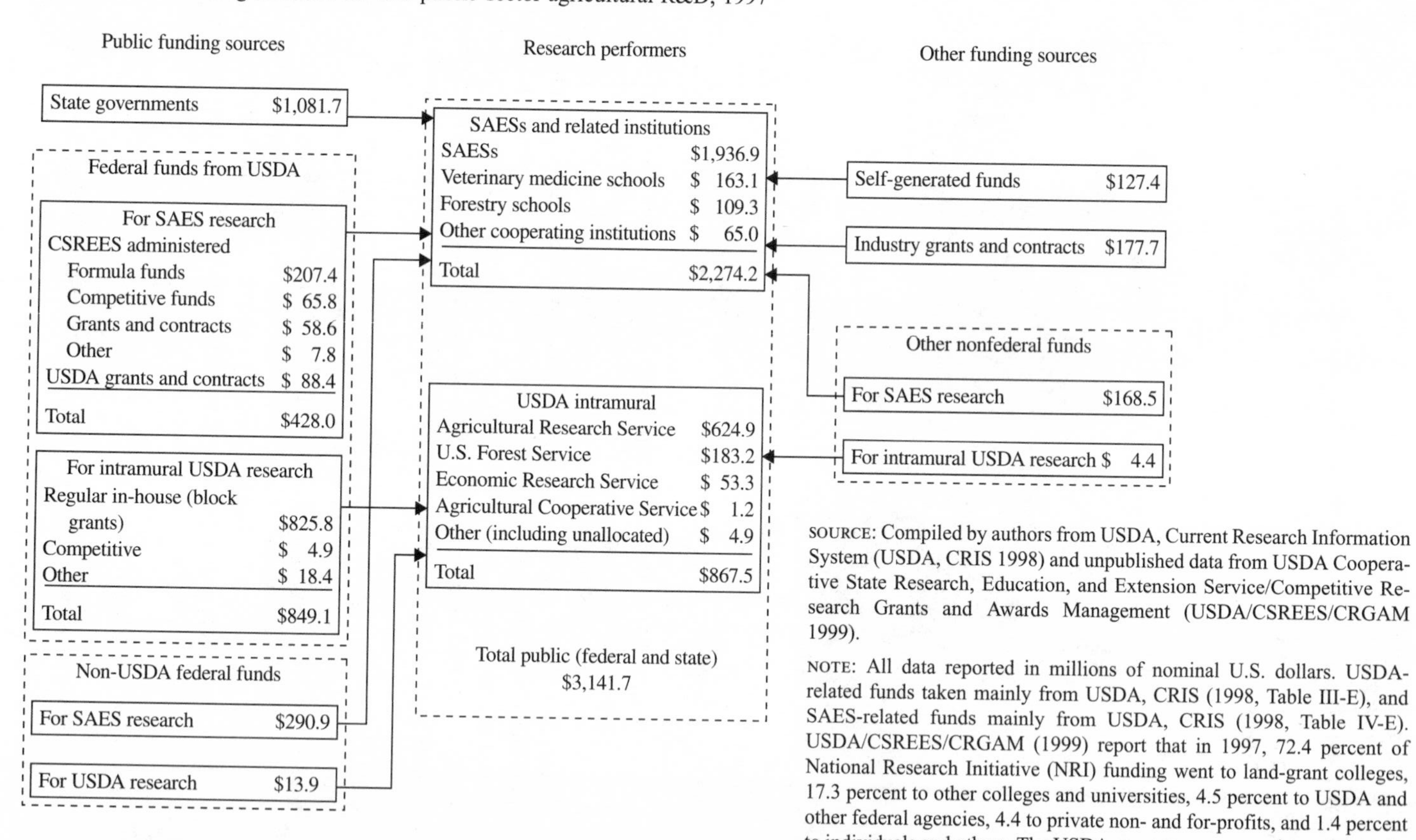

SOURCE: Compiled by authors from USDA, Current Research Information System (USDA, CRIS 1998) and unpublished data from USDA Cooperative State Research, Education, and Extension Service/Competitive Research Grants and Awards Management (USDA/CSREES/CRGAM 1999).

NOTE: All data reported in millions of nominal U.S. dollars. USDA-related funds taken mainly from USDA, CRIS (1998, Table III-E), and SAES-related funds mainly from USDA, CRIS (1998, Table IV-E). USDA/CSREES/CRGAM (1999) report that in 1997, 72.4 percent of National Research Initiative (NRI) funding went to land-grant colleges, 17.3 percent to other colleges and universities, 4.5 percent to USDA and other federal agencies, 4.4 to private non- and for-profits, and 1.4 percent to individuals and others. The USDA percentage was used to estimate the amount of 1997 NRI funding reported in USDA, CRIS (1998, Table III-E) used for intramural USDA research. NRI expenditures incurred by the SAESs were taken directly from USDA, CRIS (1998, Table IV-E). Totals in boxes may not sum because of rounding.

TABLE 4.2 Extension expenditures by source of funds

Fiscal Year	Funds from State Sources: State Government Appropriation	County Appropriation	Nontax	Total	Federal Appropriations	U.S. Total
			(thousands U.S. dollars)			
1915	1,044	780	287	2,111	1,486	3,597
1920	5,229	2,866	627	8,767	5,891	14,658
1940	6,427	6,665	1,087	14,179	18,585	32,764
1960	53,583	31,231	1,542	86,356	53,715	140,071
1980	304,883	130,630	16,356	451,878	230,820	682,698
1990	620,471	235,742	53,294	909,507	354,557	1,264,046
1994	659,921	274,272	75,563	1,000,756	419,400	1,429,156
Share of U.S. total						
			(percentages)			
1915	29.0	21.7	8.0	58.7	41.3	100
1920	35.7	19.6	4.6	59.8	40.2	100
1940	19.6	20.3	3.3	43.3	56.7	100
1960	38.3	22.3	1.1	61.7	38.3	100
1980	44.7	19.1	2.4	66.2	33.8	100
1990	49.1	18.6	4.2	72.0	28.0	100
1994	46.2	19.2	5.3	70.7	29.3	100

SOURCE: Alston and Pardey (1996, Table 2-9).

available for cooperative extension between the USDA and various state extension agencies. Approximately US$36.3 million (1993 dollars) of federal funds were appropriated under the act during its first year, roughly matching the amount of federal funds then allocated to the SAESs for agricultural research (Table 4.2).[18] A further US$50.9 million was made available for extension from various state and local government sources, resulting in a total of US$87.2 million. These funds grew at an annual rate of 7.90 percent between 1914 and 1994, until by 1994 they amounted to US$1.5 billion. As with R&D, however, in recent years the average annual growth rates of funding for public extension have been considerably lower than the long-term average growth rate and have been declining. Since 1990 nominal extension expenditures have grown at an annual rate of only 3.30 percent and by only 0.18 percent in real terms.

The public provision of extension services in the United States is largely a state or local activity. In 1994 funds from within-state sources accounted for

18. As a share of the total, not just federal, funds going to the SAESs, these sources of funds collectively accounted for 9.7 percent of the total in 1972 and 15.7 percent in 1993.

71 percent of total extension funding, while federal funds accounted for the remaining 29 percent. In cooperative extension activities a substantial share of funds is provided by county governments. In 1915 county appropriations accounted for over one-third of within-state funding. Their share rose to a peak of around 47 percent during the 1930s, then declined to 30 percent in 1970 and has been about 27 percent during the 1990s.

SAES Funds: A Geographical Analysis

Although aggregate trends help provide a perspective on the comparative developments in state and federal R&D agencies, sources of support and patterns of spending have varied considerably among the SAESs and within states over time. An understanding of these state-specific details is useful for designing effective and efficient policies for agricultural R&D.

Tables 4.3 and 4.4 present data that are directly comparable for 1970 and 1991. For both tables, states were first ranked according to the value of their agricultural output. A set of share statistics was then calculated to compare the distribution of the various sources of support with the distribution of production (here measured in terms of the value of output). The last column in each table shows the percentage of the value of output by successively larger cohorts (in incremental units of five states). Thus, for example, in 1991 the largest five states produced 32.9 percent of all agricultural output; the largest 10 produced 51.2 percent of output, and so on. The data in Tables 4.3 and 4.4 show that the spatial pattern of production has become only marginally more concentrated over time. In 1991 the top 10 states accounted for 51.2 percent and the smallest 10 states collectively accounted for just 2.9 percent of the value of output.[19] In 1970 the corresponding shares were 50.8 percent and 2.9 percent.

Research funding across the categories described in Tables 4.3 and 4.4 was by no means congruent with the value of production. In 1991 the top 10 states produced more than half the value of output but accounted for only 37.4 percent of total SAES spending and received only 34.8 percent of federal SAES research funds. Conversely the smallest 10 agricultural states produced 1.3 percent of the value of output but received 4.5 percent of total SAES spending and 5.3 percent of federal support for SAES research. In 1991 distributing R&D spending in accordance with the value of production would have required US$278 million of research resources to be reallocated from the smallest 38 agricultural states to the largest 10 states.

The distributions of the individual components of the funds are revealing, although the patterns are not too surprising. The pattern of federal formula

---

19. In 1991 the top 10 states according to value of output were (in rank order) California, Iowa, Texas, Nebraska, Illinois, Minnesota, Wisconsin, Kansas, North Carolina, and Missouri. The smallest 10 states were Delaware, Vermont, West Virginia, Maine, Nevada, Massachusetts, New Jersey, Connecticut, New Hampshire, and Rhode Island.

TABLE 4.3 Research expenditures by state agricultural experiment stations, ranked by size, 1970

Number of States (Ranked in Order of Value of Agricultural Production)	Federal Funds				State Funds	Total Research	Value of Agricultural Production
	Formula Funds	Competitive Grants	Project Funds	Total			
			(cumulative percentages)				
5	15.5	na	29.3	20.4	22.4	21.3	32.3
10	27.4	na	34.3	33.8	34.1	35.3	50.8
15	42.2	na	46.0	48.2	50.2	50.4	64.0
20	54.6	na	56.2	57.9	64.1	62.3	74.8
25	66.2	na	65.5	68.9	72.2	71.3	83.5
30	75.3	na	80.5	77.1	81.4	80.3	90.5
35	83.4	na	91.7	86.0	89.2	88.6	95.5
40	89.4	na	98.2	91.3	94.3	93.5	98.1
45	97.4	na	100.0	98.0	99.1	98.7	99.7
48	*100.0*	*na*	*100.0*	*100.0*	*100.0*	*100.0*	*100.0*

SOURCE: Compiled by the authors from U.S. Department of Agriculture, Current Research Information System data.

NOTE: na indicates not available. Data represent the share of the national total of the variable in each column represented by the largest *n* states, where *n* is the number in the first column, computed for each of the expenditure or value columns. Thus, the largest 25 recipients of formula funds for research accounted for 66.2 percent of formula funding in 1970, and the largest 25 states in terms of value of agricultural production accounted for 83.5 percent of the national value of agricultural production.

TABLE 4.4 Research expenditures by state agricultural experiment stations, ranked by size, 1991

Number of States (Ranked in Order of Value of Agricultural Production)	Federal Funds						
	Formula Funds	Competitive Grants	Project Funds	Total	State Funds	Total Research	Value of Agricultural Production
			(cumulative percentages)				
5	16.0	25.5	16.7	20.2	24.4	22.8	32.9
10	30.8	41.4	24.2	34.8	38.7	37.4	51.2
15	43.8	52.3	39.9	46.2	55.6	51.7	64.7
20	56.0	62.9	56.7	57.8	64.4	61.8	75.7
25	67.5	77.0	68.4	72.4	75.0	74.7	84.8
30	76.3	81.1	80.0	79.1	80.8	80.8	91.4
35	84.7	90.2	90.3	89.7	91.0	90.8	96.4
40	91.0	93.6	94.1	94.7	95.8	95.5	98.7
45	97.4	99.4	99.1	98.9	98.8	99.0	99.8
48	*100.0*	*100.0*	*100.0*	*100.0*	*100.0*	*100.0*	*100.0*

SOURCE: Compiled by the authors from U.S. Department of Agriculture, Current Research Information System data.

NOTE: See Table 4.3.

funding has changed little over the years and is distributed in ways that fail to match the spatial pattern of production. For example, in 1991 the largest five agricultural states accounted for nearly one-third of the total value of production but received only 16.0 percent of the federal funds. In contrast, at the same time competitive grants were more nearly congruent with production. The top five states, for example, produced 32.9 percent of total output and received 25.5 percent of the grants. The same is not the case for federally sourced special project funds. The top five states received less than 17.0 percent of these funds, while the smallest 10 states received 12.2 percent of federal formula funds, 7.7 percent of federally sourced special project funds, and 7.2 percent of state-sourced funds.

Individual state data are even more revealing. In terms of value of output, for example, California was the largest agricultural state in 1991, producing 9.1 percent of national agricultural output. However, this state received only 2.4 percent of federal special project funds and 3.4 percent of federal formula funds. California was awarded 12.1 percent of federally managed competitive funds, in keeping with the tendency for competitive funds to offset shortfalls in formula and special project funding. The pattern of federal funding for California's research contrasts markedly with state support to the system; the California Agricultural Experiment Station accounted for 11.8 percent of national funding from state sources, over 2.0 percentage points more than its share of the value of output. In part, this probably reflects both the fact that California agriculture has grown much faster than its share of federal support and a response by the state to the funding shortfall. Conversely it is possible that in some states with larger federal funds support, federal funding may have crowded out state funding.

California's experience may also reflect the fact that, as well as being a relatively large agricultural state, it is the dominant producer of several horticultural products for which producer benefits from research are largely confined to the state. Thus, with smaller perceived spillover effects and fewer others engaged in the relevant research, the incentives for the state to provide R&D investments are greater.

Research Spending Intensities

The data in Table 4.5 provide estimates of research intensities for agricultural R&D expenditures. These research intensities show the ratio of R&D outlays to the size of the agricultural sector and various other criteria.[20] Public-sector research spending grew from 0.31 percent of the gross value of farm marketings in 1950 to 1.38 percent in 1997, a growth rate of 3.23 percent per year.

20. Agricultural research intensity ratios (ARIs) are usually defined as the ratio of research expenditures to either the gross value of agricultural output or agricultural GDP, although other denominators, such as farm numbers or rural population, can also be used.

TABLE 4.5 U.S. public agricultural research and extension intensity ratios

	1950	1960	1970	1980	1990	1997	Annual Growth Rate (1950–1997)
			(percentages)				(percentages)
Relative to agricultural output[a]							
Research	0.31	0.63	0.89	0.87	1.40	1.38	3.23
Extension	0.26	0.41	0.58	0.49	0.75	0.79[b]	2.58[c]
			(1993 U.S. dollars)[d]				
Relative to farm numbers							
Research	150.95	329.65	564.07	997.29	1,218.48	1,244.20	4.59
Extension	125.25	214.25	366.32	560.71	647.11	673.22[b]	3.90[c]
Relative to total population							
Research	5.60	7.23	8.11	10.71	10.48	9.57	1.15
Extension	4.65	4.70	5.27	6.02	5.57	5.34[b]	0.32[c]

SOURCE: Public agricultural R&D, see Table 4.1; agricultural extension, see Table 4.2; agricultural output, USDA, Economic Research Service (1998); farm numbers, USDA, National Agricultural Statistical Service (1997); population, U.S. Bureau of the Census (1998).

[a]Agricultural output represents gross value of agricultural production (specifically farm cash receipts).

[b]1994 figure.

[c]Annual growth rate for 1950–1994 period.

[d]Research and extension series deflated with revised and updated version of the R&D deflator from Pardey, Craig, and Hallaway (1989).

When private-sector R&D expenditures are also included, the contemporary agricultural research intensity ratio increases from 1.38 percent to 3.51 percent of the value of output. There is US$1.92 spent on research for every public dollar spent on extension. Thus extension expenditures presently represent around 0.79 percent of the gross value of agricultural output, about half the corresponding research intensity ratio. In terms of 1993 dollars, the U.S. public research system now spends US$1,244 per farmer compared with just US$151 four decades ago. Research spending per capita has risen too. Nearly US$9.60 is now being spent per capita on public-sector agricultural R&D, an increase of 1.15 percent per year since 1950.

The USDA's Current Research Information System (CRIS) data allow us to consider state- and region-specific agricultural research intensity ratios. Figures 4.5 and 4.6 plot state-specific SAES research intensities—that is, SAES research expenditure expressed as a percentage of each state's value of agricultural production, pennies of research spending per dollar of value produced by agriculture—for 1970 and 1991. A comparison of Figures 4.5 and 4.6 shows that the agricultural research intensities for all SAESs (except Rhode

Island) rose between 1970 and 1991, with some changes in the rankings. The average SAES agricultural research intensity, which was 0.61 percent in 1970, had increased to 1.37 percent in 1991. Research intensities may change because of changes in either the numerator (research spending), the denominator (value of production), or both. In the United States, both forces have been at work. Although it was not the case in 1970, the states with the highest agricultural research intensities in 1991 were all New England states, perhaps more as a consequence of declining agricultural production than of rising expenditures on research relative to other states. In 1991 even Maine, which was near the middle of the distribution in 1970, had joined the other New England states and New York with SAES agricultural research intensities of more than 3.00 percent in 1991. These states with high agricultural research intensities tend to have high population densities and low and declining agricultural production.

Most of the states at the lower end of the agricultural research intensity scale in 1991 are specialized in agricultural production, have relatively small nonagricultural sectors and relatively low population densities, and are specialized within agriculture, especially in grain production. Many of these states also had relatively low agricultural research intensities in 1970. These states' low agricultural research intensities could reflect relatively low nonagricultural tax bases for financing agricultural research or lower perceived research potential relative to nongrain crops or livestock. Because much grain-related R&D is likely to have large interstate spillovers, low agricultural research intensities in the grain belt may also simply be a reflection of the disincentive effect of the free-rider problem. This explanation could also justify efforts to create multistate, regional research programs to cover crops with the potential for large interstate technology spillovers. As can be seen in Figure 4.6, much of the variation in state-specific agricultural research intensities can be attributed to the variation in state funding for the SAESs, but the other sources of funds also show similar patterns. Some clear outliers (for example, Colorado and Massachusetts) have disproportionately large shares of federal funding for their SAESs, whereas some others (for example, Connecticut, Florida, Georgia, California, and Kentucky) have disproportionately low shares of federal funding for their SAESs.

There is no target agricultural research intensity that can serve as a guide for policy. However, the existence of a few major states with high levels of state support, such as Florida and California, gives some hints about what determines the direction of policy. In these two states fairly unique production environments may have allowed local producers to capture large shares of the producer benefits from research, thus enhancing local political support for continued investments in agricultural technology. Other states do not have the distinctive geographies of California and Florida, in that each of those political jurisdictions approximately coincides with the relevant agricultural production

FIGURE 4.5 State-specific agricultural research intensities for state agricultural experiment station spending, 1970

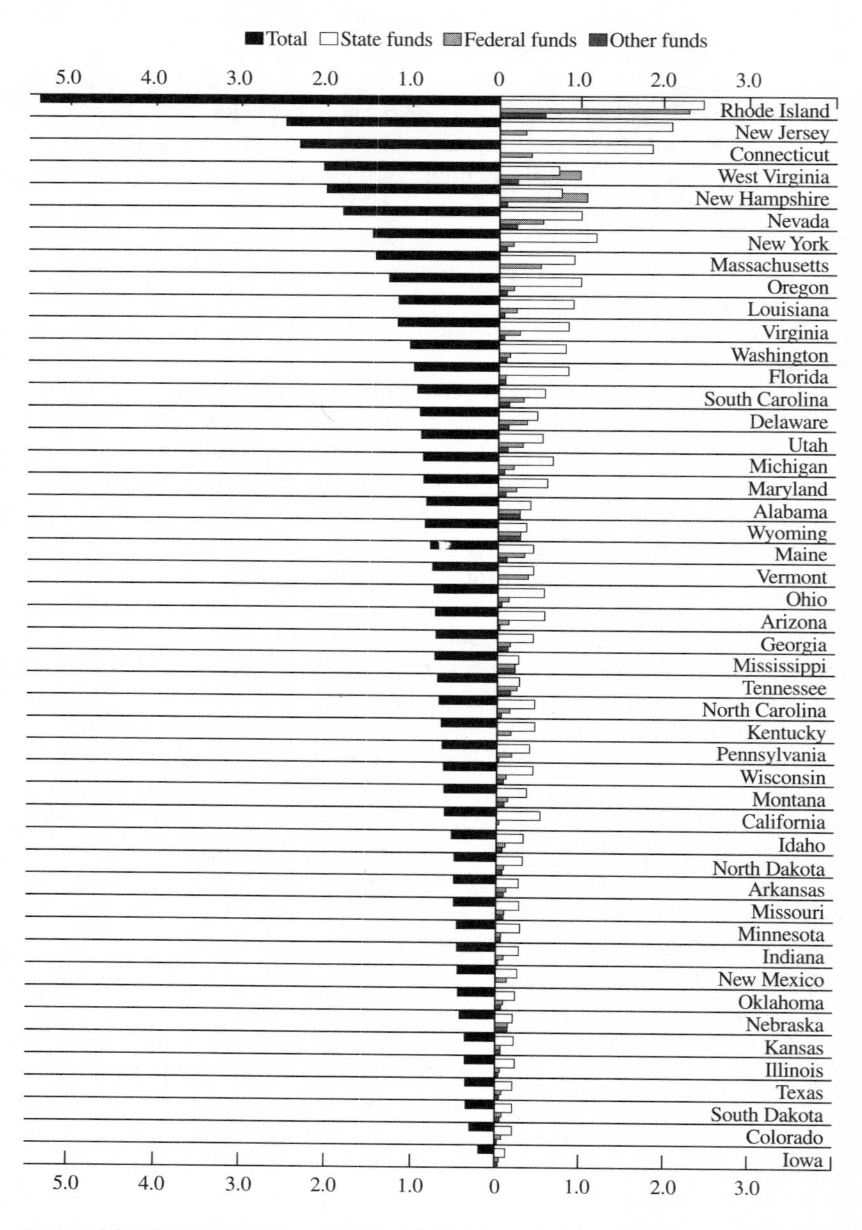

SOURCE: R&D data compiled by authors from USDA Current Research Information System data tapes. Gross value of agricultural production from Craig and Pardey (1996b).

FIGURE 4.6 State-specific agricultural research intensities for state agricultural experiment station spending, 1993

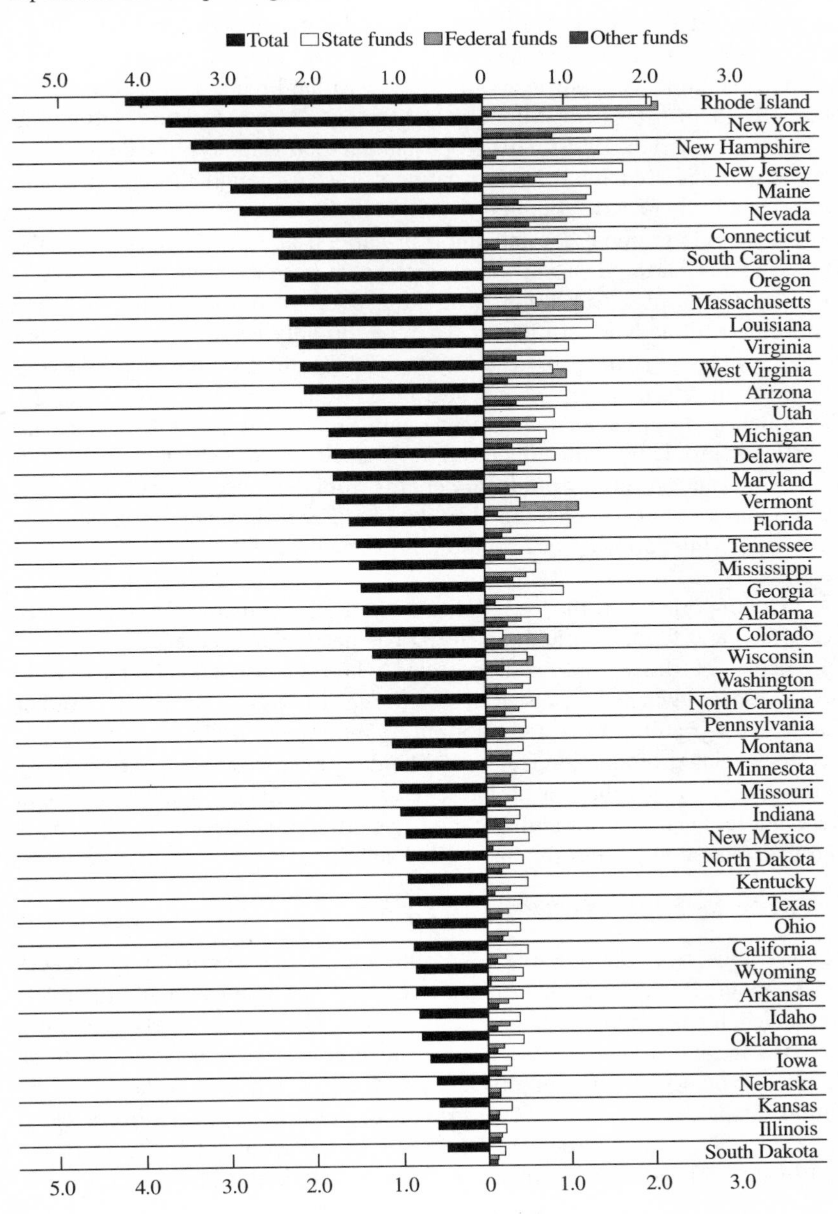

SOURCE: See Figure 4.5.

zone. Relatively lower rates of investment in R&D, especially from nonfederal sources, in the major agricultural states of the United States may be a result of the failure of federal policy to respond to the difference between agroecologically defined production regions and state political jurisdictions.

One example is the diffusion of semidwarf hard red spring wheats in the Northern Plains states of Minnesota, Montana, North Dakota, and South Dakota (Pardey et al. 1996). This crop's annual total value is quite large but its agricultural research intensity is relatively low. Although the end product, wheat, is a fairly generic commodity sold in highly competitive markets, there is considerable regional specialization in varietal and class selection: in the United States, hard red spring wheats are grown in substantial quantities only in those states, plus California, with little overlap in varietal use between California and the Northern Plains. There is virtually no spillover of relevant superior wheat varieties from the Northern Plains to other regions. The very high rates of return observed for wheat-breeding R&D in that region indicate that much higher rates of investment might be appropriate, unless there are significant problems with either public finance or cost sharing, especially among the states receiving producer benefits from the successful research. The development of instruments that facilitate both cost and benefit sharing appears to be a high priority for agricultural R&D policy at the federal or regional level.

The importance of the availability of spill-ins as an alternative to locally performed and funded agricultural R&D has been addressed econometrically by Khanna, Huffman, and Sandler (1994), who find that states prefer to fund SAES research when some research benefits can be appropriated locally rather than when agricultural research is a pure public good. States specializing in livestock-related production (for example, dairy) showed less appropriability of research than did other states. These results are consistent with the view that state and regional appropriability conditions influence state funding decisions and that federal and other funding policy should respond to these conditions.

Private-Sector R&D Expenditures

The private sector committed US$3.4 billion to in-house agricultural R&D in 1992, about 20 percent more than the amount spent on agricultural R&D conducted by the public sector (Table 4.6).[21] There was nearly a 17-fold increase in the amount of privately conducted research in the past three decades, which is a substantially faster rate of growth than occurred in the public sector. As a result, for every dollar of publicly conducted research in 1992 the private sector spent US$1.31, compared with just US$0.94 in the early 1960s. This dramatic expansion of the private-sector investment in agri-

21. The private R&D estimates are from USDA, ERS (1995) as reported in Alston and Pardey (1996). As the latter noted, it is problematic to measure privately conducted agricultural R&D in ways that can be meaningfully compared with the public-sector figures reported here.

cultural R&D coincided with the substantial strengthening of intellectual property rights, as discussed earlier in this chapter, associated with the enactment of and amendments to the Plant Variety Protection Act and with the extension to living organisms of utility-patent protection. However, much of the increase in private spending occurred in the agricultural chemical industry (Table 4.6), which, in common with the other chemical industries, has long had strong intellectual property protection. Although private spending on breeding has also increased in real terms, Venner (1997) argues that changes in the intellectual property regime have had little impact on private activity, at least in the wheat-breeding industry. However, as in the public sector, the growth in private spending on agricultural R&D has slowed considerably in recent years. The annual rate of growth in real spending on private agricultural R&D was 3.3 percent throughout the 1960s and 1970s, but dropped to only 1.5 percent after 1980.

The focus of private agricultural research has also shifted (Figure 4.7). In 1960 agricultural machinery and post-harvest and food-processing research accounted for about 81 percent of total private agricultural R&D. By 1992 these areas of research accounted for only 42 percent of the total, with the share of total private-sector research directed toward agricultural machinery dropping from 36 percent in 1960 to around 12 percent. Two of the more significant growth areas were plant breeding and veterinary and pharmaceutical research. Spending on agricultural chemicals research grew most quickly and now accounts for about one-third of total private agricultural R&D.

These data point to a dramatic shift in the pattern of publicly and privately conducted agricultural R&D in the United States over the past three decades. If these patterns of change continue, as seems likely, they will have major implications for the structure, conduct, and content of public research at both the state and federal levels.

Summary on Funding Trends

In recent years there has been a significant slowdown in the growth rate of public-sector support for agricultural research and extension. For public agencies, the most dramatic decline has been in the growth rate of state-government support for the SAESs. Private-sector spending growth has also slowed, even eclipsing the public-sector slowdown during the 1990s. However, except in the 1990s, the overall trend over the past few decades has been for increases in both the private-sector share of total R&D spending and the private-sector's contributions to public-sector research. In the 1960s, on average, the public sector spent US$1.06 for every dollar of private agricultural R&D; by 1992 the public sector spent only US$0.76 for every dollar of private agricultural R&D.

In a global context, although U.S. agricultural R&D investments have continued to grow, the U.S. share of global agricultural R&D has declined because agricultural R&D growth rates have been greater in other countries,

TABLE 4.6 Private-sector agricultural R&D spending

Year	Input-Oriented				Post-Harvest and Food Processing	Total	
	Agricultural Chemicals	Machinery	Veterinary and Pharmaceutical	Plant Breeding		Current	Real
	(millions U.S. dollars)						(millions 1993 U.S. dollars)[a]
1960	27	75	6	6	92	206	1,249
1965	64	96	23	9	131	323	1,611
1970	98	89	45	26	206	464	1,724
1975	169	138	79	50	273	709	1,944
1980	395	363	111	97	488	1,453	2,912
1985	683	304	159	179	842	2,167	3,076
1990	1,127	360	245	314	965	3,012	3,309
1991	1,227	382	276	342	946	3,173	3,320
1992	1,279	394	306	400	1,038	3,416	3,479
Annual growth rates							
	(percentages)						
1960–1992	12.8	5.3	13.1	14.0	7.9	9.2	3.3
1960–1970	13.8	1.7	22.3	15.8	8.4	8.5	3.3
1970–1980	15.0	15.1	9.4	14.1	9.0	12.1	5.4
1980–1992	10.3	0.7	8.8	12.5	6.5	7.4	1.5

SOURCE: USDA, ERS (1995, Table 2).

[a]Deflated with a revised and updated version of the U.S. agricultural R&D deflator from Pardey, Craig, and Hallaway (1989).

FIGURE 4.7 Composition of private-sector R&D

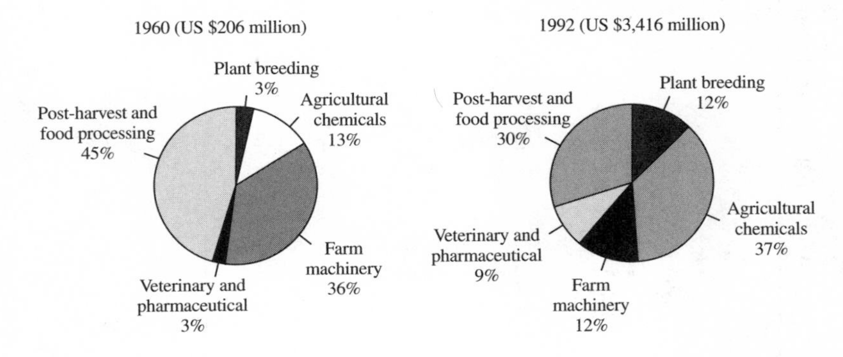

SOURCE: USDA, ERS (1995, Table 2).

especially less-developed countries. Moreover, the pattern of public spending has become increasingly subject to political intervention, as earmarked funds and special grants have been more important. Competitive grants based on peer review should be less vulnerable to political factors, but to date these have accounted for a very small fraction of the total R&D effort. Table 4.7 summarizes these major trends.

Performance of U.S. Public Agricultural R&D

Spending patterns provide evidence about the choices made by research managers in response, at least partially, to funding structures. A number of elements of research choice may be of concern: (1) the character of work (development and basic and applied research) and the related disciplinary orientation; (2) whether research should focus on outputs from production, resources used in production, or other areas altogether; (3) commodity-oriented or more generic research; and (4) types of expenditure, especially between direct expenditures on research personnel, buildings, equipment, and materials.

Character and Orientation of R&D

A critically important question is whether federal agricultural R&D policy changes have affected the nature of science being undertaken in the USDA or the SAESs. The following tables show the changing balance of research effort among fields of science and the basic-applied-developmental orientation of R&D between 1970 and 1993.

FIELD OF SCIENCE. As can be seen in Table 4.8, the biological sciences are the most important area within agricultural science. Since 1970 the share of resources devoted to biological sciences has risen from 58 to 68 percent of the

TABLE 4.7 Summary of funds available for U.S. agricultural R&D

	1950–1959	1960–1969	1970–1979	1980–1989	1990	1997	Annual Growth Rate (1950–1997)
Funds available			(millions U.S. dollars per year)				(percentages)
SAES research	89.55	200.18	483.68	1,212.89	1,766.07	2,234.55	8.06
USDA research	46.08	109.32	258.58	500.37	614.08	653.68	6.78
Total public research	135.63	309.50	742.26	1,713.25	2,380.15	2,888.23	7.70
Total private research	na	300.30	746.70	2,053.40	3,012.00	3,416.00[a]	9.17[b]
Extension	99.62	186.70	415.33	955.88	1,264.05	1,429.15[c]	6.98[d]
Real funds available			(millions 1993 U.S. dollars per year)				
SAES research	679.19	983.21	1,331.92	1,752.52	1,940.22	1,982.00	2.71
USDA research[e]	347.75	539.23	711.77	732.69	674.63	579.80	1.49
Total public research	1,026.95	1,522.44	2,043.68	2,485.20	2,614.85	2,561.80	2.37
Total private research	na	1,472.77	2,044.86	2,964.94	3,309.01	3,478.71[a]	3.25[b]
Extension	760.93	924.47	1,168.21	1,390.08	1,388.69	1,390.2[c]	1.55[d]
Funding ratios			(dollars per dollar)				
SAES research/total public research	0.66	0.65	0.65	0.70	0.74	0.77	0.34
SAES research/USDA research	1.97	1.83	1.89	2.40	2.88	3.42	1.20
Public research/private research	na	1.04	1.01	0.84	0.79	0.76[a]	–0.98[b]
Total public research/extension	1.34	1.64	1.83	1.79	1.88	1.92[c]	1.06[d]

SOURCE: SAES and USDA agricultural R&D series, see Table 4.1; private agricultural R&D, see Table 4.6; extension series, see Table 4.2.

[a]1992 figure.
[b]Growth rate for 1960–1992 period.
[c]1994 figure.
[d]Growth rate for 1950–1994 period.
[e]Deflated with a revised and updated version of the U.S. agricultural R&D deflator from Pardey, Craig, and Hallaway (1989).

TABLE 4.8 R&D expenditures by field of science, U.S. Department of Agriculture (USDA) and state agricultural experiment stations (SAESs)

	USDA			SAESs		
	1970	1980	1983	1970	1980	1993
	(percentages)					
Biological sciences	58.2	62.9	67.8	77.4	76.8	76.1
Biochemistry and biophysics	5.2	5.5	5.0	7.8	6.5	6.4
Environmental, systematic, and applied biology	13.1	14.5	16.5	18.0	19.0	17.3
Genetics and breeding	7.7	6.8	7.5	11.0	10.8	9.8
Other[a]	32.2	36.1	38.8	40.6	40.5	42.6
Physical sciences	32.9	27.2	23.0	13.7	14.0	14.6
Chemistry	12.8	10.0	8.2	5.4	5.0	5.2
Engineering	13.3	9.0	6.8	4.9	5.2	5.1
Other	6.8	8.2	8.0	3.4	3.8	4.3
Economics	8.2	8.6	8.3	6.2	5.9	5.8
Other fields	0.7	1.3	0.9	2.7	3.3	3.5
Total	*100.0*	*100.0*	*100.0*	*100.0*	*100.0*	*100.0*

SOURCE: Compiled by the authors from USDA, Current Research Information System data tapes.

[a]Includes molecular biology, entomology, immunology, microbiology, nematology, nutrition and metabolism, parasitology, pathology, pharmacology, physiology, and virology.

USDA's intramural R&D; in the SAESs this share has essentially held steady, declining slightly from 77 to 76 percent. Correspondingly, between 1970 and 1993 the shares accounted for by the physical sciences—especially chemistry and engineering—have declined markedly from 33 to 23 percent. During the same period, the SAES share of physical science R&D increased slightly, from about 14 to 15 percent. The share of economics in R&D expenditures has remained stable, accounting for about 8 percent of USDA research and about 6 percent of SAES research. In general terms, the research portfolios of the SAESs and the USDA's intramural agencies have become more similar over time.

BASIC, APPLIED, AND DEVELOPMENTAL RESEARCH. The classification of research activities into basic, applied, or developmental research is generally subjective and usually difficult.[22] Legitimate questions may be asked about whether the data reflect genuine differences rather than simply different perceptions of the distinctions between these classes of R&D. Perhaps worse, the data could reflect cynical or self-serving responses to changing impressions

22. The definitions of different types of research that are meaningful for scientists may not be as meaningful for the economic analysis of science. In any event, there is little evidence of a widespread, shared view of the real meanings of the terms.

TABLE 4.9 Basic, applied, and development R&D by U.S. Department of Agriculture (USDA) and state agricultural experiment stations (SAESs), selected years

	USDA			SAESs		
	1970	1980	1993	1970	1980	1993
			(percentages)			
Basic	40.9	40.5	48.4	36.7	37.2	47.0
Applied	55.3	51.8	43.1	54.7	54.8	46.3
Developmental	3.8	7.7	8.5	8.6	8.0	6.7
Total	*100.0*	*100.0*	*100.0*	*100.0*	*100.0*	*100.0*

SOURCE: Compiled by the authors from USDA, Current Research Information System data tapes.

about which types of R&D would be more heavily rewarded or supported; that is, the description of the work changes more than the nature of the work actually being undertaken. In any event, Table 4.9 shows some interesting similarities between USDA intramural R&D programs and SAES R&D programs.

In 1970, the USDA spent 41 percent of its budget on basic research and 55 percent on applied research. The proportions for the SAES budgets were similar: 37 percent and 55 percent. Between 1970 and 1993—and more particularly after 1980—the share allocated to basic research rose to 48 percent in the USDA budget and 47 percent in the SAES budget. At the same time, applied research fell to 43 percent in the USDA and 46 percent in the SAESs. Between 1970 and 1993, developmental research rose from 4 to more than 8 percent in the USDA's intramural R&D; it fell from about 9 to about 7 percent in the SAESs.

The differential patterns of disciplinary orientation provide somewhat more support for a distinctive and substantial intramural presence in agricultural R&D than does the comparison of the character of funded research. The large, self-identified federal intramural program in "basic research" seems anomalous. First, it is not clear what is basic research in the context of agricultural research. Basic research is usually defined as work for which no expected application can be identified, while by its very nature agricultural R&D is intended to provide results that can be used in the agricultural industry. To some extent the same critique applies to the apparent emphasis on basic research in the SAES system, although there are certainly many scientists in the SAESs conducting basic research, especially in fields related to modern biology. Yet a description of any commodity-oriented research as "basic" appears to be a misuse of the term. Second, if there is such a thing as basic agricultural research, one may question whether it can generally be performed well within the administrative and managerial styles of a government agency.

Although there have been exceptions, historically in the United States basic research has been the domain of university departments, which have traditions of self-directed, investigator-initiated research; management by peer review; extensive publication; and direct and informal dissemination of results through graduate students and postdoctoral research associates.

COMMODITY ORIENTATION. The commodity orientation of public-sector agricultural R&D and the balance between commodity-oriented and noncommodity R&D have varied among states and over time, as well as between the SAESs and the intramural USDA research programs. Table 4.10 summarizes R&D expenditures in 1970, 1980, and 1993 for each major commodity group for both SAES and intramural USDA programs. On the whole, over the 23-year period commodity-group shares have been fairly stable and similar between the two sets of institutions. The USDA intramural research program and the SAESs both spent about one-third of their budgets on crop-related R&D, about one-fifth on livestock research, and about one-half on other areas (including fish, forestry, and other resource-related areas). More important differences are found within these categories.

Within the USDA intramural programs the share of total expenditures allocated to crop research in 1993 (36 percent) is very close to its value in 1970. Within crop research, however, some important changes have occurred. In particular the share of total USDA R&D resources devoted to corn has doubled, rising from under 2 to nearly 4 percent, while USDA cotton and tobacco research programs have declined in relative terms. In contrast to crop research, over the same period the share of USDA's intramural research resources devoted to livestock research has fallen from about 19 to about 15 percent. This overall decline in livestock's share of USDA R&D performance is mirrored in declines in each category except swine. Considering the small size of the U.S. sheep industry (with marketings of US$489 million in 1993, which is about 0.5 percent of total livestock production), it is notable that the share of sheep and wool research was almost as large as that of swine in 1970 (a large industry with 1993 farm marketings of US$10.9 billion) and remains relatively large.

The slight decline in the share of livestock research allowed modest increases in the remaining categories, which generally concerned natural resources and related food and fiber industries. The share of revenues allocated to these industries increased from 45.3 percent in 1970 to 48.6 percent in 1993. Thus, there is not much evidence of a changing commodity orientation (or commodity versus noncommodity orientation) in USDA intramural agricultural research.

R&D spending patterns within the SAESs have been remarkably similar to those in the USDA intramural programs but with some differences in the patterns of change (see also Figure 4.8). In the SAESs the shares of both crop

TABLE 4.10 R&D expenditures by commodity, U.S. Department of Agriculture (USDA) and state agricultural experiment stations (SAESs)

	USDA			SAESs		
	1970	1980	1993	1970	1980	1993
			(percentages)			
Crops	35.3	30.9	36.0	37.6	35.3	31.8
Corn	1.7	2.3	3.9	3.3	2.9	2.8
Rice	0.5	0.4	0.6	0.4	0.5	0.8
Wheat	2.3	2.1	3.0	2.5	2.0	2.2
Soybeans	1.0	2.6	3.1	2.0	3.2	2.4
Fruit	5.5	4.5	5.5	7.4	6.1	4.9
Vegetables[a]	4.1	3.5	4.7	6.4	6.5	6.5
Cotton[b]	8.0	4.7	4.5	2.7	1.8	1.3
Tobacco	2.6	0.9	0.7	1.3	0.8	0.7
Range, pasture, and forage	3.7	4.7	4.6	4.5	4.9	3.6
Other	6.0	5.3	5.3	7.1	6.5	6.5
Livestock	19.4	16.4	15.4	26.3	24.8	23.3
Poultry	3.3	2.2	2.7	5.2	3.7	3.1
Beef cattle	5.3	5.4	4.3	7.6	8.7	6.8
Dairy cattle	5.1	3.5	3.3	6.9	6.0	5.2
Swine	2.5	2.8	3.0	3.1	3.4	3.1
Sheep and wool	2.1	1.5	1.2	1.9	1.4	1.5
Other	1.0	1.0	1.0	1.5	1.6	3.5
Natural and recreational resources	13.3	12.1	13.0	8.5	8.8	9.1
Trees, forest, and forest products	17.3	16.9	15.9	4.6	6.4	6.9
Fish and game	0.7	1.6	2.3	2.3	2.9	4.4
Human resources, organizations, and institutes	6.3	7.5	9.5	5.1	5.6	5.6
Other	7.7	14.5	7.9	15.5	16.1	18.8
Total	*100.0*	*100.0*	*100.0*	*100.0*	*100.0*	*100.0*

SOURCE: Compiled by the authors from USDA, Current Research Information System data tapes.

[a]Includes potatoes.

[b]Includes cottonseed.

and livestock research in total expenditure declined between 1970 and 1993. The share of R&D resources allocated to crops decreased from 37.6 to 31.8 percent. The livestock share declined from 26.3 to 23.3 percent.

Within crops the pattern of spending by the SAESs is quite similar to that of the USDA's intramural research (including a decline of tobacco and cotton research). Within livestock research SAES support for poultry research has declined, perhaps reflecting a response to the rising concentration of the indus-

FIGURE 4.8 R&D expenditures in state agricultural experiment station system by research program group, 1970–1993

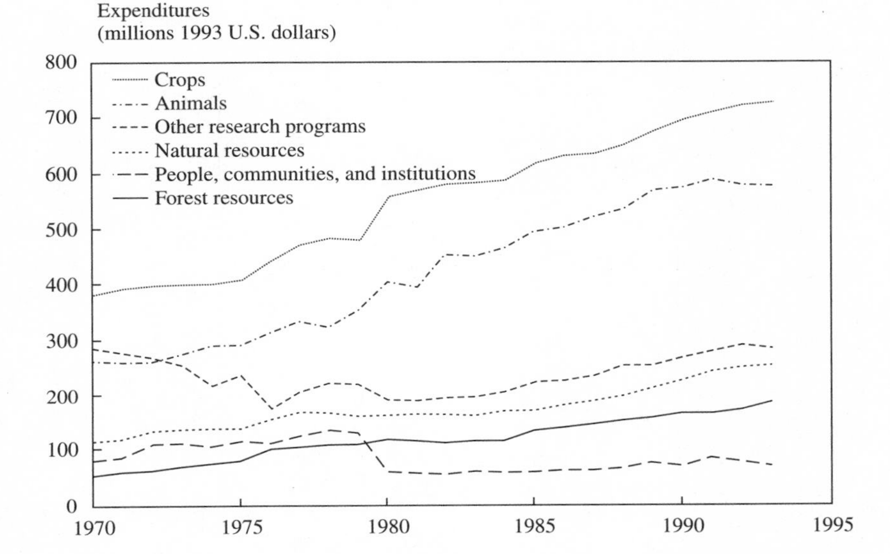

SOURCE: Compiled by the authors from USDA, Current Research Information System data tapes.

try and the increased role of the private sector in poultry R&D.[23] The SAESs still spend 1.5 percent of their R&D budget on sheep and wool.

The larger declines in the shares of crop and livestock research were coupled with larger increases in the remaining categories' share of SAES research. Thus SAES research on noncommodity areas increased from 36.0 percent in 1970 to 44.8 percent in 1993, which is much closer to, but still below, the share of USDA intramural research allocated to these topics. It is not clear whether the adjustments in R&D resource allocations by the SAESs over the 1970–1993 period reflect a greater responsiveness on the part of the SAESs to changing research demands or a delayed response to trends recognized and incorporated more quickly into USDA's intramural programs. It is noteworthy that the mix of expenditures within the noncommodity areas differs more substantially between SAES and USDA intramural programs than in the cases of crops and livestock.[24]

---

23. This is not to say that past public-sector poultry research has not been socially profitable. Peterson (1967) used both an index-number approach to estimate the research-induced shift in poultry supply functions and an econometric estimate of a research-augmented production function to estimate rates of return to SAES poultry research of between 20 and 30 percent per year.

24. This may not reflect reality so much as a desire to be seen as responsive to emerging priorities for "greener" research.

TABLE 4.11 State agricultural experiment stations' (SAESs') R&D expenditures on selected commodities, grouped by farm production regions, 1993

Region	Crops					Livestock		
	Corn	Rice	Wheat	Fruits and Vegetables	Cotton	Beef	Dairy	Swine
				(percentages)				
Northeast	2.2	0.3	0.5	13.8	0.0	1.1	6.8	1.3
Great Lake	2.9	0.2	0.7	9.4	0.0	3.1	9.3	5.2
Corn Belt	7.9	0.2	1.2	5.0	0.2	6.5	5.7	9.3
Northern Plains	4.9	0.2	9.6	3.9	0.0	15.2	3.6	5.7
Appalachian	4.0	0.7	0.9	8.6	0.8	5.4	6.3	4.1
Southeast	1.1	0.5	1.0	18.2	1.4	7.5	3.9	2.5
Mississippi Delta	0.6	6.5	1.6	8.2	7.0	6.7	4.9	1.6
Southern Plains	2.2	1.6	4.0	6.7	5.6	12.1	3.3	1.5
Rocky Mountain	1.5	0.0	4.2	7.8	2.0	16.7	4.9	0.5
Pacific	0.6	0.4	2.3	18.8	0.6	3.1	2.9	0.6
Total SAESs	*2.8*	*0.8*	*2.2*	*11.0*	*1.3*	*6.9*	*5.2*	*3.2*

SOURCE: Compiled by the authors from U.S. Department of Agriculture, Current Research Information System data tapes.

Among the major production regions there are some even greater differences in the SAESs' R&D investment patterns (Table 4.11). Primarily, these patterns reflect the geographic distribution of agricultural production (for example, corn research is carried out in Corn Belt states while cotton research is carried out in California and the Mississippi Delta states).

The pattern of concentration of livestock research is perhaps even more striking. For example, beef research ranges from 1 percent of the total R&D budget in the Northeast to 17 percent in the Mountain states. Reflecting the fact that every state produces some milk, the dairy industry's share of total R&D budgets is generally above 3 percent and is about 6 percent or more of research expenditure in the major dairy regions in the Great Lake states, Corn Belt, and Northeast. The dairy industry's share of the total R&D budget is also above 6 percent in the Appalachian region, which is not known for dairy production, and lowest of all in the Pacific region, which is dominated by California, the fastest growing and most efficient milk producer. However, there are many competing demands in California's agricultural R&D budget.

The geographic patterns of research specialization are interesting in relation to questions about the regional distribution of federal commodity-specific research resources. As discussed by Alston and Pardey (1995a, 1996), there seems to be potential for a greater degree of regional specialization in research, perhaps through coordination of individual SAESs, federal initiatives, or the use of funds raised by industry checkoff arrangements. For some commodities (for example, rice or corn) the potential for regional specialization in research may be greater than others (for example, dairy). The extent of this potential is not fully revealed by the data because they do not show whether research programs conducted in individual SAESs are broadly applicable and duplicative or region-specific, adaptive research. Nor do the data provide insights about the extent to which economies of scale, size, or scope in research might be exploited. However, the data do provide insights about potential directions for further research on the allocation of R&D funds.

Type of Expenditure

A fundamental distinction is made between expenditure to support the employment of research personnel and expenditure on other costs of research, such as equipment, buildings, and administrative support. The quantity and quality of research results is directly related to the numbers of scientists and engineers engaged in research. We present here data on the changing stock of researchers.

AGRICULTURAL RESEARCH PERSONNEL. As research expenditures have increased, so too have the numbers of research personnel. In 1980 the SAESs employed approximately 13,500 scientists, which provided about 8,300 full-time equivalent years of effort when allowance was made for teaching and extension functions (Table 4.12). In addition to SAES staff, significant numbers of researchers employed by USDA or funded by other federal agencies are located at the SAESs and work in cooperation with SAES staff. In 1980, the latest year for which we have compiled the necessary figures, there were about 1,600 USDA staff and about 760 additional staff working on various cooperative agreements and located at the SAESs. Table 4.13 gives regional details of the evolving pattern of SAES and cooperating scientists working on experiment station research throughout the various USDA farm regions.

The composition and deployment of SAES personnel has changed substantially over time (Table 4.14). When the SAES system began, more than two-thirds of its scientists were trained to either B.S. or M.S. levels and about one-half of these scientists worked full time on research. By the 1970s over 70 percent of SAES professional staff held Ph.D.s. Immediately after World War II, there was a substantial shift away from full-time research appointments so that by the mid-1980s only one-quarter of SAES scientists worked full time on research. More than half of all SAES professionals had joint research and teaching appointments and about 15 percent of SAES personnel who did some research were also involved in extension work.

SPENDING PER SCIENTIST. Research into animal production has traditionally been regarded as more expensive than plant research because of the need for larger and more expensive experimental units. This perception is borne out by the data in Figure 4.9, which shows real spending (1993 prices) per scientist in the SAESs as differentiated by research areas. Spending per livestock scientist, however, has been flat since the late 1970s, whereas spending per plant scientist and per natural resource scientist has continued to climb throughout the period 1970–1993. The distinction between livestock and plant scientists has narrowed considerably. This may reflect the growth of more-basic science, such as molecular biology, with its attendant relatively heavy capital costs. The science of livestock and crop research has certainly changed. For example, animal scientists used to work on whole and frequently large animals. Now they often work on parts of cells, in ways that are similar to those used by their colleagues in other disciplines.

Conclusion

The U.S. agricultural economy is supported by a complex system of R&D agencies, which has been historically dominated by publicly funded institutions but with a growing and now significant private-sector presence. The two main components of this system, the research agencies of the USDA and the SAESs, have developed in parallel since their beginning in the second half of the nineteenth century and today perform a wide variety of R&D activities. The size and nature of these activities and their effects on the agricultural sector and the rest of the economy are influenced by the funding arrangements that support and to some extent direct agricultural R&D. The changing nature of

TABLE 4.12 Research personnel trends in the state agricultural experiment stations (SAESs)

	Number of Research Personnel				Full-Time Equivalents[a]			
Fiscal year	USDA Co-ops	Other Co-ops	SAESs	Total	USDA Co-ops	Other Co-ops	SAESs	Total
1890	0	0	325	325	0	0	244	244
1900	0	0	650	650	0	0	484	484
1910	0	0	1,321	1,321	0	0	966	966
1920	55	0	1,752	1,807	55	0	1,217	1,272
1930	123	0	3,129	3,252	123	0	2,322	2,445
1940	350	1	4,179	4,530	350	1	3,000	3,351
1950	497	3	6,069	6,536	497	3	4,275	4,775
1960	1,162	59	9,021	10,242	1,162	59	6,171	7,392
1970	1,534	276	11,068	12,878	1,534	276	7,071	8,881
1980	1,627	762	13,483	15,872	1,627	762	8,347	10,736

SOURCE: Alston and Pardey (1996, Table 2-6).

NOTE: Data are for SAESs and other cooperating institutions.

[a]Counts of SAES researchers converted to full-time equivalents using the following scaling factors: 1.0 for personnel with research-only appointments, 0.5 for personnel with research-teaching or research-extension appointments, 0.3 for personnel with research-teaching-extension appointments.

TABLE 4.13 Number of state agricultural experiment stations (SAESs) and cooperator research staff by farm production region

	1890–1899		1910–1919		1930–1939		1950–1959		1970–1979		1985	
Region	SAES	Co-op	SAES	Co-op	SAES	Co-op	SAES	Co-op	SAES	Co-op	SAES	Co-op
Northeast	94	0	287	2	680	7	1,245	11	1,814	78	2,039	252
Corn Belt	42	0	237	6	543	16	876	81	1,474	192	1,656	197
Great Lake	30	0	130	3	349	15	539	29	963	84	1,066	104
Northern Plains	28	0	138	10	218	21	441	54	854	123	1,067	250
Appalachian	30	0	120	15	266	11	604	42	1,185	192	1,519	244
Southeast	24	0	83	5	192	9	439	62	956	261	1,443	308
Mississippi Delta	29	0	61	3	141	6	331	23	708	111	1,156	169
Southern Plains	5	0	33	2	135	4	351	55	561	211	912	268
Rocky Mountain	22	0	121	7	316	20	530	73	1,105	317	1,140	440
Pacific	20	0	110	5	291	16	681	72	1,449	243	1,750	268
Total United States[a]	*325*	*0*	*1,321*	*58*	*3,129*	*123*	*6,036*	*500*	*11,068*	*1,810*	*13,748*	*2,500*

SOURCE: Alston and Pardey (1996, Table 2-7).

[a]Includes the 48 contiguous states (that is, excludes Alaska and Hawaii). Totals subject to rounding error.

TABLE 4.14 Degree and research appointment status of state agricultural experiment station professional staff

	1890	1910	1930	1950	1970	1985
			(percentages)			
Degree status						
Ph.D.	29.3	19.7	32.7	47.1	72.5	na
M.S.	31.3	29.2	42.1	33.0	19.2	na
B.S.	39.4	51.1	25.2	20.0	8.3	na
Research appointment status						
Research only	50.1	46.2	49.8	43.4	30.4	25.7
Research and teaching	49.9	53.8	44.0	50.6	57.4	58.6
Research and extension	0.0	0.0	2.2	2.1	4.2	6.1
Research, teaching, and extension	0.0	0.0	4.0	3.9	8.0	9.6

SOURCE: Alston and Pardey (1996, Table 2-8).

NOTE: na indicates not available.

FIGURE 4.9 State agricultural experiment station R&D spending per scientist by research program group, 1970–1993

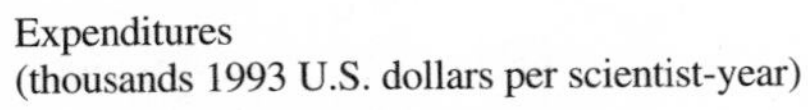

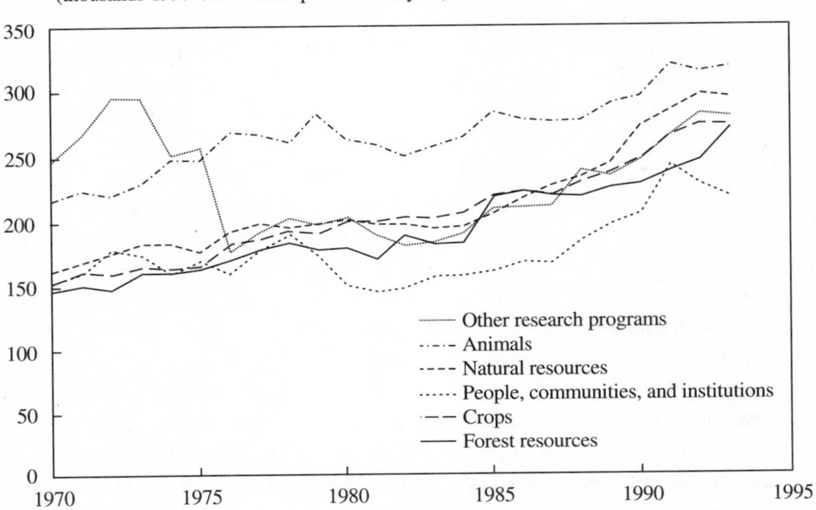

SOURCE: Compiled by the authors from USDA, Current Research Information System data tapes.

these funding arrangements and the concurrent evolution of U.S. public agricultural R&D institutions have been the subject of this chapter.

From the beginnings of the U.S. agricultural research system, policymakers have recognized the essentially regional nature of many agricultural problems and much agricultural research, along with the national interest in resolving such regional problems. A primary benefit of agricultural research—a stable and low-cost food supply—is shared by the entire nation, although the achievement of that result requires local efforts. At the same time, some part of the benefits from successful R&D is realized by the agricultural sectors and by rural communities. There are possibly substantial local benefits as well.

The balancing of national and local benefits and mostly local effort continues to be a major challenge of federal science policy for agriculture. The national interest in local research has been expressed clearly through formula funds and grants and contracts administered by the USDA and disbursed to the SAESs. The recognition of local benefits from research is seen in the requirement that federal funds be matched by state resources. The states themselves have found sufficient local benefits from research to make major expenditures, frequently in excess of matching requirements, to support their local agricultural research endeavors.

Of course not all agricultural research is local or regional. The progress of commodity-specific research programs depends in part upon a strong foundation of basic science, which by its nature has no predictable field or region of application. To some extent agricultural research has benefited from the achievements of the R&D activities financed by the NSF and NIH, which have supported many of the revolutionary advances in the biological sciences in the last quarter century. However, between the most fundamental work in immunology and biochemistry and application in breeding and pest control there lies a vast gray area where the work is essential for commodity-specific applications and is often of identifiably agricultural interest but is also potentially applicable to many commodities and regions.

The support of the national infrastructure of more fundamental agricultural science is the second major challenge. To a large extent this objective, the development of nationally applicable agricultural knowledge, has been a central mission of the intramural federal research of the USDA, and grants and contracts to the SAESs have in part supported this program; more recently, competitive-grants programs have supported the same mission. The SAESs have carried out much of this work as well.

Although the existing funding structures and institutions have in large measure succeeded at meeting the broad set of challenges facing U.S. agricultural science policy, as evidenced by the size and achievements of the U.S. agricultural research system and the outstanding productivity of U.S. agriculture, these structures are under strain. Federal support for agricultural R&D, along with other federal spending activities, faces tightening fiscal constraints.

Within the system there are other sources of strain. First, the formulae that served for the past half century as the principal expressions of the national interest in local research appear to be out of date and to direct substantial research funds to states that have ceased significant agricultural production. Second, although matching-fund provisions have worked fairly well in eliciting state-supported agricultural R&D in some states, particularly California and Florida, which have distinctive climates and growing conditions, they seem to have been far less effective at eliciting state support in regions where many neighboring states compete in the production of a commodity, such as the corn- and wheat-producing states of the Midwest. Third, the articulation of performance and activity between an intramural federal effort that ostensibly concentrates on problems of national application and an SAES system that is devoted to commodity-specific and local research programs seems to be largely obsolete. Moreover, the slow growth of competitive-grants programs and the continued prominence of federal research facilities indicates that institutional arrangements have been slow to reflect this reality. Finally, congressional involvement in the details of research management has led to increasing appropriations of funds for reasons that seem to fall outside the bounds of traditional science policy and interfere with the central mission of achieving a stable, low-cost, and internationally competitive feed, food, and fiber supply while preserving environmental resources through the efficient conduct of agricultural R&D.

APPENDIX TABLE A4.1 Source of funds for state agricultural experiment stations (SAESs) and other cooperating institutions

	Federal				Nonfederal					Grand Total	
Fiscal Year	CSREES Administered[a]	USDA[b]	Other[c]	Total	State Government	Self-generated Funds	Industry[d]	Other	Total	Nominal	Real[e]
					(millions U.S. dollars)						(millions 1993 U.S. dollars)
1972	71.5	7.0	28.2	106.7	205.5	23.2	16.6	11.0	256.3	363.0	1,213.9
1973	78.2	7.7	29.6	115.4	222.1	28.1	17.7	11.7	279.6	395.1	1,253.1
1974	83.2	8.8	32.0	124.0	247.4	32.4	21.0	12.2	313.0	437.0	1,293.6
1975	92.0	11.1	35.3	138.4	284.7	37.3	24.0	15.0	361.0	499.4	1,369.6
1976	104.8	10.5	40.8	156.1	309.7	30.7	28.3	16.4	385.1	541.3	1,388.0
TQ 1976[f]	26.2	2.6	10.2	39.0	77.4	7.7	7.1	4.1	96.3	135.3	347.0
1977	118.9	12.6	55.6	187.0	321.2	39.1	32.7	21.9	414.8	601.8	1,483.1
1978	134.5	16.5	57.9	208.8	374.9	40.1	34.7	22.4	472.1	680.9	1,582.5
1979	156.3	21.1	64.6	242.1	413.5	46.7	37.1	27.2	524.5	766.6	1,663.0
1980	162.8	27.5	71.6	261.9	456.4	55.9	48.4	30.5	591.3	853.1	1,709.7
1981	174.3	33.3	83.0	290.6	501.2	59.1	53.5	38.2	652.1	942.7	1,728.2
1982	199.2	36.2	107.6	343.0	545.2	62.5	61.3	45.5	714.6	1,057.6	1,790.3
1983	204.9	38.9	95.2	339.0	576.5	65.4	66.7	49.1	757.7	1,096.7	1,702.5
1984	210.5	38.5	103.2	352.3	621.8	66.3	71.0	54.4	813.5	1,165.7	1,748.6
1985	221.0	35.9	112.4	369.4	678.3	70.5	79.1	61.5	889.3	1,258.7	1,786.7
1986	222.7	35.8	140.6	399.1	741.7	69.4	85.1	70.2	966.5	1,365.6	1,829.5
1987	230.8	36.8	148.1	415.7	778.8	75.4	93.8	85.1	1,033.1	1,448.8	1,831.7

1988	247.8	42.2	153.5	443.5	823.4	84.8	99.1	91.1	1,098.3	1,541.8	1,874.8
1989	261.0	48.9	169.7	479.6	894.4	92.4	111.3	102.1	1,200.2	1,679.8	1,932.7
1990	272.8	54.1	188.6	515.5	950.1	102.4	126.6	112.4	1,291.5	1,807.0	1,985.2
1991	290.8	57.8	199.4	548.0	985.9	113.6	134.0	114.8	1,348.4	1,896.3	1,984.5
1992	316.6	60.7	221.3	598.7	981.5	116.1	143.4	121.0	1,362.1	1,960.7	1,996.7
1993	331.0	68.6	249.0	648.5	985.4	110.0	146.1	134.8	1,376.3	2,024.8	2,024.8
1994	353.2	79.8	270.0	703.0	1,010.9	116.7	152.9	148.2	1,428.7	2,131.7	2,073.6
1995	356.5	83.8	285.8	726.1	1,048.9	114.2	155.3	148.4	1,466.8	2,192.9	2,060.6
1996	344.1	83.1	288.3	715.6	1,049.5	120.5	162.5	164.2	1,496.8	2,212.4	2,022.3
1997	339.6	88.4	290.9	718.9	1,081.7	127.4	177.7	168.5	1,555.3	2,274.2	2,017.2
Annual growth rates											
						(percentages)					
1972–1997	6.4	10.7	9.8	7.9	6.9	7.1	10.0	11.5	7.5	7.6	2.1
1990–1997	3.2	7.3	6.4	4.9	1.9	3.2	5.0	5.9	2.7	3.3	0.3

SOURCE: 1972–1993, Alston and Pardey (1996, Table 2-2); 1994–1997, USDA, Current Research Information System (various years, Table IV-E).

NOTE: Includes all 50 SAESs, forestry schools, 1890/Tuskegee institutions, veterinary schools, and other cooperating institutions.

[a]Includes formula funds, special grants, and competitive grants.
[b]Includes monies received from USDA grants, contracts, and cooperative agreements.
[c]Includes contracts and grants received from agencies such as the National Science Foundation, Energy Research and Development Administration, Department of Defense, National Institutes of Health, Public Health Service, National Aeronautics and Space Administration, and Tennessee Valley Authority.
[d]Includes funds received through industry grants and agreements.
[e]Deflated with a revised and updated version of the U.S. agricultural R&D deflator from Pardey, Craig, and Hallaway (1989).
[f]Includes appropriations for the period from July 1, 1976, to September 30, 1976, the so-called transition quarter (TQ) whereby the federal government moved from a fiscal year ending June 30 to one ending September 30.

APPENDIX TABLE A4.2 U.S. Department of Agriculture, appropriations for research and education

Year	CSREES Administered			Agricultural Research Service	Forest Service	Economics and Statistics			Total Research[a]	National Agricultural Library	Education			Total Research and Education[b]
	Competitive	Others	Total			Economic Research Service	Statistical Service	Total			Extension Service	Other	Total	
						(millions U.S. dollars)								
1970	0.0	62.7	62.7	160.1	45.6	na	na	17.0	290.3	0.0	146.2	4.8	151.1	441.4
1971	0.0	69.6	69.6	178.6	48.8	na	na	18.4	320.3	0.0	165.6	5.5	171.0	491.4
1972	0.0	83.0	83.0	191.7	54.4	na	na	18.8	358.0	0.0	182.2	6.1	188.3	546.3
1973	0.0	91.5	91.5	208.1	57.8	na	na	20.6	387.9	0.0	197.9	6.5	204.3	592.3
1974	0.0	90.1	90.1	205.0	64.7	na	na	22.0	386.9	0.0	206.7	6.8	213.6	600.4
1975	0.0	101.7	101.7	224.4	77.6	na	na	24.9	433.5	0.0	217.2	7.9	225.0	658.6
1976	0.0	114.5	114.5	282.8	82.3	na	na	28.9	516.0	0.0	230.2	8.3	238.4	754.4
TQ 1976[c]	0.0	28.6	28.6	64.4	22.3	na	na	7.4	124.6	0.0	56.0	2.1	58.1	182.8
1977	0.0	129.0	129.0	282.9	89.8	24.5	4.7	29.2	539.7	0.0	232.7	9.2	241.9	781.6
1978	15.0	142.9	157.9	313.9	90.6	26.0	5.0	31.0	609.8	6.6	257.5	20.8	278.3	894.7
1979	15.0	159.3	174.3	328.0	95.0	28.2	5.4	33.6	648.2	7.0	263.8	21.2	285.0	940.2
1980	15.5	170.4	185.9	358.0	95.9	26.1	5.0	31.1	687.5	7.3	274.0	21.5	295.5	990.3
1981	16.0	184.7	200.7	404.1	108.4	39.5	7.5	47.0	778.0	8.2	292.2	22.4	314.6	1,100.8
1982	16.3	204.3	220.6	423.2	112.1	39.4	7.0	46.4	816.3	8.2	315.7	11.0	326.7	1,151.2
1983	17.0	215.3	232.3	451.9	107.7	38.8	7.6	46.4	856.6	9.1	328.6	11.9	340.5	1,206.2
1984	17.0	220.7	237.7	471.1	108.7	44.3	8.2	52.5	886.7	10.4	334.3	18.2	352.5	1,249.6
1985	46.0	230.6	276.6	491.0	113.8	46.6	8.4	55.0	956.2	11.5	341.2	21.3	362.5	1,330.2
1986	42.3	227.3	269.6	483.2	113.6	44.1	8.0	52.1	945.1	10.8	328.0	18.4	346.4	1,302.3
1987	40.7	253.0	293.7	511.4	126.7	44.9	3.4	48.3	1,002.2	11.1	339.0	18.7	357.7	1,371.0

1988	42.4	260.7	303.1	544.1	132.5	48.3	3.6	51.9	1,054.6	12.2	358.0	19.8	377.8	1,444.6
1989	39.7	270.9	310.6	569.4	138.3	49.6	2.9	52.5	1,097.9	14.3	361.4	21.9	383.3	1,495.5
1990	38.6	288.0	326.6	593.3	150.9	51.0	2.8	53.8	1,155.3	14.7	369.3	28.6	397.9	1,567.9
1991	73.0	300.3	373.3	631.0	167.6	54.4	3.2	57.6	1,265.4	16.8	398.5	34.8	433.3	1,715.5
1992	97.5	316.9	414.4	670.6	180.5	59.0	3.6	62.6	1,371.4	17.8	419.3	35.8	455.1	1,844.3
1993	97.5	317.5	415.0	671.7	182.7	58.9	3.9	62.8	1,373.0	17.7	428.4	35.7	464.1	1,854.8
1994	112.1	325.2	437.3	679.2	193.1	55.2	3.5	58.7	1,413.6	18.2	434.6	37.5	472.1	1,903.9
1995	101.0	318.0	419.0	708.0	194.0	54.0	4.0	58.0	1,409.0	19.0	432.4	37.6	470.0	2,001.0
1996	94.0	307.0	401.0	700.0	178.0	53.0	4.0	57.0	1,371.0	20.0	419.0	44.0	463.0	1,933.0
1997	94.0	307.0	401.0	710.0	181.0	54.0	3.0	57.0	1,388.0	20.0	418.0	45.0	463.0	2,028.0
1998	97.0	312.0	409.0	740.0	189.0	72.0	3.0	75.0	1,452.0	19.0	415.0	45.0	460.0	2,002.0
Annual growth rates							(percentages)							
1970–1998	9.8[d]	5.9	6.9	5.6	5.2	5.3[e]	−2.1[e]	5.4	5.9	5.4[d]	3.8	8.3	4.1	5.5
1970–1980	8.2[d]	10.5	11.5	8.4	7.7	5.8[e]	−3.9[e]	6.2	9.0	6.9[d]	6.5	16.1	6.9	8.4
1980–1990	9.6	5.4	5.8	5.2	4.6	6.9	−5.6	5.6	5.3	7.3	3.0	2.9	3.0	4.7
1990–1998	12.2	1.0	2.9	2.8	2.9	4.4	0.9	4.2	2.9	3.3	1.5	5.8	1.8	3.1

SOURCE: 1970–1994, Alston and Pardey (1996, Table 2-3); 1995–98, unpublished USDA budget tables (1996, Table 8).

NOTE: na indicates not available.

[a]Also includes research undertaken by Agricultural Cooperative Service, Agricultural Marketing Service, Alternative Agricultural Research and Commercialization, Animal and Plant Health Inspection Service, Federal Grain Inspection Service, Nutrition Research and Education Service, and Foreign Agricultural Service.
[b]Includes columns designated as Total Research, National Agricultural Library, and Total Education.
[c]Includes appropriations for the transition quarter (TQ), which covers the period from July 1, 1976, to September 30, 1976.
[d]Growth rate for the periods 1978–1998 and 1978–1990 respectively.
[e]Growth rate for the periods 1977–1998 and 1977–1990 respectively.

5 Agricultural R&D Policy in Australia

JULIAN M. ALSTON, MICHAEL S. HARRIS,
JOHN D. MULLEN, AND PHILIP G. PARDEY

This chapter reviews the agricultural R&D institutions in Australia, particularly those in the public sector, and recent policy changes that affect them. Agricultural R&D institutions and policies have been subject to periodic public review in Australia. Documentation of arguments and data have appeared in the government green paper titled *Rural Policy in Australia* (Harris et al. 1974), the report following the Industries Assistance Commission inquiry in 1976, and many other subsequent reports including, most recently, the Industry Commission (1994b, 1995) reports on an inquiry into R&D generally. In addition to these public inquiries, agricultural R&D and policy changes have been studied in government, universities, and industry.[1] We draw heavily on those previous studies, integrate their findings with our own arguments and analysis, and use some new data.

The Australian public-sector agricultural R&D system is particularly interesting for several related reasons. First, Australia invests relatively heavily in public-sector agricultural R&D (agricultural research intensities are higher there than in most other industrialized countries). Although Australia provides little direct assistance to its agricultural sector, it provides more support than most countries for public-sector agricultural R&D; most developed countries provide more total assistance but mainly through price supports and other direct interventions in commodity markets that have become quite unimportant in Australian agriculture. Second, an important element of the Australian agricultural R&D system has been a system of research and development corporations (RDCs), which are funded by commodity taxes (or checkoffs) combined with matching government grants.

The recent rise in importance of the RDCs, especially in the past decade, has been an important change in the institutional structure and the nature of research funding and management in Australia, increasing the roles of primary producers in these areas. More recently a group of cooperative research centers (CRCs) has also formed and the structure and strategies of the RDCs have been

1. In addition, see Baker, Baklien, and Watson (1990), ABARE (1995b), and DIST (1995).

evolving. Understanding the implications of these institutional developments for efficiency and equity in public-sector agricultural R&D is an important objective of this chapter.

The remainder of this chapter comprises three main parts. The first part reviews Australia's public-sector R&D institutions, including the events leading to the recent policy changes, and draws some contrasts between Australia and other countries. The second part presents empirical information on the roles of the various research providers and sources of funds and attempts to show the influences of institutional changes on total funding for public-sector R&D, sources of funds, and disposition of funds. Finally, the third part draws inferences for policy.

Australia's Public-Sector R&D Institutions

Background and Introduction

Like most other industrialized countries Australia has a well-developed system for providing agricultural R&D services, with a high degree of public-sector involvement in both the financing and the performance of R&D. This in turn reflects the perception that although social returns to investment in rural R&D are high on average, private returns are often lower, which leads to an underinvestment problem when R&D is left to the private sector (see, for example, Harris and Lloyd 1991; Alston and Pardey 1996). Some, however (Mullen and Cox 1995; Alston, Craig, and Pardey 1998), question the magnitude of the problem.[2]

Government involvement in the Australian system of agricultural research began in the 1850s when the Victorian Board of Agriculture took over the first experimental farm that had been established by a farmers' committee (for details see Baker, Baklien, and Watson 1990). Subsequent developments in Australia's public agricultural R&D in many ways paralleled those in the United States, including the post–World War II boom that ended in the 1970s. As in most other countries, government involvement has been mainly in the form of government-funded R&D conducted by public agencies.[3]

From the 1930s to the present, mechanisms have progressively developed to facilitate a growing role for industry in providing funds and, perhaps to a

2. Rural research is the term commonly used in Australia to describe agricultural R&D (for example, Harris et al. 1974; IAC 1976; and Industry Commission 1995).

3. A distinction between "private" and "public" R&D may be drawn in terms of who pays for the R&D, who does the work, who uses the research results, or who benefits. Generally, here it refers to where the research is done, in the sense of which sector (private or public) has provided the facilities and expertise to perform the research. Some privately performed R&D can benefit from public funds in the form of subsidies or tax concessions; similarly public R&D can be partly financed by industry funds, as is commonly the case in Australia and somewhat the case in the United States.

lesser extent, in setting research directions. In the beginning industry R&D funding arrangements were relatively informal. These arrangements were partial, having evolved in a fragmented manner, and they lacked a coherent rationale (see, for example, Gleeson and Lascelles 1992; Smith 1992). The past 10 years have seen further changes in Australia's rural R&D system with the intent of formalizing and strengthening the private-sector's role in R&D, both as a source of finance and in determining where the R&D effort should be directed.

This evolution has culminated in the creation of the current RDCs, which now fund about one-third of all publicly performed agricultural R&D in Australia. The RDC model is a mechanism by which the factors that lead to underinvestment (that is, public-good characteristics of research, the difficulty of excluding free-riders, and the nonrival use of research findings) can be ameliorated to allow industry, a principal beneficiary, to assume greater responsibility for the funding and direction of research.

The critical step was taken in 1985, when commodity-specific R&D councils were established to help fund and direct agricultural R&D with levies on agricultural products. These councils were later turned into corporations with greater independence than the semigovernment councils, and, later, some new corporations with natural resource rather than commodity orientations were also established. Smith (1992) lists the essential features of this arrangement as follows:

- a defined industry (that is, defining who should contribute),
- a production levy on all producers based on the number of units of production of stipulated commodities,
- an agreement by government to match producers' contributions,[4]
- independent appointment of a competent board to manage the funds,
- nomination of an industry organization with which the board managing the funds can negotiate agreed goals and priorities,
- negotiation of research activity against these goals and priorities, and
- a high awareness by all parties of the need to adopt research findings into commercial practice.

The objectives of this arrangement were spelled out in the relevant legislation as[5]

4. Until recently the government typically matched industry funds dollar-for-dollar up to a limit of 0.5 percent of the gross value of production in the industry, although there were exceptions. A revision in these matching arrangements is under discussion following the Industry Commission's (1995) recommendations.

5. This is taken from Section 3 of the Primary Industries and Energy Research and Development Act of 1989. Note that the original wording of the first subpoint referred to increasing "the economic, environmental, or social benefits"; the "or" was changed to "and" in a 1993 amendment.

- increasing the economic, environmental, and social benefits to members of primary industries and to the community in general by improving the production, processing, storage, transport, or marketing of the products of primary industries;
- achieving the sustainable use and sustainable management of natural resources;
- making more effective use of the resources and skills of the community in general and the scientific community in particular; and
- improving accountability for expenditure on R&D activities in relation to primary industries.

Since this change was implemented, the RDCs have increased their influence over Australia's rural research effort and broadened their scope beyond concerns about specific commodities to include agencies focused on natural resource issues. In this chapter we attempt to evaluate the effects of this change. In particular we consider the effects of the development of the RDC system on total funding for rural R&D, the share of the funding burden borne by the private and public sectors (as well as the appropriateness of the allocation of activities between the private and public sectors), and changes in the types of R&D undertaken in aggregate.

Government and the Rural Sector in Australia

Australia has a federal system of government. That is, like Canada and the United States, a number of preexisting state authorities (established as British colonies) agreed to establish a national government. The state (or provincial) governments continue to act as autonomous entities that are not subordinate to the federal government (unlike unitary government systems such as Great Britain's). As in the United States, the Australian constitution established a structure including both a federal government (locally labeled the Commonwealth) and state governments; six major states are self-governing. In addition two Commonwealth territories enjoy limited self-government, but they are not important for the purposes of this discussion.

The Australian constitution delegates a set of powers to the Commonwealth (that is, federal) government and defines, explicitly or implicitly, the powers lodged at the state level. Constitutionally, the Commonwealth government has a monopoly on customs and excise duties and, owing to legislation and to High Court interpretations of the constitution, it has acquired effective monopolies of revenue devices such as income and sales taxes. Thus, the federal government possesses the bulk of the revenue-raising powers, whereas the state governments carry much of the spending responsibilities.[6] A major

6. More detail on public finance in Australia may be found in Groenewegen (1990) and in Mathews and Jay (1972).

source of revenues for state government programs is income tax collected by the federal government and returned to the states. For instance, the International Monetary Fund (IMF 1995) reported that in 1994 total spending by all governments in Australia was A$200.9 billion, including Commonwealth outlays of A$123.8 billion and state government spending of A$77.1 billion. Intergovernmental transfers from Commonwealth to state and local governments totaled A$32.3 billion.

Government intervention in the Australian rural sector has historically taken many forms, including both federal and state arrangements for marketing or for protecting producers of particular commodities. Examples of past interventions include input subsidies, output subsidies, infrastructure support (including subsidized irrigation schemes, transport and handling facilities, and research facilities), quotas, buffer stocks, buffer funds, and home consumption price schemes. However, especially since the late 1970s, Australian agriculture has been much less protected than in most developed countries. Indeed the major industries, that is, extensive livestock, grazing, and cropping (described locally as broadacre agriculture), have for the most part received little if any direct assistance through commodity market programs. Government assistance has been concentrated in a few industries (especially milk, dairy, eggs, tobacco, and sugar), along with various industry marketing arrangements such as the buffer stock scheme for wool and cost pooling and price stabilization practices for wheat (Piggot 1990; Kenwood 1995:Chapter 3).[7]

In many agricultural industries in Australia, government-funded R&D (including extension) has been the main form of assistance to agriculture. According to the Organization for Economic Cooperation and Development (OECD 1995a), the level of assistance to primary producers as measured by the average producer subsidy equivalent applying to Australian agriculture was about one-quarter of the OECD average during the early 1990s. In contrast, Australia had one of the highest public agricultural research intensities in the world, with total public spending on agricultural R&D greater than 3.7 percent of agricultural gross domestic product (GDP) since 1991. These points suggest an interesting stylized fact regarding government support for Australian agriculture: there is relatively little support given to agriculture, and a higher proportion of that support is in the form of agricultural R&D.

The Provision of Rural R&D in Australia

The structure of Australia's public R&D infrastructure has reflected Australia's dual governmental structure. In Australia the major federal research providers are the Commonwealth Scientific and Industrial Research Organization

7. Sieper (1982) provides a review of the longer history of agricultural policy in Australia. See also Mauldon (1990) and Piggott (1990). OECD (1995a) provides details on agricultural producer- and consumer-subsidy equivalents.

(CSIRO), the higher education sector, and some smaller research institutions (for example, the Australian Bureau of Agricultural and Resource Economics). CSIRO came into existence in 1916 as the Advisory Council on Science and Technology, and it became the Council for Scientific and Industrial Research in 1926. Since then it has developed into its current form as CSIRO, the largest R&D body in the nation. At first the predominant focus of CSIRO was on the rural sector. However, that has changed such that in 1994 expenditure on production R&D in agriculture accounted for 20 percent of the CSIRO budget (Mullen, Lee, and Wrigley 1996). Traditionally CSIRO has carried out much of the basic scientific research on which more-applied rural research has been built (see, for example, Gleeson and Lascelles 1992; Industry Commission 1995). Increasingly research organizations supported by the Commonwealth are being pressured to find outside sources of funds. CSIRO is currently under direction to generate 30 percent of its finances from nongovernment sources, and as a consequence the research undertaken by CSIRO has an important applied component. Federal funds have also financed most of the university research in Australia.

Much of the nation's applied R&D that is targeted toward specific agricultural industries is carried out by state governments in departments of agriculture or their equivalents.[8] State departments of agriculture also have responsibilities for extension and regulation, sometimes jointly with or as an agent of the Commonwealth (as in plant and animal quarantine regulation). The multiple roles have some virtues: they may increase the stability of appropriation funding and enable some sharing of overheads among functions. Perhaps more important, there are clear and strong complementarities among the three functions. Only occasionally does a conflict arise between one role and another (for instance, where the regulatory role leads to a disincentive for industry to provide information to the government). Some state departments have had other, related responsibilities including running agricultural colleges, overseeing government business enterprises such as abattoirs and cool stores, administering rural finance schemes, administering welfare relief schemes, and providing community services such as botanic gardens.

Separate state government departments dealing with conservation and natural resources, for example, may also conduct research on agriculture-related issues such as soil erosion and land quality, state rivers and water supply, and irrigation. All of these departments receive funding from state governments—much of which is derived from the Commonwealth government in the form of grants to the states. The states also receive some project funding from outside the government, particularly from RDCs.

In May 1990 the federal government announced its intention to fund a planned 62 CRCs, which are joint agreements between different research pro-

8. State departments also undertake a significant amount of basic research.

viders to undertake R&D into particular areas. These CRCs were created in part to deal with research spillovers across agencies. As originally envisaged, CRCs were to become self-financing after seven years, or at least no longer reliant on federal funds. About one-quarter of the CRCs are concerned with rural research, including premium-quality wool, wheat quality, meat quality, cotton production, and soil and land management. They all count CSIRO as a core participant, and most list state government departments and universities as members that provide in-kind contributions of facilities and research staff.

Funding Research: The Changing Roles of Government and Industry

Government schemes for organizing industry participation in financing or directing research into and promoting the major rural commodities in Australia originated in the interwar years. In their comprehensive survey of Commonwealth rural research policy, Williams and Evans (1988:28–30) list 108 acts of Parliament relating to industry/government research schemes passed between 1936 and 1988. Many are amendments of earlier acts; but of the 32 still current in 1988, nine date from before the 1980s.

Much of the following discussion is concerned with the evolution of the statutory arrangements between government and industry for financing agricultural R&D, which culminated in the establishment of the RDC system in 1989. However, it should be noted that the debate about the role of the public sector is much broader than the evolution of the RDC system. This debate, which is reviewed in the Industry Commission Report (1995), also encompasses the extent of market failure and the competitive nature of the research industry. These issues are discussed below. It also should be noted that government had been funding research in several industries on a nonstatutory basis since early in the twentieth century (Williams and Evans 1988:73–84). Figure 5.1 shows schematically the flows of funds for rural research in Australia as it presently exists.

From the point of view of statutory arrangements, the federal government first focused its attention on the wool industry, with the Wool Publicity and Research Act of 1936 (Williams and Evans 1988). The act's provisions featured a voluntary levy and its emphasis was more on promotion than on research. Eight more wool acts were passed before any other industries were given similar attention. However, in the second half of the 1950s the tobacco, fishing, wheat, and dairy industries were the subjects of statutes that imposed levies on producers.

The Wool Use Promotion Act of 1945 was a significant landmark. The legislation introduced a one-to-one supplementation of industry levy funds by the Commonwealth government. Subsequent wool industry acts broadened the areas of government involvement to include research planning and the management of government contributions. It was not until the Wool Re-

FIGURE 5.1 Flows of funds for rural research in Australia

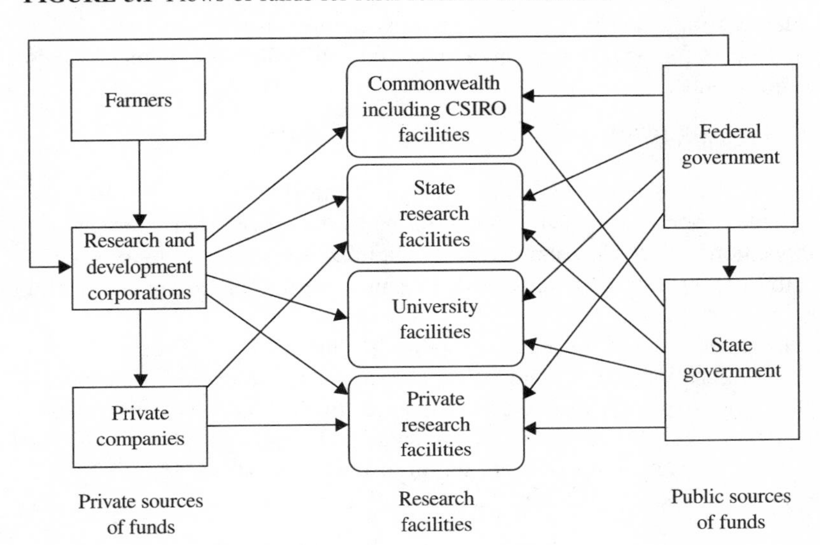

search Act of 1957, however, that government and industry money was combined and jointly managed so as to render the funds from different sources indistinguishable.

The wheat industry was granted its own arrangement in 1957 under the Wheat Research Act, which provided authority for levy collection of 0.25 pence per bushel on all wheat delivered to the Australian Wheat Board. The arrangement was similar to the 1945 Wool Act, with grower levies held in a separate account to be administered on a state basis by committees nominated by the respective state agriculture ministers. Meanwhile, expenditure of the Commonwealth funds was to be determined by a body called the Wheat Industry Research Council, which would have Commonwealth, state, university, and CSIRO representation. As with the Tobacco Industry Act, an emphasis was placed on new research—that is, there was an explicit intention that the funds raised through these schemes would not be used to fund R&D that would have been performed anyway. However, there was no clear means of determining what constituted new research in the first place. Statutory arrangements in the dairy, meat, honey, and eggs industries soon followed.

According to Smith (1992), these various Rural Industry Research Funds were subject to a variety of criticisms concerning the accountability of fund managers, the disproportionate influence of the research providers (the scientists) on research priorities, and the commercial applicability (or lack thereof) of the results of the research. These criticisms and the generally ad hoc nature of the arrangements were key factors leading to the introduction of a formal process for establishing industry councils to collect and manage funds in the mid-1980s.

The Kerin Reforms: RDCs Introduced

The 1980s was an era of dramatic change in the rural sector. In 1983 the Australian Labor Party was elected to national office after some years in opposition. With the appointment of John Kerin as primary industries' minister, Australian agriculture began a period of deregulation and rationalization.

The formalization and implementation of the RDC model came in two rounds of legislation in 1985 and in 1989. The specific rationale for the policy design is most explicitly discussed in the government's 1989 *Research, Innovation and Competitiveness* policy statement, which was released jointly by the minister for primary industries and the minister for resources and which accompanied the 1989 legislation. The main objectives can be summarized as follows:

- to increase total spending on rural R&D;
- to bring R&D priorities more into line with industry requirements, rather than being driven by research providers; and
- to increase the level of R&D funding by industry.

The following objectives might be regarded as secondary, but they are still important:

- to maintain an explicit role for government in the financing of R&D,
- to establish a formal degree of accountability to industry and to government, and
- to improve the commercial potential and the actual adoption rate of research outcomes and new technologies.

The 1985 reforms were the first part of the federal Labor government's major attempt to restructure the funding system for rural R&D. The Rural Industries Research Act of 1985 was designed to be enabling legislation that would provide for the establishment of a research fund for any industry willing to impose a research levy on its producers, with dollar-for-dollar matching of such funds by the government up to a limit of 0.5 percent of the gross value of production (GVP) in that industry. A board or council would be appointed to manage this fund, with a single government representative and industry representatives on the board. By 1986 12 such commodity-based councils had been created. Also, the Australian Special Rural Research Council had been established to nurture small industries and to set up research funds. The Australian Meat and Livestock Research and Development Corporation was also established; the Horticultural Research and Development Corporation followed in 1987. These two corporations and the one for the wool industry were dealt with under separate legislation.

The 1989 Primary Industries and Energy Research and Development Act (the PIERD Act) replaced the 1985 legislation. As a result, all the major R&D councils were to become corporations, which meant in practice a greater degree of independence from government.[9] The umbrella body for smaller industries and cross-industry research, the Australian Special Rural Research Council, was replaced by the Rural Industries Research and Development Corporation (RIRDC). Fifteen levy-funded, industry-specific organizations operated after passage of the act in 1989, although it should be noted that three were actually established under their own legislation (those relating to meat, horticulture, and wool) and five were smaller councils operating under the RIRDC umbrella. In early 1994 a sixteenth body, the Forest and Wood Products Research and Development Corporation, was introduced, with a different levy/matching-funds system based on the government's perceptions of the industry's particular circumstances.[10] Two new corporations funded almost entirely by government were also established: the Land and Water Resources Research and Development Corporation and the Energy Research and Development Corporation. The structure of the RDC system is summarized in Figure 5.2.

In 1993/94, the RDCs collectively controlled about one-third of total publicly performed agricultural R&D spending (including funds raised from levies as well as from public sources). In terms of aggregate value of levies collected and spending by the RDCs, their importance has grown fairly rapidly during the past decade. The value of industry levies collected and managed by the RDCs rose from 0.17 percent of the gross value of agricultural production in 1984/85 to 0.46 percent of the gross value in 1993/94. For several RDCs, the

---

9. More specifically the RDCs can apply for patents or otherwise manage intellectual property, employ staff and set terms and conditions for their employment, charge for services and information, enter into agreements or joint ventures for R&D activities, manage their own financial resources, and form companies (Gleeson and Lascelles 1992).

10. These perceptions, as expressed in the 1989 policy statement (Kerin and Cook 1989), were that the industry was less fragmented than many agricultural industries and that there would be less difficulty internalizing the benefits of R&D to the researcher. However, because the industry does contain a large number of small firms, the government agreed to contribute one dollar for every two dollars the industry raised, up to a ceiling of 0.25 percent of the gross value of production.

FIGURE 5.2 Structure of the research and development corporation system

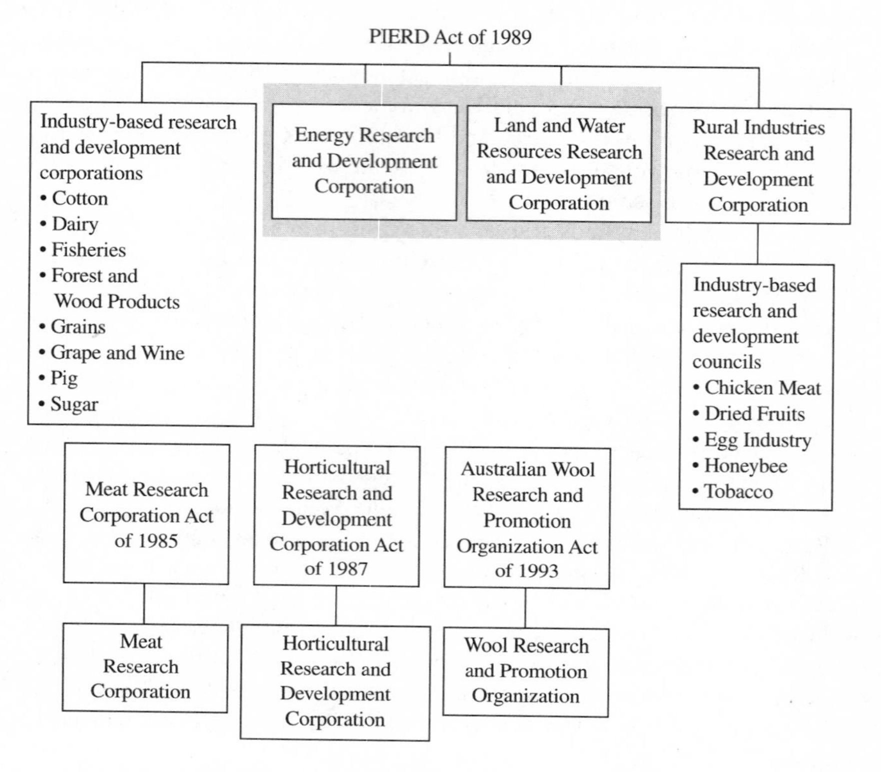

SOURCE: Adapted from Industry Commission (1995:695).

0.50 percent limit on matching contributions from the government is a binding constraint.

Issues Concerning RDCs

Several issues arise in relation to the conduct of the RDCs and their effects on the pattern of R&D. Some of these were addressed by the Industry Commission (1994a, 1995). Sufficiently detailed data are not available to address many of these issues in a concrete fashion. The first is the question of accountability. What mechanisms should be introduced to ensure that the RDCs are appropriately accountable to their constituents and to the society more broadly? What such mechanisms currently are in place, and how well do they work?

The second is the question of management. What procedures are in place to ensure that the R&D resources are allocated effectively, and how well do they work? Some of the RDCs have imposed requirements for formal ex ante cost-benefit analysis of project proposals (see, for example, GRDC 1992), although we have no information on how accurately those assessments have been made or the extent to which they have been used to allocate funds.[11] On the other hand, there is little evidence of much attention to ex post analysis of past projects that have been funded, perhaps on the basis of the ex ante claims made about them, although increasing attention has been given to such studies of late.

A third issue has been referred to as "the tail wagging the dog." How has the RDC "tail" (which is now large enough not to be considered marginal) affected the allocation of the public-sector R&D "dog"? Specifically it has been suggested that, especially in times of tight funding, competitive grants are likely to leverage several times their own value in core funds (for example, grants may provide operating money while not covering the overhead of capital costs and scientists' time); competitive grants can even cause a diversion of government funds from core to marginal funding. In any event competitive grants are likely to change both the balance of the nature of expenditure (less capital and more operating resources) and the project mixture on which other funds are spent. Details on this aspect are difficult to obtain, but anecdotal evidence is available that suggests that RDCs have led to a shift in the portfolio of public-sector agricultural R&D toward the more-applied R&D projects that are more emphasized by the RDCs.

Another relatively unexplored economic question is how the incidence of the commodity levy (checkoff) is shared between consumers and producers. That is, although the tax is paid by the industry, some of the cost of the tax is passed on to consumers, which means that consumers effectively pay part of it. If some of the tax can be passed on to foreign consumers, it means that some of the R&D bill is financed by the rest of the world (for the relevant theory and an application to the Australian wool industry, see Alston and Mullen 1992).

Finally, there are interindustry issues. The Industry Commission Report (1994a, 1995) provides some discussion of the issues that arise where certain research questions are not adequately addressed by individual RDCs. Different RDCs interact on and jointly fund some research projects, and the Rural Industry Research and Development Corporation is set up to do the overlapping and small-industry R&D. Some of this interaction is beginning to take place in the context of the CRCs. However, this question remains: are there still incentives for particular RDCs to try and free-ride on things that should be in their interest (and society's) to help fund?

---

11. For example, beginning in 1995/96, the Grains Research Development Corporation required that all new projects submitted to the corporation include an ex ante benefit-cost assessment.

Cooperative Research Centers

In 1991 the Australian government initiated the CRC Program. The program was developed in response to a number of perceived weaknesses in the conduct of R&D (including rural R&D) throughout Australia. One concern was that links between industry and public research institutions were weak and ineffective, as were the links among the research agencies themselves. It was also felt that the geographical and institutional dispersion regarding the conduct of R&D in Australia provided insufficient incentives to develop cooperative programs of research that could develop a critical mass and realize potential economies of scale and scope. These constraints were considered particularly acute with regard to competitive research grants, which are normally awarded on an individual-scientist basis with little regard to potential synergies and scale economies across research institutions.

CRCs are contractual groupings of government research agencies, universities, and industries in Australia that contribute cash and in-kind resources and collaborate in the management and conduct of R&D over a set time period, which is usually seven years. Federal government funding—up to a maximum of 50 percent of each center's total budget—is designed to draw these groups together and facilitate their collaboration. In the case of CRCs linked to the rural sector, collaborators are likely to include CSIRO, state departments of agriculture, the RDCs, and other private-sector interests. The CRCs are similar to the Interdisciplinary Research Centres in the United Kingdom, the Science and Technology Research Centers and the Engineering Research Centers in the United States, and the Centres of Excellence in Canada.

Research centers are selected for funding after a competitive bidding process against other CRC proposals managed by a secretariat within the federal Department of Industry, Science, and Technology. The centers are generally funded with cash grants from the Australian government and mainly in-kind (but some cash) contributions from the other partners. Although the partners are expected to contribute at least half of the center's resources, the federal government's contribution was only 30 percent after the first two selection rounds (held in March and December 1991) that established 34 centers, and the trend has been to even greater commitments by collaborating partners in subsequent rounds (DIST 1995, commonly known as the Myers Report). In general Australian government funding is only assured for seven years and research programs are regularly reviewed during this period. In the 1996 (fifth) round of center selection, 13 of the original 15 centers and 23 new applicants applied for funding. Only six of the existing centers received all the support they sought for a further seven years. Some received partial funding and others were denied any more public support. In 1997 there were 65 centers receiving a total of A$146 million in Commonwealth government funding (Finkel 1997).

Given the lags involved in R&D, it is premature to identify the economic impact of the CRC Program. However, the Industry Commission inquiry into research (Industry Commission 1995:848–863) suggested a number of criteria against which the program should be evaluated. The commission called for an assessment of (1) the new links developed by the program and the degree to which established links were strengthened, (2) the degree to which program funds have been used to support public R&D rather than crowd out private research that would otherwise have taken place, and (3) the extent to which benefits from building integrated research teams are being realized (there is a concern that the dispersed nature of many rural based CRCs has limited the extent of potential benefits from taking an integrated approach to R&D).

The Myers Report summarized the recent findings of the CRC Program Evaluation Steering Committee by concluding that

> the CRC Program is very well conceived and the prospects of the Government's broad objectives for the scheme being achieved are excellent. Indeed there is already clear evidence of a significant and beneficial change in research culture—especially insofar as it concerns universities and their cooperation with government research agencies and industry. The change in culture extends to industry and other research users who are showing a general enthusiasm for the program and a willingness to become actively involved with longer term and more basic research. (DIST 1995:1)

The Myers Report did not directly address issues related to market failures and the provision of public goods. However, it did note "a shift toward universities and CSIRO approaching industry to become involved in research focused on long term outcomes . . . a valuable addition to the more common practice of industry approaching universities or CSIRO to solve short term problems" (DIST 1995:6). The report also noted that it was unrealistic to expect CRCs with heavy commitments to basic research to become self-financing in seven years.

During the first several rounds, 24 of the 62 approved CRCs were related to agriculture (Table 5.1 and Appendix Table A5.1). In 1994/95 the agricultural CRCs undertook A$158.4 million of research, of which A$49.4 million was provided by the government through the CRC Program.

The Industry Commission Review

In September 1993, the government instructed the Industry Commission to undertake a review of the nation's R&D policies, including the arrangements relating to rural research. A draft report was circulated late in 1994, in which the arrangements for rural R&D were assessed and recommendations were made (Industry Commission 1994a,b). In making these recommendations the Industry Commission sought to encourage

TABLE 5.1 Cooperative research centers (CRCs) for agriculture, rural-based manufacturing, and the environment

		Personnel		Funding in 1994/95	
Name	Date Established	Postgraduate	Full-Time Equivalent Researchers	CRC Program	Total Resources
				(millions Australian dollars)	
Plant Science	July 1991	26	63	2.7	9.0
Soil and Land Management	July 1991	51	42	2.7	6.9
Temperate Hardwood Forestry	July 1991	37	20	1.8	5.8
Tissue Growth and Repair	July 1991	30	50	1.8	4.0
Tropical Pest Management	July 1991	44	48	1.8	7.7
Hardwood Fibre and Paper Science	April 1992	19	17	2.2	5.9
Biological Control of Vertebrate Pest Populations	July 1992	17	19	2.2	6.8
Catchment Hydrology	July 1992	43	31	2.2	5.3
Legumes in Mediterranean Agriculture	July 1992	46	105	2.2	6.9
Tropical Plant Pathology	July 1992	35	39	1.6	5.9
Viticulture	July 1992	18	49	2.2	6.7

Ecologically Sustainable Development of the Great Barrier Reef	April 1993	25	26	2.1	7.2
Cattle and Beef Industry	July 1993	12	31	3.0	9.6
Food Industry Innovation	July 1993	21	30	1.8	5.1
Freshwater Ecology	July 1993	48	53	2.1	7.5
Premium Quality Wool	July 1993	16	29	2.6	9.1
Sustainable Cotton Production	July 1993	43	40	2.1	7.7
Aquaculture	October 1993	18	37	2.3	7.8
Quality Wheat Products and Processes	July 1995	14	46	2.2	11.1
Sustainable Development of Tropical Savanas	July 1995	5	48	2.5	8.8
Sustainable Sugar Production	July 1995	10	35	2.2	6.3
Weed Management Systems	July 1995	19	42	2.2	7.3
Molecular Plant Breeding	July 1997	34	42	2.7[a]	10.6[a]
Sustainable Rice Production	July 1997	16	44	2.2[a]	8.0[a]

SOURCE: Cooperative Research Centres Program (1996); DIST (1999).

[a]Figures are for 1997/98.

- rural industries to take greater financial and managerial responsibility for research, the benefits from which were confined to their industry;
- government to take full responsibility for and to confine its activities to research with predominantly public-good characteristics; and
- a more competitive environment in the provision of public research services by suggesting that public authorities charge fully for their services and that there be separate authorities to plan and provide research services.

Although most economists would not have objected to the principles enunciated in the commission's draft report, many were concerned about how they would be applied in practice. The specific recommendations in the draft report elicited substantial criticism in responses submitted by a range of organizations, including research providers (departments of agriculture, CSIRO, universities, and individual scientists working therein) and research funders (the RDCs), as well as other industry interests and independent economists. It was suggested that the commission did not fully appreciate the complexity of the economic and scientific relationships in Australia's agricultural R&D institutions or the implications of the radical changes in the institutional structure for the provision of agricultural R&D services that would occur if the Industry Commission recommendations were fully implemented.

The key specific recommendation in relation to the RDCs was that the ceiling (in most cases, 0.5 percent of GVP) for matched funds should be lifted, while the one-to-one ratio of government funds to levy funds should be gradually reduced to a one-to-four ratio. It was reasoned that there was no plausible rationale for the ceiling and that it acted as a disincentive for the industry to raise funds above that limit. The commission believed that the one-to-four ratio would bring public support for agricultural R&D to approximate parity with that given to industrial R&D through the 125-percent tax deduction for eligible expenditure on R&D (reduced from 150 percent in August 1996).

The basis for this recommendation was the Industry Commission view that there was no evidence that spillovers from agriculture to the general community were higher than those from other sectors of the economy, with the extent of spillovers providing the rationale for government support. The commission also recommended that changes be made to the use of outside funding by CSIRO and the state departments; that they should undertake only public interest research and not seek outside funding; and that they should fully cost any contract research.

The counter-arguments raised were

- that the reduced level of support embodied in the proposed one-to-four funding formula could lead to lower total funding for rural R&D, with the result that a budgetary saving of a few million dollars could be achieved at the expense of many millions of dollars of forgone potential benefits;

- that agricultural R&D was likely to have considerably greater spillover benefits to society than industrial R&D and so the funding formula for one should not be used as a criterion for the other;
- economies in the provision of R&D arising from "jointness" in the supply of research, regulatory, and extension services would be lost if the arrangements for funding and operating these services were totally separated;
- regarding the commission's public interest and outside funding arguments, that it was neither practicable nor desirable to separate research projects into private-good (beneficial to industry) and public-good (beneficial to the community) research because many projects have elements of both; and
- research projects should be funded in the least-cost way and conducted by the research organization with the comparative advantage, not according to some arbitrary (and ill-defined) notion of whether it is basic research or whether the product is expected to be more or less a public good.

The final commission report was made available in June 1995. The recommendations for the rural sector were substantially modified in directions consistent with some of the counter-arguments. The recommendations for the rural sector were to make "changes to enhance the role of the RDCs in rural research" (Industry Commission 1995:1). The commission concluded, "While there has been limited experience to date with rural research corporations, the Commission judges the evidence so far to be favorable. The system has increased the financial contribution of farmers to rural R&D and the R&D that is done appears to be carefully assessed and directed to the needs of the rural sector" (Industry Commission 1995:37).

Nevertheless, the commission suggested that the system could be improved in a number of ways, mainly to do with the government contribution: the ceiling on matching contributions from the government (0.5 percent of the GVP) was arbitrary and the government share of funding was too high. In light of comments received, however, the commission substantially softened its recommendation on matching contributions, compared with the draft report.

> The Commission recommends that the present levy matching scheme through RDCs, involving dollar for dollar contributions by the Commonwealth up to 0.5 percent of GVP, be amended as follows:
>
> - the Commonwealth Government continue to provide one dollar for every industry dollar spent on R&D up to 0.25 percent of GVP; and
> - thereafter, to contribute at the rate of one dollar for every two dollars from industry, with no ceiling. (Industry Commission 1995:38)

The Industry Commission accepted arguments that spillovers from agriculture may be larger than from other sectors because of the high proportion of basic and strategic basic research done in agriculture. The other argument for the higher ratio is that the commission ultimately recognized that the research priorities of individual farmers are unlikely to coincide with the collective priorities determined by the RDCs. The commission considered that its recommendation would induce more R&D for the same government outlay than the previous mechanism (of one-to-one up to 0.5 percent) and would minimize the risk of a reduction in the amount of R&D funded by the RDCs that might arise from its one-to-four proposal. These recommendations have yet to be accepted by the federal government.

The Industry Commission (1994a, 1995) explored the issue of the "tail" of RDC funding "wagging the dog" of public-sector agricultural R&D expenditures. However, its final recommendations on the roles of state governments in relation to projects funded by industry were also much softer and more equivocal than its draft recommendations that state departments should not seek outside funding: "State departments should cost all externally commissioned research and price it to recover full costs unless additional social benefits not already subsidized are identified" (Industry Commission 1995:39).

In its final report the Industry Commission explicitly acknowledged the complementarities between research and other activities (such as regulation and extension), although relatively more emphasis was given to complementarities between university teaching and research. In addition to recommendations about funding and the roles of various research agencies, the commission recommended the adoption of institutional arrangements for managing R&D. The commission also recommended that those state governments that had not already done so consider establishing their agricultural research departments as separate corporations or institutes, as well as establishing forums for developing state priorities and setting performance indicators to assist in monitoring and evaluating the effectiveness of their research agendas.

Later, in the policy assessment section of this chapter, some of these findings and recommendations of the Industry Commission are examined in more detail. The presence of a more competitive research industry, as advocated by the commission, would be consistent with the broader Hilmer Report on National Competition Policy in Australia, which requires statutory authorities to provide services in a way that is competitively neutral with respect to the private sector. The application of the principles of "competition policy" specifically to research and extension services is presently under consideration.

Empirical Questions and Evidence

The most useful primary data sources on rural research in Australia are the figures on national R&D collected by the Australian Bureau of Statistics—but for which a long annual time series is unavailable—and the data set compiled by Mullen, Lee, and Wrigley (1996a) on annual research expenditures for agriculture carried out by CSIRO, the various state departments of agriculture, and the universities that have engaged in agricultural research.[12] The data set of Mullen, Lee, and Wrigley is for the period 1953–1994. It does not include off-farm research (for example, processing and marketing research), R&D by other state and Commonwealth bodies, or private R&D. These data, along with other data provided by the RDCs, are examined in depth below.

Agricultural Versus Total R&D Spending

Total expenditure on R&D of all kinds in Australia has typically been low by international standards. According to the Industry Commission (1995:6), Australia's comparative R&D performance is characterized by

- low gross expenditure on R&D (GERD) relative to GDP, which reflects relatively low business expenditure on R&D;
- a high ratio of public to private expenditure on R&D;
- a high ratio of basic to applied research, although this ratio has been declining recently; and
- a middling performance in terms of research publications and patents.

The commission reported that in 1991 Australia's GERD-to-GDP ratio was 1.34 percent, seventeenth of 23 high-income countries surveyed. The United States was ranked fourth, with GERD equal to 2.75 percent of GDP.

Notably, however, the government provided a particularly large share of the R&D in Australia (about 0.9 percent of GDP, fourth largest among the OECD countries), much of which was devoted to agriculture.[13] The Industry Commission (1995:Part D) also provides evidence on business expenditure on R&D (BERD), which shows that in the early 1990s Australia ranked sixteenth of the 20 high-income countries surveyed for BERD/GDP, despite a second place (to Singapore) for average annual real growth in BERD since 1991. The United States was third in 1992, even while being sixteenth in terms of real growth.

Considerable caution should be exercised in drawing conclusions based on comparative aggregate R&D figures because, among other things, the figures are likely not to have been constructed on a comparable basis. Moreover, perhaps the figures should differ among industries.[14] Industrial technologies

---

12. Summary R&D information is also contained in Charles (1994), Industry Commission (1994b, 1995), Ralph (1994), and DPIE (n.d.).

13. See also the related evidence in Chapter 3, the discussion around Table 3.10, and Gregory (1993).

14. The Industry Commission (1995:Part D) noted that although manufacturing industries perform a high proportion of business R&D in Australia, the manufacturing sector accounts for a much smaller share of Australia's GDP than many other countries.

can be applicable in many different environments and locations, and the place in which the R&D was performed is not necessarily a reliable guide to where it is eventually utilized and the benefit obtained. However, agricultural technologies are often more site specific than industrial technologies, although this notion of site specificity can be overblown. It is clear that some agricultural innovations have been successfully exported. For example, Byerlee and Traxler (1995) discuss the international impacts of wheat varieties developed by the International Center for Improvement of Maize and Wheat (CIMMYT) located in Mexico; Brennan and Fox (1995) evaluate their impacts in Australia; and Pardey et al. (1996) evaluate the U.S. impacts.

Public Agricultural R&D

The Committee of Economic Inquiry (known as the Vernon Report; Commonwealth of Australia 1965) estimated that rural research accounted for 35 percent of total public expenditure on R&D in Australia at that time. By Australian Bureau of Statistics estimates, since then this percentage has declined to about 15 percent (1991 and 1993 figures). Over the past 30 years, privately performed agricultural R&D appears to have risen strongly, perhaps by a factor of eight, so that the nominal *total* expenditure on agricultural R&D has grown faster (and declined less rapidly in recent years) than expenditure on agricultural R&D done by public agencies. Some of this privately performed R&D is funded by the RDCs. More details on private R&D are included below.

GENERAL TRENDS. Australia's total spending on publicly performed agricultural R&D between 1953 and 1994 in nominal and real (1993 dollars) terms is shown in Figure 5.3 and Appendix Table A5.2. In nominal terms, total expenditure shows generally smooth growth from A$9.0 million in 1953 to A$530.5 million in 1994, an average annual growth rate of 10.4 percent. The nominal, undeflated figures do not represent quantities of research or the purchasing power of research expenditures. Estimates of real expenditures were obtained by dividing the nominal series by the implicit GDP deflator for Australia. Real expenditure grew by 4.1 percent per year, from A$99.9 million (1993 dollars) in 1953 to A$520.2 million in 1994. These data show that real funding grew at a rapid rate during the 1950s and 1960s (6.6 and 8.5 percent per year, respectively), slowed substantially during the 1970s (2.0 percent per year), and virtually stalled during the 1980s (0.8 percent per year). The real rate of growth recovered somewhat during the early 1990s to average 2.9 percent per year from 1990 to 1994.[15]

15. A similar trend is evident from the shorter Australian Bureau of Statistics series. In general, the two data series are quite consistent. In recent years the biggest divergence occurred in 1988/89, when the Mullen, Lee, and Wrigley (1996a) estimate of R&D spending fell to 76 percent of the Australian Bureau of Statistics estimate. The Australian Bureau of Statistics estimate of expenditure appears to be abnormally large for 1988/89; it exceeds the estimate in 1990/91 and is almost as large as that in 1992/93.

FIGURE 5.3 Spending on publicly performed agricultural R&D, 1953–1994

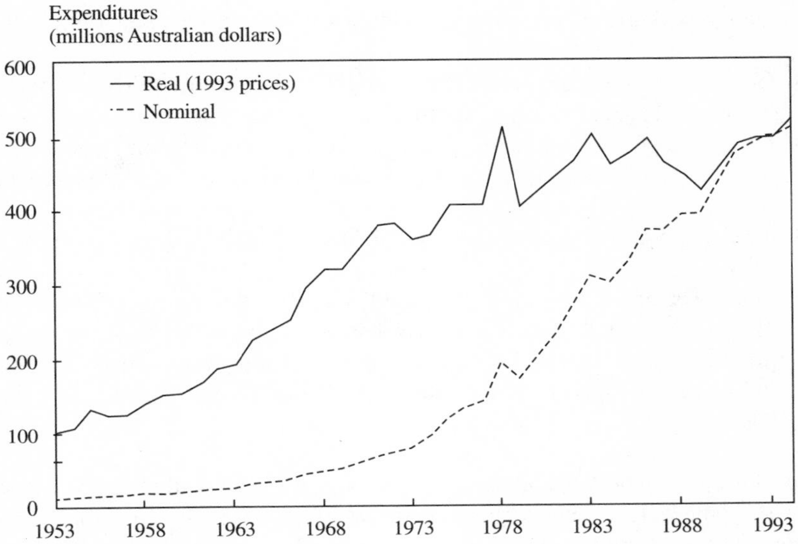

SOURCE: Appendix Table A5.2.

NOTE: The 1978 spike in the series results from abnormally high levels of capital expenditures by the Commonwealth Scientific and Industrial Research Organization (CSIRO), a reflection of the costs associated with constructing a A$200 million National Animal Health Laboratory facility in Geelong between 1978 and 1984.

The evidence of a slowdown in the growth of real support for agricultural R&D since the mid-1970s is surprising in the light of the implicit aims of the RDC model: to increase the total resources devoted to agricultural R&D and to increase the share of those resources coming from the private sector (and producers particularly). Although we cannot know what would have happened to spending in the absence of the RDCs and the CRCs, in Figure 5.3 there is no marked increase relative to trend even in the nominal, undeflated series during the recent period when the RDCs and the CRCs have risen in prominence. One possibility is that the development of the RDCs has crowded out other sources of funds for publicly performed agricultural R&D, rather than simply supplementing the existing funds. Another possibility is that without the advent of the RDCs and CRCs real public support for agricultural R&D would have slowed sooner and, perhaps, have even fallen. As always, the data refer to "before and after" rather than "with and without" the introduction of RDCs. It is probably not appropriate to use the data to directly infer the effects of the RDCs because other things were not held constant.

A variety of hypotheses could account for why public-sector agricultural R&D spending appears to have fallen (or at least has not risen) in real terms in conjunction with the rising importance of RDCs in the 1980s and 1990s:

- Government spending has fallen generally, and agricultural R&D support has been curtailed along with other forms of spending.
- Public support for agriculture has fallen, and spending on agriculture, including agricultural R&D, has fallen relative to total government spending.
- Total support for science in general has fallen, including agricultural science.
- Public-sector support for agricultural R&D has fallen for reasons specific to agricultural R&D such as reduced scientific opportunity or increased private-sector roles (for example, arising from improved property rights).
- Public-sector support for agricultural R&D has been crowded out by the RDCs.

These five hypotheses are not mutually exclusive, and all might be borne out. To some extent, anecdotal evidence would support the first four. The change in public support may have more than offset the beneficial effects of RDCs expanding the total public-sector R&D pot. Yet we cannot rule out the crowding-out hypothesis from these gross comparisons alone. Conversely, we should also ask why public support for agricultural R&D has been so high. After all, as Mullen, Lee, and Wrigley (1996a) noted, agricultural R&D spending grew faster than total government spending during the 1950s and 1960s.

Institutional Details

To shed further light on these questions, we consider some more-detailed breakdowns of the data and look at spending patterns of the private sector and various public agencies. In 1993 nearly one-third of all Australian agricultural R&D was performed by the private sector and almost 70 percent by public agencies. Most of the public research was undertaken by the state departments of agriculture (53 percent of the public-private total); CSIRO performed about one-third of the agricultural research, and the universities collectively undertook 12 percent of the 1994 total. Figure 5.4 identifies spending on public agricultural research by the state departments of agriculture, universities, and the principal federal agency, CSIRO. Real spending by all agencies grew from 1953 to 1994: 3.7 percent per year for CSIRO, 3.8 percent for the state agriculture departments, and an estimated 7 percent for the universities. These data suggest the longer-run trend was for the universities' share to grow at the expense of research carried out by both state and federal agencies.

Real agricultural R&D spending by CSIRO grew more slowly than spending by the other types of agricultural research institutions. In fact, CSIRO

FIGURE 5.4 Spending on publicly performed agricultural R&D by institution, 1953–1994

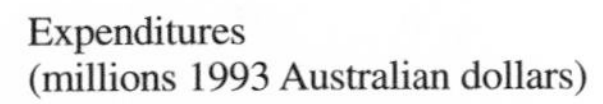

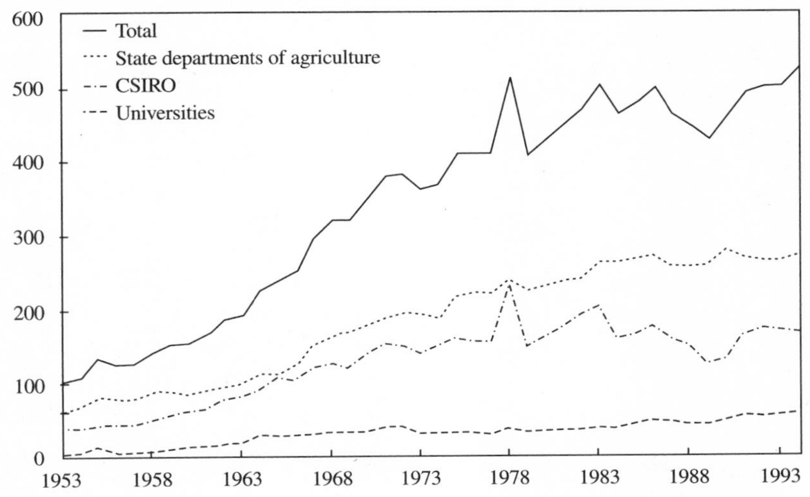

SOURCE: Appendix Table A5.2.

NOTE: See note to Figure 5.3.

research spending on agricultural R&D actually shrank in real terms by 2 percent per year throughout the 1980s. Has this reflected a decline in CSIRO overall or a shift in the direction of CSIRO R&D? It would seem to be a shift of priorities that accounts for the relative decline in CSIRO's agricultural R&D. Total research expenditure by CSIRO has not fallen in real terms, although the growth rate has slowed. Figure 5.5 shows that expenditure on agricultural production research as a proportion of CSIRO's total budget has shrunk from around 40 percent in the early 1950s to about 30 percent in the 1980s. By 1994 production agriculture accounted for only one-quarter of the agency's total budget.

Research Orientation

The Australian Bureau of Statistics breaks down research into four categories: pure basic, strategic basic, applied, and experimental development. In general government rural research (excluding universities) from the early 1980s to the early 1990s, applied and experimental development research show impressive growth—both more than doubled in nominal terms. Strategic basic research rose by about 40 percent. Pure basic research rose and then fell—the nominal spending in this category in 1992/93 was less than it was in 1981/82.

FIGURE 5.5 Commonwealth Scientific and Industrial Research Organization spending on agricultural R&D, 1953–1994

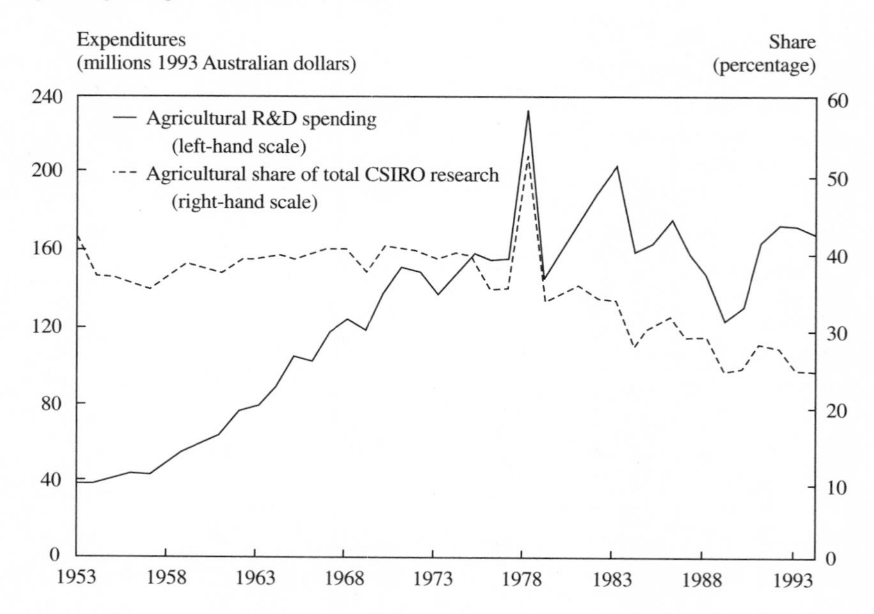

SOURCE: Adapted from Mullen, Lee, and Wrigley (1996a).

NOTE: See note to Figure 5.3.

This has meant that the share of pure basic and strategic basic rural research by government fell from about 25 percent in 1981/82 to 15 percent in 1992/93. In a submission to the Industry Commission (1995:762), the RDCs suggested that 11 percent of their expenditure supported basic research and 27 percent supported strategic research, with the remaining 62 percent spent on applied research. These data indicate a much larger share of RDC expenditure on basic and strategic research than is indicated by the Australian Bureau of Statistics. Perhaps strategic research is not being defined in the same way.

By either estimate, however, more basic research is conducted in the rural sector by public institutions than by business enterprises in the economy in general. This provided the rationale for the Industry Commission's view in its final report that the spillovers from agricultural research to the broader community were larger than the spillovers from research in other sectors and justified a higher level of public support for rural R&D than for R&D in other sectors. Basic research is likely to result in greater spillovers than applied research.

Private-Sector Roles

Considerable difficulties are encountered in constructing meaningful indicators of private-sector R&D. Identifying the private contribution to agricultural research separate from the R&D conducted by business enterprises is useful, but often difficult (see Chapter 3). Deciding what is appropriately reported as agricultural R&D and distinguishing that from the other types of R&D that firms perform is also problematic.

Table 5.2 represents our best efforts to gain a quantitative sense of contemporary trends in privately performed agricultural research in Australia, including research related to production agriculture, food processing, and agriculturally related chemical and pharmaceutical research. Privately performed research grew at an annual rate of about 21 percent in nominal terms and 15 percent in real terms from 1982 to 1993, well above the corresponding rate of growth of 5.6 percent for publicly performed research.[16] Some of the growth in privately conducted R&D has been financed by the RDCs. In addition, the introduction of a 125-percent tax concession for R&D in 1985 also contributed to the increase in rural R&D carried out by business enterprises. Nonetheless, the rapid growth in private spending was from a small base. By 1993 the private share of overall agricultural R&D was still only about 30 percent, which is well up on the 9 percent that prevailed a decade earlier, but still well below the private share for agricultural R&D performed in the United Kingdom and the United States.[17]

Taking the data in Table 5.2 at face value, food-processing research accounted for about two-thirds of the privately performed agricultural R&D in Australia throughout the 1980s and 1990s. The share of private chemical and pharmaceutical R&D directed toward agriculture has fallen (from about 24 percent in 1982 to 9 percent in 1993), which suggests that an increasing proportion of that technology is imported from elsewhere.

Agricultural Research Intensities and Spending Ratios

Australia's agricultural research intensity, which is defined as spending on public research as a proportion of agricultural output, is shown in Table 5.3. The notion that overall public-sector spending on rural R&D did not increase with the inception of the RDCs is given further support here. R&D spending as a percentage of AgGDP (a value-added measure of agricultural output) was

16. This estimate of research by the private sector includes a small portion of the levy payments by primary producers but does not include research carried out by the producers themselves.

17. Including pre- and post-harvest R&D done by the public sector in the agricultural R&D total used to calculate this private share (see note to Appendix Table A5.1 for elaboration) would lower the 1993 share of privately performed agricultural R&D from 30.3 to 26.2 percent.

TABLE 5.2 Privately performed agricultural R&D in Australia

Fiscal Year	Agriculture, Forestry, Fishing, and Hunting	Food-Processing Enterprises	Chemical and Pharmaceutical Enterprises	Total Private R&D		Private Share of Total
				Nominal	Real	
		(millions Australian dollars)			(millions 1993 Australian dollars)	(percentages)
1977	1.48	na	na	na	na	na
1979	2.38	na	na	na	na	na
1982	1.32	19.63	6.91	27.86	47.86	9.2
1985	4.65	43.92	10.40	58.97	84.24	14.9
1987	12.32	71.79	13.68	97.79	121.21	20.6
1989	23.59	88.59	18.13	130.31	140.14	24.5
1991	32.28	111.00	20.11	163.39	167.25	25.3
1993	58.00	139.80	21.45	219.25	219.25	30.3
Annual rate of growth						
			(percentages)			
1982–1993	41.1	19.5	10.8	20.6	14.8	11.5

SOURCE: Agriculture, forestry, fishing, and hunting series: 1977, Mullen, Lee, and Wrigley (1996a:Table 3); 1979–1993, ABS (1993b). Research by food-processing and chemical and pharmaceutical enterprises: OECD (1991, 1996a,b) and unpublished OECD data files (February 1997).

NOTE: na indicates not available. The agriculture, forestry, fishing, and hunting series is the agricultural R&D expenditure by business enterprise by product field data reported by the Australian Bureau of Statistics. It includes R&D directed toward products and processes related to poultry, grains, sheep, cattle, pigs, and "other agriculture, forestry, fishing and hunting." The series apparently excludes processing R&D and research into many purchased inputs, including farm chemicals and pharmaceuticals (see Mullen, Lee, and Wrigley 1996a:19). The OECD data include intramural business enterprise expenditures on R&D classified by socioeconomic objective. Textile-related R&D is not included here, so some wool-processing R&D is unavoidably missing from the series. See notes to Appendix Table A3.2 for additional details.

TABLE 5.3 Public agricultural research intensities and spending ratios

	1953	1961	1971	1981	1991	1994
			(percentages)			
Research intensity ratios						
R&D/agricultural GDP	na	1.23[a]	2.88	3.02	4.09	3.99
R&D/GVAP	0.39	0.74	1.83	2.03	2.28	2.26
			(1993 Australian dollars)			
Spending ratios						
Per capita	na	15.97	29.70	30.52	28.87	29.48
Per economically active agricultural population	na	357	863	975	1,234	1,319[b]
Per agricultural establishment[c]	489	826	2,008	2,568	3,950	4,186[b]

SOURCE: Appendix Table A5.1 for research expenditures; the World Bank (1997) for agricultural GDP; FAO (1997) for total population and economically active agricultural population; and ABARE (1995a) for agricultural establishments.

NOTE: na indicates not available. Agricultural GDP denotes agricultural gross domestic product and GVAP denotes gross value of agricultural production.

[a]1962 figure.
[b]1993 figure.
[c]Agricultural establishments include operations with an estimated value of agricultural operations of A$2,500 or more prior to 1986/87. From 1986/87 the threshold shifted to A$20,000, from 1991/92 to A$22,500, and from 1993/94 to A$5,000, thereby causing this ratio to rise faster than otherwise, at least through 1993/94.

reasonably high around the middle of the 1980s (for example, reaching 3.6 percent), but fell approaching the 1990s (to 2.7 percent) before beginning to increase again in the more recent years.

We also computed state-specific agricultural research intensities for the state departments of agriculture (noting that this represents only a subset of the R&D spending within and by the individual states) by dividing the agricultural R&D spending by the gross value of agricultural production for each state. These intensities are reported in Table 5.4, with a comparable intensity ratio covering all public agricultural R&D in Australia given in Table 5.3. Current and past research intensities differ dramatically among the states. In 1994 the state-specific research intensities ranged from as little as 0.71 percent in Victoria to as much as 1.91 percent in South Australia. Also, the intensity with which other state governments invest in agricultural R&D did not decline as far or as fast as those in New South Wales and, particularly, Victoria. Chapter 4 noted similar differences in intensity ratios among the states within the United States.

TABLE 5.4 State departments of agriculture research intensities

	New South Wales	Victoria	Queensland	South Australia	Western Australia	Five-state Average
			(percentages)			
1953	0.30	0.14	0.25	0.19	0.27	0.23
1961	0.48	0.21	0.61	0.25	0.46	0.41
1971	1.36	0.48	0.93	0.77	1.03	0.94
1981	1.45	0.62	1.13	1.21	1.15	1.10
1991	1.33	0.90	1.17	1.37	2.05	1.29
1994	1.29	0.71	1.13	1.91	1.69	1.23

SOURCE: Mullen, Lee, and Wrigley (1996a).

NOTE: See Table 5.3 for relevant details.

Table 5.3 also reports some public research spending ratios. In 1994 Australia spent A$30.1 (1993 dollars) per capita on public agricultural R&D, almost double the A$16.0 per capita that was spent in 1961. Real spending per farm (or more precisely, per agricultural establishment) and per agricultural worker (as measured by the economically active agricultural population) appears to have risen much faster than spending per capita. Australia spent A$1,304 per agricultural worker and A$4,139 per farm on public research directed toward agriculture in 1993, substantially more than was spent in 1961.

R&D Corporations

RDCs raise, allocate, and manage R&D funds but perform little of their own research. Most is done under contract to the corporations. Where they get those funds, who and what they fund, and how they administer the research are all important policy issues. Here we assemble and interpret the available evidence.

SPENDING PATTERNS. According to data provided by the Commonwealth Department of Primary Industry and Energy, total expenditure by the agricultural RDCs (that is, excluding the Energy Research and Development Corporation) rose from just A$47.5 million in 1985 to A$239.2 million in 1995, an annual rate of growth of 17.5 percent—or 13.2 percent per year if spending is measured in constant prices (Figure 5.6). This reflects strong growth by most of the RDCs as well as the commencement of a number of new corporations, such as the Land and Water Resources Research and Development Corporation and the Forest and Wood Products Corporation. RDC spending grew faster than public agricultural R&D or total agricultural R&D.[18]

18. Some RDC spending goes to administration and research areas and agencies not included in the overall figure for public agricultural R&D reported by Mullen, Lee, and Wrigley (1996a).

FIGURE 5.6 Overall spending trends of the agriculturally related research and development corporations (RDCs), 1985–1995

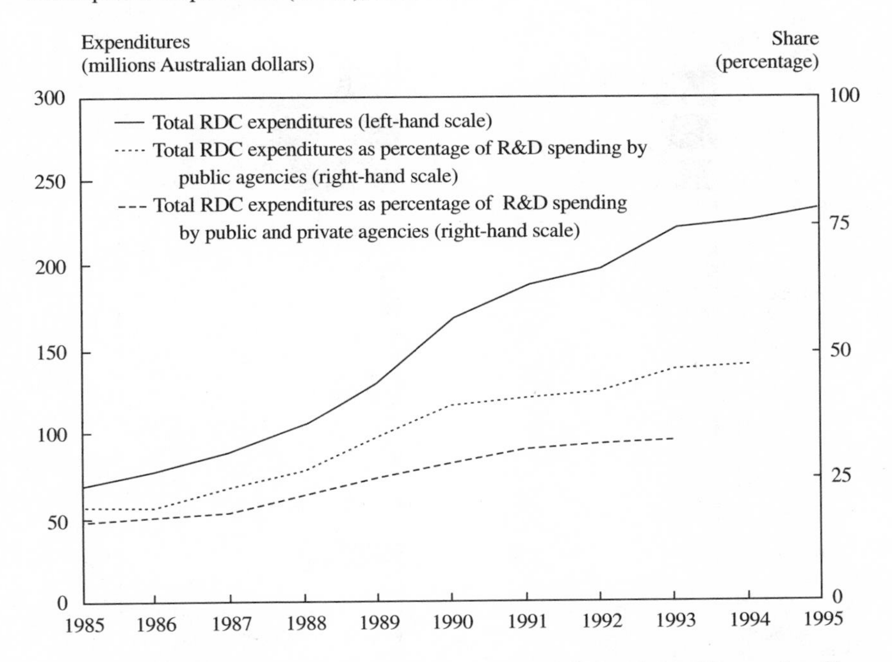

SOURCE: RDC data from unpublished Department of Primary Industries and Energy data files. Public agency data from Mullen, Lee, and Wrigley (1996a). Private agency data, see Table 5.2.

The big three, with annual expenditures in the A$25–55 million range in 1995, are the RDCs for meat, grains, and wool, in decreasing order (Figure 5.7). In 1995 these three corporations accounted for nearly half the total spending by the RDCs. Six RDCs spent between A$10 and A$25 million; in decreasing order of importance they were the RDCs for horticulture, land and water, dairy, the Rural Industry Research and Development Corporation, energy, and fisheries. The RDCs for sugar, pigs, cotton, grapes and wine, and forest and wood products each spent less than A$10 million annually.

FUNDING SOURCES. Figure 5.7 also provides an indication of the RDC funds derived from industry levies compared with the total expenditures of each corporation. In 1995 the revenues from industry levies totaled A$102.6 million and contributions from the Commonwealth government were A$126.19 million.[19] Thus the levy share of total spending in 1995 averaged 43 percent, down somewhat on the 47 percent reported for 1994, but less than the 50 percent

19. The RDC totals reported in this subsection relate to agricultural research, excluding the Energy Research and Development Corporation.

FIGURE 5.7 Spending patterns among the research and development corporations

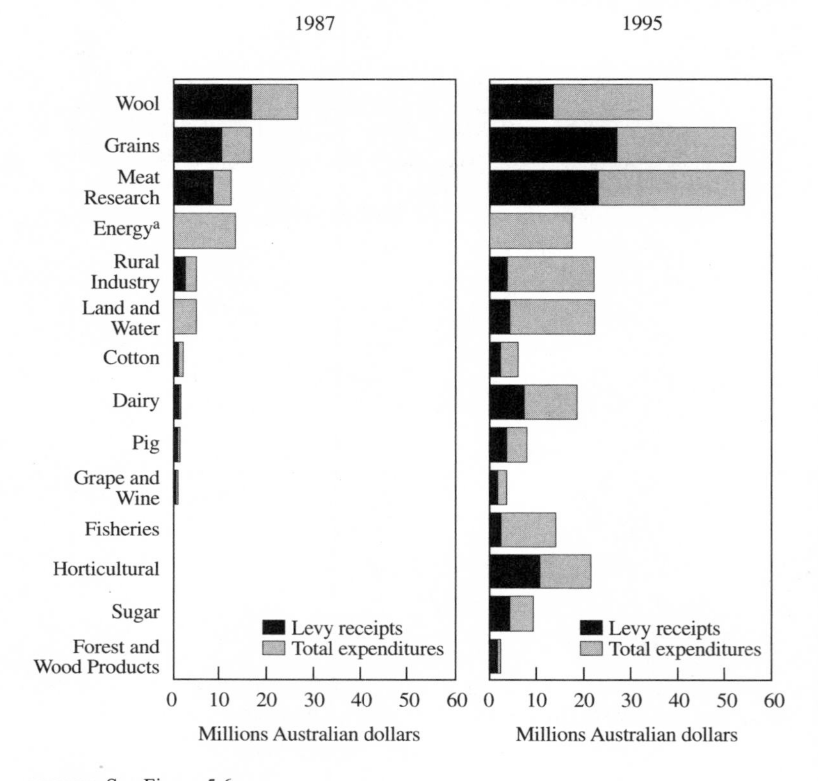

SOURCE: See Figure 5.6.

[a]Excluded from most of the data reported elsewhere in this chapter because research undertaken by this corporation does not directly relate to agriculture.

share reported for 1996. These changes are part of a longer-term pattern of fluctuations in the levy contributions due, in part, to changes in market conditions that affect levy receipts. Over time the corporations have increasingly relied on various sources of own-income (for example, from royalties or license income on patented technologies and from interest income on accumulated savings) to smooth their pattern of spending or to top up income earned from industry levies and government contributions. For example, in 1995 RDC expenditures totaled A$239.2 million, some A$10.4 million more than the combined contribution from the Commonwealth government and the levy receipts.

Table 5.5 shows that although levy funds grew by 14.5 percent per year since 1985 (10.3 percent in inflation-adjusted terms), contributions from the

Commonwealth government grew even faster (18.2 percent annually in nominal terms, 13.8 percent after adjusting for inflation). In addition, the public dollars directed toward the RDCs grew faster than the total amount of publicly performed agricultural R&D in Australia. This suggests that public R&D funds that were previously disbursed on an institutional basis directly from government treasury departments to state departments of agriculture, CSIRO, and the universities are increasingly being channeled through the research corporations. In other words, the RDCs may have crowded out other, more-direct forms of providing public funds to public research performers rather than having generated any additional public funding for public research agencies.

Industry and government (matching) contributions to the RDCs have grown, not only in absolute terms but also relative to the gross value of agricultural production. By 1994 RDC spending was 1.0 percent of the value of agricultural output, compared with just 0.3 percent in 1985, with the industry-funded share of RDC spending growing from 0.2 percent of the GVAP to 0.5 percent (Table 5.5) as new RDCs were established and others increased their levy rates toward the 0.5 percent ceiling for matching federal funds.

INSTITUTIONAL ORIENTATION. In 1993 about 31 percent of total RDC spending went to the state departments of agriculture, 26 percent to CSIRO, and 17 percent to the universities.[20] About 26 percent of the corporations' funds were used to cover the costs of operating the corporations or to fund various other agencies, including private consulting firms and providers of specialist R&D services (notably, research related to marketing; Ralph 1994). Combining these institutional shares with the estimated share of RDC funding directed to off-farm research (40 percent) suggests that in 1994 almost 22 percent of the public agricultural R&D total reported by Mullen, Lee, and Wrigley (1996a) was funded by the RDCs—similar to the estimate by Mullen, Lee, and Wrigley (1996a:22). It seems the corporations now play a pivotal role in the funding and, thereby, the performance of Australian agricultural R&D. This view is bolstered further by the evidence presented in Figure 5.8. This figure tracks the share of agricultural R&D spending by the CSIRO and four state departments of agriculture (New South Wales, South Australia, Victoria, and Western Australia) derived from "industry funds" including both the producer levy and the government component of RDC expenditures.

The establishment of the RDCs in the mid-1980s corresponds to a dramatic reversal in the previously declining importance of industry funds as a source

---

20. Institutional shares are based on data provided in Industry Commission (1995:Table F1, Appendix F.5) and unpublished DPIE data and exclude the Energy Research and Development Corporation. Ralph (1994) provides similar data that are presumed to be for 1993/94, wherein 38 percent of RDC funds go to state government agencies, 23 percent to CSIRO, 22 percent to universities, 1 percent to Commonwealth government agencies (other than CSIRO), and 16 percent to various other uses.

TABLE 5.5 Structure of agriculturally related research and development corporation (RDC) finances

	1985	1987	1989	1991	1993	1995
			(millions Australian dollars)			
Revenue						
Levy receipts	26.52	40.82	48.52	60.91	95.26	102.60
Commonwealth contribution	23.71	41.14	68.51	83.05	105.75	126.19
Total expenditures	47.50	71.29	121.20	184.32	222.04	239.21
			(percentages)			
Research intensities						
Levy receipts/GVAP	0.17	0.24	0.21	0.29	0.43	na
Commonwealth/GVAP	0.15	0.24	0.30	0.39	0.48	na
Expenditures/GVAP	0.30	0.41	0.52	0.87	1.01	na
Spending ratios						
Levy receipts/RDC expenditures	55.83	57.26	40.03	33.05	42.90	42.89
RDC expenditures/total public agricultural R&D	14.14	18.89	30.25	38.22	44.46	na
RDC expenditures/total public and private agricultural R&D	12.03	15.03	22.83	28.55	30.90	na

SOURCE: Unpublished Department of Primary Industries and Energy files for RDC expenditure data.

NOTE: na indicates not available. Excludes the Energy Research and Development Corporation.

of revenue for both CSIRO and the state departments of agriculture. In the case of CSIRO, trust funds had been a shrinking source of revenue since the 1950s. For the state departments, the decline set in around the late 1960s. By 1995 data from Mullen, Lee, and Wrigley (1996a) suggest that trust (largely RDC) funds accounted for 31 percent of CSIRO's agricultural research expenditures and 21 percent of the research undertaken by the state departments of agriculture. Mullen, Lee, and Wrigley (1996a) reported difficulties in estimating R&D expenditures by universities and the university share funded by RDCs. They assumed that the RDCs and other external funding bodies accounted for half of university research, whereas incomplete ABS data suggest the RDC share was lower. Alternatively, Ralph (1994) estimated that 22 percent of RDC spending went to universities in 1993, suggesting that almost 60 percent of the agricultural R&D undertaken by the universities that year was funded by the RDCs.

RESEARCH DIRECTIONS. It has been argued that RDCs would spend a smaller proportion of their research funds on more-basic R&D, which has

FIGURE 5.8 Industry funds as a percentage of total research expenditures, 1953–1994

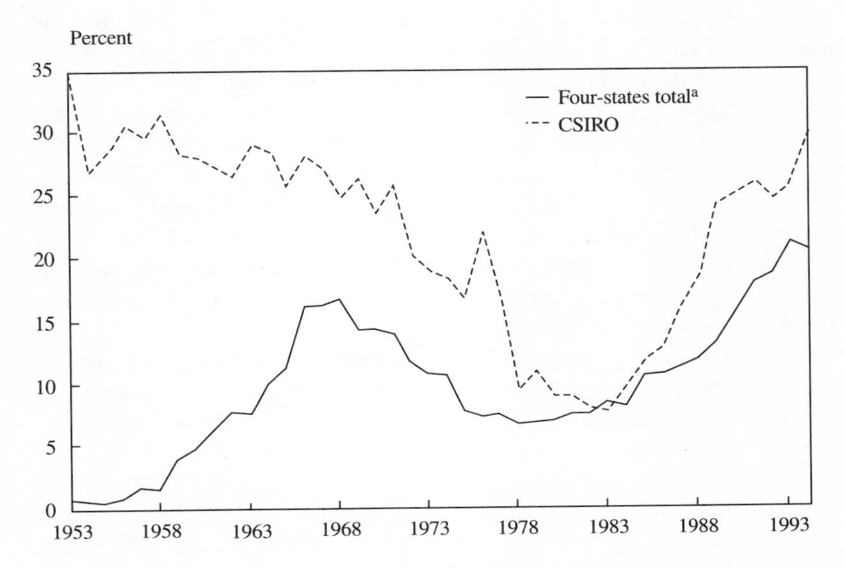

SOURCE: Constructed by the authors from data in Mullen, Lee, and Wrigley (1996a).

[a]New South Wales, South Australia, Victoria, and Western Australia.

correspondingly smaller spillover benefits, than would the public sector in general. Thus an expanding role for RDCs in financing and directing the public-sector R&D effort would lead to a relative, if not absolute, decline in the importance of research with greater public-good benefits. Has the RDC system led to a greater emphasis on applied R&D, or commercialization, at the expense of more-basic research? Aggregate time-series data on research directions in agricultural R&D are not available, although the questions are important ones. The Industry Commission (1995:Figure E4.4) reports a large variation in the mix of basic, strategic, and applied research funded by the various RDCs; the reported 1994/95 average was about 12 percent basic, 25 percent strategic, and 63 percent applied research.

Another issue of interest is the balance of research between on-farm and off-farm problems. Although time-series data on this aspect are not available, Department of Primary Industries and Energy estimates suggest that in 1994 production research accounted for 60 percent of R&D expenditure by the RDCs, although some considerable variation among RDCs was observed: five RDCs invested over half their R&D budgets in off-farm (usually postproduction) research.

ADMINISTRATIVE COSTS. Alston and Pardey (1996) point out that every method of allocating research resources is bound to involve four types of costs (1) information costs, (2) various other transaction costs, (3) opportunity costs of inefficient research resource allocation, and (4) rent-seeking costs. A possible economic justification for the RDCs is that the efficiency gains they generate through the improved allocation of research resources more than offsets any additional costs they may incur compared with, for example, the prior methods of allocating research resources. Is the substantial and growing investment by RDCs in managment and analysis evidence of a sensible reaction to the complex tasks facing them, or is it a sign of the buildup of a wasteful bureaucracy? Are the procedures used to account for the impact of past research and to judge the potential payoffs to current and future research cost effective? What is the appropriate balance between competitive and contract research given the alternative information, administrative, and other costs associated with each approach?

Unfortunately these questions are easier to pose than to answer. However, we do have some limited data on the administrative, program management, evaluation, and communication costs of the RDCs' and the CRCs' operations. The 1993 Department of Primary Industry and Energy survey shows that administrative costs averaged 6 percent of total RDC budget expenses (Ralph 1994). Program management, evaluation, and communication costs were 7 percent of the overall expenses, so that "overhead" charges totaled 13 percent. Table 5.6 provides some reported overhead charges for selected agencies for selected years. Although it is difficult to detect any clear trends, there seems to have been a tendency for the share of overhead expenses in total RDC commitments to increase over time. Although these overhead rates appear much lower than the rates commonly charged by universities and various other agencies, most of the RDCs' funds are channeled to these research performing agencies, leaving such funds exposed to a "double-overhead" charge.

There are some suggestions of economies of size or at least some considerable structural differences across RDCs in their overhead rates: overhead rates for the comparatively larger RDCs (that is, the Meat Research Corporation and the Grains Research and Development Corporation) are substantially lower than for the Rural Industries Research and Development Corporation. This is perhaps a reflection of the comparative size of the respective RDC operations, but it may also reflect the substantial differences in the organizational structures of the agencies (the Rural Industries Research and Development Corporation, for instance, supports a number of research councils, making it quite distinct from the other RDCs). The data also point to the lower overhead rates of the CRCs' secretariat operation compared with the RDCs'. This may partly reflect differences in the coverage of the cost data between the two groups of agencies, but it probably also reflects the substantial operational differences between the CRCs and the RDCs. It should also be borne in mind that each CRC has its own administrative unit.

TABLE 5.6 Overhead costs of selected research and development corporations and cooperative research centers

	1991	1992	1993	1994	1995	1996
Rural Industries Research and Development Corporation (RIRDC)						
a. Program support and administration (millions Australian dollars)	na	1.98	1.93	na	2.44	3.42
b. R&D programs (millions Australian dollars)	na	11.64	15.77	na	17.30	16.25
(a) divided by (a + b) (percentage)	na	14.5	10.9	na	12.4	17.4
Meat Research Corporation (MRC)						
a. Program support and administration (millions Australian dollars)	na	na	na	na	3.56	4.73
b. R&D programs (millions Australian dollars)	na	na	na	na	42.33	40.34
(a) divided by (a + b) (percentage)	na	na	na	na	7.8	10.5
Grains Research and Development Corporation (GRDC)						
a. Program support and administration (millions Australian dollars)	na	na	na	na	4.44	5.56
b. R&D programs (millions Australian dollars)	na	na	na	na	45.33	49.04
(a) divided by (a + b) (percentage)	na	na	na	na	8.9	10.2
Cooperative Research Centres (CRC)						
a. Program support and administration (millions Australian dollars)	0.549	0.648	0.694	0.542	0.859	na
b. R&D programs (millions Australian dollars)	0.00	18.28	45.37	90.60	114.00	na
(a) divided by (a + b) (percentage)	na	3.4	1.5	0.6	0.7	na

SOURCE: RIRDC (1993, 1996); MRC (1996); GRDC (1996); and DIST (1995).

NOTE: na indicates not available.

Policy Assessment

In this chapter we set out to

- review public-sector agricultural R&D institutions in Australia, including a summary and interpretation of new data on investments in rural research;
- evaluate the consequences of the recent policy changes, while emphasizing the roles of the RDC-type institutions; and
- draw inferences about the potential for such arrangements in other countries.

Here we summarize the key elements of the previous sections in relation to the institutional developments and their consequences and draw some broad inferences.

It is important to note that industry R&D funding arrangements, which involve the use of levies combined with matching grants from the government, have had a long history in Australia. The recent events associated with the creation of the entities explicitly called RDCs cannot be fully appreciated without some understanding of that history. The recent rise of RDCs followed a previous period of *relative* decline in their importance. It is also notable that the recent rise of RDCs has been associated with a decline of general government support for agriculture and research and a rise of privately funded and executed agricultural R&D. As well as understanding these events, it would be helpful to understand why private-sector R&D has been and continues to be so small; why general support for agriculture has been so low; and why support for public-sector agricultural R&D has been so high in Australia compared with most OECD countries.

The contribution of the RDCs to agricultural research by public and private institutions in Australia needs to be evaluated within the context of the changing attitudes to the role of government arising at this time. Partly as a result of these other concurrent changes, it is not possible to precisely identify the unique contribution of the RDCs and their development.

Expenditure on agricultural research grew rapidly in nominal terms from the 1950s to the 1970s and continued to grow thereafter at a slightly slower rate. In real terms, in the 1980s and 1990s there was little or no growth. Although the RDCs were not constituted as such until 1985, their forerunners in the major agricultural industries such as wool, wheat, and meat were created much earlier. Most notably, the Wool Research Trust Fund had contributed around 40 percent of CSIRO's rural research funds in 1953. By the end of the 1960s rural industry research funds were contributing about 20 percent of research budgets in the state departments of agriculture. This was a time of strong public and industry support for public agricultural institutions.

We have not been able to ascertain why public support for agricultural research increased so strongly in the 1950s and 1960s relative to both the size of the agricultural sector and the growth in government spending in general. It is perhaps tempting to see the growth in the departments and CSIRO as an example of successful rent seeking by the agricultural sector. However, few research and extension programs administered by state departments had the political profiles of rural adjustment and finance schemes, statutory marketing issues, and input subsidy schemes, presumably because their impact on farm profitability was less direct.

Alternatively we have tried to infer why support grew from general policies toward the rural sector, from science policy, and from views about the role of government at that time. Some appreciation of these issues can be gained from the Vernon Report (Commonwealth of Australia 1965). In the mid-1960s the rural sector was still a much larger part of the economy than it is now. The value of rural production as a proportion of gross national product had declined from 21 percent in 1948/49 to 13 percent in 1961/62, whereas the share of rural exports in merchandise exports only declined from 86 percent to 77 percent in the same years. The Vernon Report (Commonwealth of Australia 1965:157), quoting a 1952 speech by the federal minister of commerce and agriculture, noted that an expansionist farm policy (which was still in place when the Vernon Report was written) was also based on concerns about defense requirements, food security, and the "dollar problem." The report noted that an expansion of research programs was one of several measures used to achieve these policy ends.

With respect to research policy the Vernon Report noted that "the relationship between R&D and the growth of productivity was self-evident" (Commonwealth of Australia 1965:418) and that Australia imported much new technology embodied in inputs. The report argued for an increase in Australian R&D. With respect to rural research the report noted that "Australia spends about three times as high a proportion of its gross national product on research in primary industry as does the United States. Since the primary industry contribution to gross national product in Australia is about three times as great as in the United States, it appears that Australian research in primary industry is roughly comparable in scale with that in the United States" (Commonwealth of Australia 1965:424). At that time about 35 percent of public research funds went to primary industries, which probably amounted to about 25 percent of all research expenditure.

There was no direct discussion of the roles of the public and private sector in financing rural research in the report. The writers of the Vernon Report seemed to accept that rural research should largely be funded by the public sector, perhaps reflecting a view that it was appropriate for governments to make investments in science and technology that would increase productivity and add to wealth. This central role of government also seemed to be accepted by Donaldson (1964), who reviewed the use of levies on producers to fund research. He argued for continuing public funding on the grounds that consumers benefited from rural research in the form of lower prices. This rationale for public funding is broader than that enunciated in more recent reviews of the role of government in agricultural research, such as that by the Industry Commission. In particular, Donaldson overlooked the fact that producers and consumers share not only the benefits from research, but also the incidence of levies imposed to fund research, and thus levies ameliorate the "free-rider" problem.

In the 1970s the share of research budgets funded by the precursors to RDCs fell sharply. The contributions of the RDCs did not start to increase again until after 1985. During the 1980s a more conservative view about the role of government developed. Smaller government and "the user pays" became politically important ideas. At the same time the rural industries themselves were demanding greater control over the direction of research, which may have suited governments wishing for a more arms-length relationship. The establishment of the new RDCs was consistent with these developments.

We need to assess the contribution of the RDCs in terms of the rationale for their establishment. This rationale has been detailed above, but includes (1) an increase in resources available for agricultural research, (2) an increase in industry support for agricultural research, and (3) greater opportunities for industry to influence the direction of research. According to Smith (1992), the new system introduced in 1985 represented a great improvement over the previous arrangements. Research expenditures rose both through more industries becoming involved and through increases in the agreed levies. However, there were still perceived problems with inadequate planning, slow transference of research into practice, and an (allegedly) excessive emphasis on on-farm production research. There were also questions about efficiency in management and about whether interindustry problems were being neglected. These concerns were used to justify the reforms in 1989, which mainly shifted control toward industry.

Smith (1992) also pointed out that the new arrangements have led to their own set of questions. In particular, does the composition of the boards of the respective corporations result in an appropriate range of skills being made available, without loss of contact with the producer constituency? Given that RDCs contribute only a fraction—albeit a sizable one, now—of total funds devoted to rural R&D, is their influence on research directions somehow excessive? Also, has there been an overemphasis on sharply defined and time-bounded projects, which may act to the detriment of long-term capability?

It is also interesting to compare the way in which CRCs operate with the RDCs. Like the CRCs, the RDCs have also sought to reduce unnecessarily duplicative research and to increase the degree of cooperation among researchers and institutions. However, an increasing share of RDC funds are going to fund large cooperative projects in areas of research identified and closely defined by the RDC as being of direct relevance to their industry rather than by the research institutions, as is the case in the CRCs. Given the mix of public and private outcomes from many research activities, it is perhaps reasonable that these two broad approaches to fostering links between research institutions coexist, although it would also be reasonable to expect a much larger proportion of the budget for a CRC to be devoted to research having broad community spillovers than would be the case for a commissioned RDC project.

AMOUNTS OF FUNDING. The RDCs have not succeeded in markedly increasing the total resources available for agricultural research. Although nominal expenditure has continued to rise since 1985, real expenditure has remained constant. Research intensities stagnated throughout the 1980s and have only begun to rise again, slowly, in the 1990s. This undoubtedly reflects, in part, a drive for smaller government, but total government outlays as a percentage of Australian GDP only declined from 42.3 percent in 1985/86 to 38.5 percent in 1993/94.

Anecdotal evidence (including the authors' personal experiences) indicates that the pressures for reductions in government agricultural R&D budgets arose independently of the rise of the RDCs. Indeed, the decline in real public-sector support has occurred at different times and at different rates in the different states and the Commonwealth, which suggests separate influences (perhaps state-level budgetary pressures and party politics as much as anything else), rather than a single influence, the RDCs. Thus there is no clear evidence that the RDCs have crowded out other sources of funds for public-sector agricultural R&D. It seems clear that without increased contributions from the RDCs, total funding for agricultural R&D would have fallen faster in nominal and real terms.

INDUSTRY SUPPORT. Clearly the RDCs have been successful in increasing industry support for research. Expenditure by the agricultural RDCs rose from A$47.5 million in 1984/85 to more than A$239.2 million in 1994/95 in nominal terms. Because levies are related to the quantity of production or the value of output, whereas the cost of research services is more closely related to the general price level, the real contribution of the RDCs has not risen so quickly. R&D expenditure by business has also risen markedly. Some of this can be attributed to the RDCs, but, in part, it is a response to the 150 percent tax concession for research expenditure. The increase in research funded by the RDCs has not been enough to maintain the share of agricultural research in the total research budget.

MIX OF RESEARCH. There seems to have been a marked shift away from basic research toward applied rural research. Undoubtedly the RDCs are responsible for some of this shift, but some state departments have encouraged more-applied research under the slogan of being "market focused." The fact that RDC funding attracts additional public funding, because state agencies and universities (unlike federal agencies such as CSIRO) in general do not fully charge RDCs for contract research, adds to the RDC influence and prompts the question of whether "the tail is wagging the dog." However, some of these developments may reflect a widespread view that government should be more responsive to industry needs. The user-pays or, more appropriately, beneficiary-pays principle was seen not only as a means of raising revenue, but also as a means of identifying and responding to industry needs. Until recently state departments were active in seeking RDC funds and in investigating ways of "commercializing" their services. At the federal level arbitrary external funding quotas were set so that CSIRO now has to raise 30 percent of its funds externally. As the Industry Commission (1995) has recently pointed out, such developments reflect a fundamental misunderstanding of the appropriate role of public institutions in supplying public rather than private goods.

In principle, economists generally argue that public institutions should concentrate on providing public goods from the public purse. Private goods should be fully funded by their beneficiaries. These principles are not widely accepted at present. As evidenced by their submissions to the Industry Commission, some RDCs have little understanding about the appropriate role of public institutions. Additionally, as the RDCs appropriately fund more-applied research, the extent of public subsidy should decline. The Industry Commission has recommended that public institutions should fully cost contract research as a first step down this path. Although that view is perhaps an extreme one, the development of a more competitive research industry is a reasonable objective of government and rules for interactions between the private sector and the public sector should be devised with this in mind.

Because RDC activities are still heavily subsidized, it is not clear that the change in the balance between basic and applied research is appropriate. A more-applied emphasis in rural research in public institutions presumably has come at the expense of more-basic research, which has a higher public-good component. Little is known about the appropriate amount of provision of public goods through rural research. Because producers bear such a small share of the cost, there is a significant divergence between the social and industry returns from rural research. Therefore it is difficult to assess whether the amount of investment in rural research is appropriate. In addition, this potential conflict of interest raises further issues about the governance structures for RDCs.

On the whole, like the Industry Commission, we would conclude that the rising role of RDCs has been beneficial for Australian agriculture, but that it has some drawbacks. If similar arrangements were to be developed and adopted elsewhere, they should be designed with a view to minimizing the crowding out of other public-sector funds for agricultural R&D and with regard to the potential for the total research agenda to be distorted in directions that suit the private interests of the RDCs, and their constituents, but at the expense of the public interest they are also meant to serve. As elsewhere, the debate about the appropriate role of the public sector in providing research services to agriculture is continuing. The central issue at present is the extent of competition in the research industry and pressure for public institutions to act in a competitively neutral way with respect to the private sector.

APPENDIX TABLE A5.1 Cooperative research centers for agriculture, rural-based manufacturing, and the environment: core participants

Name	Core Participants			
	Industry	Universities	Commonwealth	State
Plant Science	• Groupe Limagrain Pacific Pty. Ltd.	• Australian National University	• CSIRO, Division of Plant Industry[a]	
Soil and Land Management		• University of Adelaide	• CSIRO, Division of Soils	• South Australian Research and Development Institute • Primary Industries South Australia • Agriculture Victoria • Forestry Tasmania
Temperate Hardwood Forestry	• Boral Timber Tasmania Pty. Ltd. • Amcor Plantations Pty. Ltd. • Bunnings Forest Products Pty. Ltd. • Australian Newsprint Mills • North Forest Products	• University of Tasmania Departments of Plant Science Agricultural Science	• CSIRO, Division of Forestry and Forest Products	
Tissue Growth and Repair[b]	• Dairy Research and Development Corporation	• University of Adelaide	• CSIRO, Division of Human Nutrition	
Tropical Pest Management	• Boral Timber Tasmania Pty. Ltd.	• University of Queensland	• CSIRO, Division of Entomology	• Queensland Departments of Lands Primary Industries

Hardwood Fibre and Paper Science	• Pulp and Paper Manufacturers Federation of Australia	• Monash University, Australian Pulp and Paper Institute, Department of Chemical Engineering • University of Melbourne School of Agriculture and Forestry	• CSIRO, Division of Forestry and Forest Products	
Biological Control of Vertebrate Pest Populations		• Australian National University John Curtain School of Medical Research, Divisions of Biochemistry and Molecular Biology Faculty of Science Cell Biology	• CSIRO, Division of Wildlife and Ecology	• Wildlife Research Center • Western Australian Department of Conservation and Land Management • Agricultural Protection Board of Western Australia
Catchment Hydrology		• Monash University • University of Melbourne	• Bureau of Meteorology • CSIRO, Division of Water Resources	• Victorian Department of Conservation and Natural Resources • Melbourne Water • Rural Water Authorities
Legumes in Mediterranean Agriculture		• University of Western Australia School of Agriculture, Department of Botany • Murdoch University School of Biological and Environmental Sciences	• CSIRO, Center for Mediterranean Agricultural Research	• Agriculture Western Australia

(*continued*)

APPENDIX TABLE A5.1 *Continued*

Name	Core Participants			
	Industry	Universities	Commonwealth	State
Tropical Plant Pathology		• University of Queensland Departments of Botany, Biochemistry, Microbiology • Queensland University of Technology	• CSIRO, Division of Tropical Crops and Pastures	• Queensland Department of Primary Industries • Bureau of Sugar Experiment Station
Viticulture[c]		• University of Adelaide • Charles Sturt University	• CSIRO, Division of Horticulture	• New South Wales Agriculture • Agriculture Victoria • South Australian Research and Development Institute • Primary Industries South Australia
Cattle and Beef Industry (Meat Quality)		• University of New England	• CSIRO, Divisions of Animal Production, Animal Health, Food Science and Technology, Tropical Animal Production	• New South Wales Agriculture • Queensland Department of Primary Industries

Food Industry Innovation	• Burns Philp Group Burns Philp and Company Ltd. Burns Philp R&D Pty. Ltd. Mauri Laboratories Pty. Ltd. • Goodman Fielder Ingredients Ltd.	• University of New South Wales Departments of Biotechnology Food Science Technology Schools of Biochemistry and Molecular Genetics Microbiology and Immunology	• CSIRO, Divisions of Food Science and Technology Human Nutrition	
Freshwater Ecology	• Albury-Wodonga Development Corporation	• Monash University • University of Canberra • La Trobe University	• CSIRO • Murray-Darling Basin Commission	• Australian Capital Territory Government • Melbourne Water • New South Wales Fisheries • Australian Capital Territory Electricity and Water • Sydney Water • Gippsland and Southern Rural Water • Goulburn-Murray Rural Water • Wimmera-Malle Rural Water

(*continued*)

APPENDIX TABLE A5.1 *Continued*

Name	Core Participants			
	Industry	Universities	Commonwealth	State
Premium Quality Wool[d]	• Australian Wool Research and Promotion Organisation	• University of New England • University of Western Australia • University of New South Wales • University of Adelaide	• CSIRO, Divisions of Animal Production Wool Technology	• Agriculture Western Australia
Sustainable Cotton Production	• Cotton Research Development Corporation	• University of New England • University of Sydney	• CSIRO, Divisions of Plant Industry Entomolgy	• Queensland Department of Primary Industries • New South Wales Agriculture
Aquaculture		• University of Tasmania • University of Technology, Sydney • James Cook University of North Queensland	• CSIRO, Division of Fisheries • Australian Institute of Marine Science	• Queensland Department of Primary Industries • Department of Primary Industry and Fisheries, Tasmania • New South Wales Fisheries • South Australian Research and Development Institute

Quality Wheat Products and Processes[d]	• Arnott's Biscuits Ltd. • Australian Wheat Board • Bread Institute of Australia Inc. • Defiance Mills Ltd. • George Weston Foods Ltd. • Goodman Fielder Ltd. • Grains Research and Development Corporation	• University of Sydney	• CSIRO, Division of Plant Industry	• New South Wales Agriculture • Agriculture Western Australia
Sustainable Development of Tropical Savannas		• Northern Territory University • James Cook University of North Queensland • Australian National University	• CSIRO, Divisions of Tropical Crops and Pastures Wildlife and Ecology Soils • Australian Nature Conservation Agency	• Northern Territory Parks and Wildlife Commission Power and Water Authority Departments of Primary Industries and Fisheries Mines and Energy Lands, Planning and Environment • Queensland Departments of Lands Primary Industries

(*continued*)

APPENDIX TABLE A5.1 *Continued*

Name	Core Participants			
	Industry	Universities	Commonwealth	State
				• Western Australian Departments of Agriculture Conservation and Land Management
Sustainable Sugar Production	• Bundaberg Sugar Company Ltd. • CSR Ltd. • Mackay Sugar Cooperative Association Ltd. • New South Wales Sugar Milling Cooperative Ltd. • Sugar North Ltd. • Sugar Research and Development Corporation • The Australian Canegrowers Council Ltd.	• James Cook University of North Queensland Department of Botany and Tropical Agriculture • University of Queensland Department of Agriculture • Central Queensland University Department of Social Sciences	• CSIRO, Divisions of Soils Tropical Crops and Pastures	• Queensland Department of Primary Industries • Bureau of Sugar Experiment Stations
Weed Management Systems	• Grains Research and Development Corporation • Avcare Ltd.	• University of Adelaide • Charles Sturt University • University of New England	• CSIRO, Divisions of Entomology Plant Industry	• New South Wales Agriculture • Agriculture Western Australia

				• Department of Conservation and Natural Resources, Victoria
Ecologically Sustainable Development of the Great Barrier Reef	• Association of Marine Park Tourism Operators	• James Cook University of North Queensland	• Australian Institute of Marine Science • Great Barrier Reef Marine Park Authority	• Queensland Department of Primary Industries
Agricultural and Rural-Based Manufacturing	• Forestry Tasmania • North Forest Products • Amcor Plantations • Australian Newsprint Mills • Bunnings Treefarms	• University of Tasmania • Southern Cross University • Griffith University	• CSIRO, Divisions of Forestry and Forest Products Entomology	• Queensland Department of Primary Industries
Molecular Plant Breeding[e]	• Grains Research and Development Corporation • Dairy Research and Development Corporation • Vanderhave Sudwestdeutsche Saatzucht • Australian Barley Board • Turf Grass Technology	• University of Adelaide • Southern Cross University • University of Western Australia • Flinders University of South Australia		• Victoria Department of Natural Resources and Environment • South Australian Research and Development Institute

(*continued*)

APPENDIX TABLE A5.1 *Continued*

Name	Core Participants			
	Industry	Universities	Commonwealth	State
Sustainable Rice Production[f]	• Ricegrowers' Cooperative Ltd.	• University of Sydney • Charles Sturt University	• CSIRO, Divisions of Plant Industry Water Resources	• New South Wales Departments of Land and Water Conservation Agriculture

SOURCE: Cooperative Research Centers Program Compendium (1996).

[a]CSIRO, Commonwealth Scientific and Industrial Research Organization.
[b]Also funded by the Child Health Research Institute.
[c]Also funded by the Australian Wine Research Institute.
[d]Also funded by the New Zealand Institute for Crop and Food Research International Ltd.
[e]Also funded by the International Maize and Wheat Improvement Center (CIMMYT).
[f]Also funded by the Rural Industries Research and Development Corporation.

APPENDIX TABLE A5.2 Public expenditures on rural research in Australia

Year[a]	Nominal Spending				Real Spending			
	State Departments of Agriculture[b]	Universities[c]	CSIRO[d]	Total	State Departments of Agriculture[b]	Universities[c]	CSIRO[d]	Total
	(millions Australian dollars)				(millions 1993 Australian dollars)			
1953	5.30	0.33	3.41	9.03	58.57	3.61	37.68	99.85
1954	6.09	0.45	3.55	10.10	65.10	4.83	37.96	107.89
1955	7.54	1.29	3.97	12.80	77.76	13.30	40.98	132.04
1956	7.81	0.58	4.47	12.87	75.33	5.62	43.12	124.07
1957	8.16	0.74	4.61	13.51	75.93	6.86	42.87	125.66
1958	9.32	0.81	5.40	15.54	85.15	7.42	49.31	141.87
1959	9.50	1.13	6.19	16.82	85.92	10.18	55.98	152.08
1960	9.67	1.56	7.13	18.35	82.10	13.27	60.51	155.88
1961	10.77	1.78	7.85	20.40	88.49	14.61	64.47	167.57
1962	11.25	2.11	9.37	22.74	91.23	17.14	76.03	184.40
1963	12.41	2.33	10.14	24.89	97.03	18.25	79.31	194.59
1964	14.22	3.69	11.75	29.66	108.16	28.08	89.32	225.57
1965	14.81	3.67	14.24	32.72	109.44	27.14	105.29	241.87
1966	17.07	4.16	14.41	35.65	122.01	29.74	102.99	254.74
1967	21.43	4.48	17.01	42.93	148.53	31.07	117.90	297.50
1968	24.00	5.00	18.57	47.57	161.46	33.65	124.90	320.01
1969	26.35	5.30	18.61	50.25	168.95	33.95	119.32	322.22
1970	29.30	5.59	22.47	57.36	179.36	34.21	137.57	351.13
1971	32.83	7.06	26.43	66.32	188.28	40.51	151.58	380.38
1972	37.18	7.75	28.58	73.51	194.22	40.48	149.32	384.02
1973	42.41	6.92	30.23	79.56	193.73	31.61	138.09	363.43

(continued)

APPENDIX TABLE A5.2 *Continued*

Year[a]	Nominal Spending				Real Spending			
	State Departments of Agriculture[b]	Universities[c]	CSIRO[d]	Total	State Departments of Agriculture[b]	Universities[c]	CSIRO[d]	Total
	(millions Australian dollars)				(millions 1993 Australian dollars)			
1974	48.96	8.18	38.71	95.85	188.67	31.53	149.18	369.37
1975	64.75	9.62	47.61	121.98	216.98	32.23	159.54	408.74
1976	73.48	11.22	51.74	136.44	221.71	33.84	156.11	411.66
1977	79.67	11.02	56.24	146.93	222.75	30.82	157.24	410.81
1978	92.85	14.83	89.82	197.50	241.26	38.53	233.41	513.20
1979	97.42	14.84	63.35	175.61	226.76	34.54	147.46	408.76
1980	110.60	17.04	76.39	204.03	232.94	35.88	160.88	429.70
1981	124.46	19.26	92.19	235.91	238.12	36.85	176.37	451.34
1982	141.45	21.11	111.71	274.27	242.97	36.26	191.89	471.12
1983	163.62	24.54	126.91	315.07	263.37	39.50	204.27	507.13
1984	173.25	26.42	105.08	304.74	264.29	40.30	160.29	464.88
1985	188.04	32.07	115.73	335.84	268.62	45.82	165.31	479.75
1986	205.28	38.61	132.96	376.86	273.49	51.44	177.14	502.08
1987	208.97	39.04	129.37	377.37	259.04	48.39	160.36	467.79
1988	226.55	39.06	131.56	397.17	258.08	44.49	149.87	452.44
1989	241.80	43.09	115.79	400.68	260.05	46.34	124.53	430.92
1990	270.30	47.64	126.45	444.39	251.89	49.68	131.87	463.44
1991	264.94	55.54	161.85	482.32	271.19	56.85	165.67	493.71
1992	266.20	55.20	172.89	496.70	269.14	55.81	174.80	499.76
1993	266.96	58.52	173.89	505.10	266.96	58.52	173.89	499.37
1994	280.49	61.84	173.73	530.50	275.06	60.65	170.37	506.07

				(percentages)				
Annual growth rates								
1953–1960	9.0	25.1	11.1	10.7	4.9	20.5	7.0	6.6
1960–1970	11.7	13.6	12.2	12.1	8.1	9.9	8.6	8.5
1970–1980	14.2	11.8	13.0	13.5	2.6	0.5	1.6	2.0
1980–1990	9.3	10.8	5.2	8.1	1.9	3.3	–2.0	0.8
1990–1994	0.9	6.7	8.3	3.8	–0.6	5.1	6.6	2.2
1953–1994	10.2	13.6	10.1	10.4	3.9	7.1	3.7	4.0

SOURCE: Mullen, Lee, and Wrigley (1996a,b), updated by authors for 1992–1994 university data.

NOTE: For 1961–1994, deflated with a rebased implicit gross domestic product (GDP) deflator taken from the World Bank (1997); for years prior to 1961, the implicit GDP deflator was backcast using the deflator reported in Mullen, Lee, and Wrigley (1996a:22).

[a]Fiscal year (for example, 1994 refers to 1993/94).
[b]Includes New South Wales, Queensland, South Australia, Victoria, and Western Australia.
[c]Includes Universities of Sydney, New South Wales, New England, Queensland, Melbourne, and Western Australia; and Monash, LaTrobe, Adelaide, and Murdoch Universities.
[d]Commonwealth Scientific and Industrial Research Organization.

6 Agricultural R&D Policy in the United Kingdom

COLIN THIRTLE, JENIFER PIESSE, AND
VINCENT H. SMITH

This chapter examines the post–World War II history of agricultural science policy and agricultural research in the United Kingdom, with particular emphasis on the policy innovations and shifts in direction that have occurred in the 1980s and 1990s. Six major developments are identified. First, since 1980 the government has redefined the roles of the public and private sectors, with productivity-enhancing research, near-market research, and technology-transfer activities becoming the responsibility of the private sector. Public R&D has become concentrated on broadly based, basic research and public-interest research. Second, this has led to reductions in the expenditures of the public system and cuts in core funding from general government revenues. Third, fees for service have been introduced where possible; institutions have been privatized, closed, and amalgamated; and new institutions have been created to raise funds from commodity groups. Fourth, to create a competitive market in research, competitive bidding for research projects has, to an extent, replaced automatic program funding. Fifth, to extend market incentives to within public organizations, accountability has been increased by the formation of executive agencies, monitoring and evaluation procedures have been strengthened, and increasing numbers of staff have been hired on short-term contracts. Sixth, the government has committed itself to providing clear policy leadership and improving information flows by means of publications, improved management structures, and increased collaborative research.

The chapter is organized as follows. In the first section we describe the current organization of agricultural research in the United Kingdom and discuss flows of funds and expenditures in the U.K. National Agricultural Research Service (NARS). Next we discuss the institutional history of public agricultural R&D and R&D policy in the United Kingdom since World War II. In the third section we provide a quantitative review of public funding and expenditures in the postwar period. More recent changes are investigated in the fourth section by comparing public and private R&D funding and expenditures in 1987 with those in 1994. In the final section we summarize the findings.

An Overview of the Current Research System

The current organization of the U.K. NARS is illustrated in Figure 6.1, which shows the major groups of private and public institutions involved in agricultural and food research, flows of funds between these institutions (the numbers attached to arrows), and their expenditures (the numbers in the boxes) for the 1993/94 fiscal year.[1]

Flows of Funds

Total agricultural and food research expenditures in 1993/94 were about £893 million. Public institutions accounted for £339 million, or 38 percent of these expenditures, whereas the private sector accounted for £554 million or 62 percent.

Core funding from general tax revenues supported about 80 percent (£270 million) of the agricultural research expenditures of the major, nonuniversity public institutions. The remaining public outlays were funded either through government revenues channeled through the Higher Education Funding Council (HEFC) for research at universities or through contracts for research funded by the private sector or other government departments. Recently a modest but increasing proportion of the public research budget (£38 million in 1993/94) has been allocated to food and food-safety research. The private sector invests heavily in food industry and agricultural input research. In 1993/94 estimated food industry outlays on R&D were £256 million and estimated agricultural input industry expenditures were £284 million (slightly less than the £304 million spent in private laboratories because part of these expenditures were funded from boards, foundations, and other sources of funds).

The above data indicate that in the United Kingdom the private sector accounts for the lion's share of food research (about 90 percent in 1993/94) and that food research in the United Kingdom currently represents 35 percent (£310 million) of total agricultural and food R&D expenditures (£897 million). The public sector's role in relation to more traditional agricultural R&D is much more substantial. In 1993/94 51 percent of all nonfood agricultural R&D expenditures (£587 million) consisted of R&D expenditures by core public institutions (£301 million), and general tax revenues (about £232 million) funded about 40 percent of total agricultural R&D outlays. Considerable funding for research at public institutions is also provided by the private sector through contracts or grants funded by checkoffs or commissioned directly by

1. The U.K. public agricultural research sector has recently undergone substantial changes, with major institutions receiving new names and functions. On April 1, 1994, for example, the Agricultural Food and Research Council was incorporated in the Biotechnology and Biological Sciences Research Council (BBSRC). New names of institutions are used in this section to enable readers to understand the current system.

FIGURE 6.1 Funding and performance of agricultural and food research, 1994

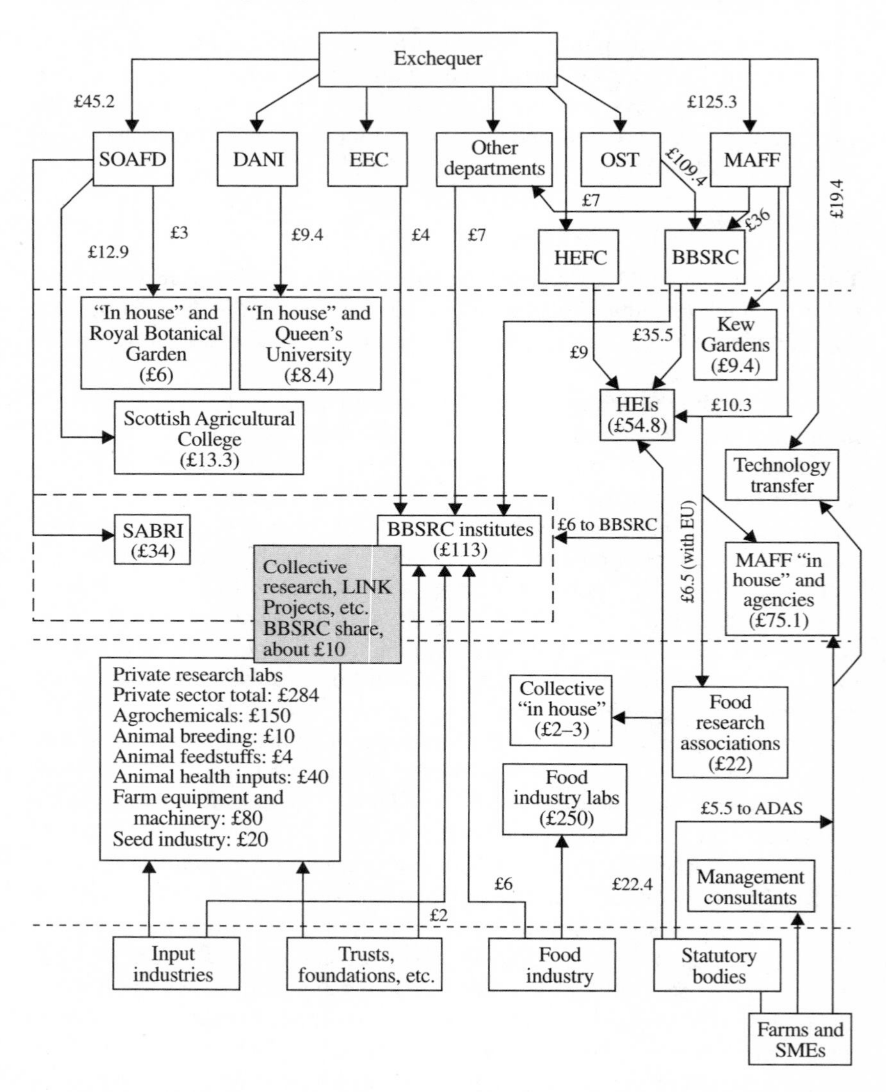

NOTE: Monetary figures in millions. ADAS, Agricultural Development Advisory Service; BBSRC, Biotechnology and Biological Sciences Research Council; DANI, Department of Agriculture for Northern Ireland; EEC, European Economic Community; EU, European Union; HEFC, Higher Education Funding Council; HEIs, higher education institutions; MAFF, Ministry of Agriculture, Fisheries and Food; OST, Office of Science and Technology; SABRI, Scottish Agricultural and Biological Research Institute; SMEs, small and medium enterprises; SOAFD, Scottish Office of the Agricultural and Fisheries Department.

individual private firms and charitable foundations. In 1993/94 these private expenditures amounted to about £48 million, almost 15 percent of the budgets of public research institutions.

The general interdependency between the public and private systems in R&D is clearly illustrated in Figure 6.1. To understand the current U.K. NARS and the nature of these interdependencies, we begin by examining the current organization and structure of public-sector agricultural research.

The Public Sector

The U.K. public agricultural research system encompasses four countries and has evolved over a period of almost 90 years.[2] In this section, we first examine the major elements of the U.K. public agricultural research system and then consider the roles in U.K. agricultural research of the regional departments with responsibilities for Scotland and Northern Ireland and the European Union (EU).

MAJOR ELEMENTS. In the United Kingdom, most domestic public-sector agricultural research is now funded through the Office of Science and Technology (OST) and the Ministry of Agriculture, Fisheries and Food (MAFF).[3] OST's functions in the United Kingdom roughly correspond to those of the National Science Foundation in the United States, whereas MAFF's functions roughly correspond to those of the U.S. Department of Agriculture. The OST funds research in all areas of science, including agriculture. In 1993/94 its total budget for all science research councils was about £1.15 billion, of which the agricultural research council, the Biotechnology and Biological Sciences Research Council (BBSRC), received about £109 million. The BBSRC also received £36 million from MAFF for commissioned research and about £11 million from industry, other government organizations, and miscellaneous sales, which resulted in an income of about £156 million (BBSRC 1994:14–15). The BBSRC used more than 60 percent of this income, £98 million, to support eight BBSRC agricultural research institutes. Most of the rest of this income was used to fund university research (about £36 million) and pensions for retired BBSRC institute staff (£20 million). The remainder (over £2 million) was allocated to other councils and government departments.

The BBSRC institutes, which are listed in Table 6.1, deal with a wide array of research areas, including animal physiology and genetics, arable crops, plant science, food, and engineering. These institutes now focus more on more-basic than on more-applied research issues. In 1993/94 the total operating

2. Provision for a national agricultural research system was made under the 1909 Development and Road Improvement Act.

3. The Office of Science and Technology was established in 1992, combining parts of the Department of Education and Science (DES) and the Cabinet Office. The name change reflects the Conservative government's increased focus on the role of science and technology in economic growth.

TABLE 6.1 Grants to Biotechnology and Biological Sciences Research Council and Scottish Agricultural and Biological Research Institutes, 1994

Biotechnology and Biological Sciences Research Council Institute	Recurrent Grant	Capital Grant
	(millions pounds sterling)	
Brabaham Institute (animal physiology and genetics)	8.2	2.6
Institute of Arable Crops Research	12.5	1.3
Institute for Animal Health	11.0	1.8
Institute of Food Research	11.6	1.2
Institute of Grassland and Environmental Research	8.0	1.2
Institute of Plant Science Research	9.2	1.4
Roslin Institute (animal physiology and genetics)	5.2	0.7
Silsoe Research Institute (engineering)	4.9	1.2
Scottish Agricultural and Biological Research Institutes (SABRI)	*SOAFD Grant*	*Other Income*
Hannah Research Institute (biological and food research)	3.6	0.6
Moredun Research Institute (animal health)	4.2	1.1
Macaulay Land Use Research Institute	6.5	0.8
Scottish Crop Research Institute	8.0	2.0
Rowett Research Institute (nutrition)	6.0	1.2

SOURCE: AFRC (1991), BBSRC (1994), and annual reports of the Scottish institutes (Scottish Agricultural Science Agency 1994; Scottish Crop Research Institute 1994).

budget for the eight BBSRC institutes was about £113 million. Of this total, £98 million was funded by the BBSRC and £15 million was income from contracts with government departments, the EU, private industry (including the food industry and statutory bodies), foundations, and trusts. About £10 million of the institutes' budgets was allocated to collaborative research projects with private institutions and universities.

The other major component of the U.K. agricultural research system is university agricultural research performed in higher education institutions. In 1993/94 expenditures on university research amounted to almost £55 million, of which more than 60 percent (£35.5 million) was funded by the BBSRC, about 19 percent (£10.3 million) by MAFF, and about 16 percent (£9.0 million) from general tax revenues from the HEFC.

MAFF supports substantial intramural research operations through the facilities of four executive agencies.[4] The ministry has a mandate to perform

4. Between 1990 and 1992, four components of MAFF with research responsibilities were

statutory duties and research to support government policy; more than 60 percent of its R&D budget is allocated to policy support (that is, to protect the public, farm animals, and the rural and marine environment). The rest of MAFF's R&D budget funds more-applied, productivity-oriented research (MAFF 1994). The Agricultural Development Advisory Service (ADAS) is the largest of MAFF's executive agencies. In 1993/94 ADAS's income was £76 million (ADAS 1994), of which two-thirds consisted of public funds and the rest was income from fees for service and grants and contracts from producer groups and individual firms and farms. ADAS has an outreach or extension mission (for which charges have been imposed since 1987), as well as applied research functions on its experimental farms.

In 1993/94 MAFF's total R&D budget amounted to about £165 million, including £125 million in core funding from general government revenues, £20 million in fees for service and contract income, and a technology-transfer budget of about £19 million. Of this income, MAFF allocated £36 million to the BBSRC, about £10 million to university research, £7 million to other departments, about £9 million for botanical research at Kew Gardens, and about £7 million to private research institutions (much of which supported food research associations).[5] The remaining £95 million was allocated to MAFF's intramural R&D program (£75 million) and for technology transfer (about £19 million).

SCOTLAND. Most public agricultural research funding in Scotland is provided through the Scottish Office of the Agricultural and Fisheries Department (SOAFD). In 1993/94, SOAFD's research budget amounted to about £45 million. Like MAFF, SOAFD allocates research resources to in-house activities, including the Scottish Agricultural Science Agency (formed in 1992) and the Royal Botanical Gardens in Edinburgh. However, only about 7 percent of SOAFD's budget was used to support in-house research, as compared with 52 percent for MAFF. In contrast, 75 percent of SOAFD's budget (£34 million) was allocated to the five Scottish Agricultural and Biological Research Institutes (SABRIs) listed in Table 6.1, whereas MAFF allocated only about 25 percent of its budget to the BBSRC and university research. The remaining 18 percent of SOAFD's budget (£13 million) supported the Scottish Agricultural College system, which consists of a central office in Edinburgh and three regional colleges.

---

given executive agency status: the Veterinary Medical Directive (1990), the Central Veterinary Laboratory (1990), the Central Scientific Laboratory (1992), and the Agricultural Development and Advisory Service (1992).

5. The food research associations were the Camden Food and Drink Research Association (CFDRA), the Flour Milling and Baking Research Association (FMBRA), the Leatherhead Food Research Association, and the British Industrial Biological Research Association. In 1994 CFDRA and FMBRA merged to form the Camden and Chorleywood Food Research Association.

NORTHERN IRELAND. In Northern Ireland, the Department of Agriculture for Northern Ireland (DANI) manages public agricultural research funds. All of its research budget is allocated either to in-house research activities or to support research at Queen's University, Belfast. Northern Ireland has no independent research institutes, partly because it is small and because of close links between DANI and Queen's University.

EUROPEAN UNION FUNDING. The United Kingdom also funds and participates in EU research projects. In 1993/94 this system provided about £8 million in food and agricultural research funding to the United Kingdom.[6] The influence of the EU on U.K. agricultural research institutions extends well beyond its funding role. EU regulations are partly responsible for the abandonment of the U.K. system of marketing boards (established the late 1940s) that functioned as commodity-based marketing, promotion, and research organizations. The promotional and research functions that the marketing boards performed are now the responsibility of a new set of statutory bodies. The functions of these bodies are discussed below in the section on public agricultural science policy in the post–World War II period.

The Private Sector

The private sector, which is illustrated in the lower part of Figure 6.1, consists of agricultural input industries, nonprofit trusts and foundations, food industry companies, statutory bodies, farms, and small and medium enterprises (mostly connected with the food industry). In 1993/94 input industries spent almost £300 million on R&D, largely in private research facilities, but they also funded a small amount of research at collective (commodity-level) research institutions. The food industry's estimated R&D budget in 1993/94 was about £272 million, of which £250 million was allocated to private facilities and the rest to food research associations. Through fees for services, checkoffs, and other contracts, the private sector also funded £50 million of research by MAFF agencies, BBSRC institutes, and the universities. Thus private funds support about 17 percent of the agricultural research budgets of public research institutions.

About 40 percent of the private sector's support for public research, £22.4 million, is provided by commodity groups (called statutory bodies) that collect levies on producers of particular commodities and also have some limited in-house research facilities (with budgets of only £2–3 million in 1993/94). The statutory bodies are quasi-government organizations in that they operate under statutory authority and derive their resources from legislated mandatory levies

6. MAFF has shown a special interest in ensuring that U.K. agriculture benefits from a new initiative, Framework Programme IV (described in OST 1995a), which will distribute 684 million ECUs (£560 million at the June 1995 exchange rate of £0.82 to the ECU) to all areas of research from 1994 to 1998.

on agricultural output, but, through their boards of directors, are largely responsible to the industries that fund them. In several cases, the statutory bodies' strong ties to commodity groups and food processors are reinforced by mandates to implement marketing and promotion activities on behalf of their client groups. These organizations almost always support commodity-specific, near-market research.

Agricultural input industries allocate most of their research funds to their own in-house research laboratories, some of which are well-established institutions with strong research records. In 1993/94 estimated private-sector expenditures at in-house facilities were about £284 million. Of this total, about £150 million was spent by the agricultural chemicals industry; £100 million by farm equipment and machinery companies and pharmaceutical companies in relation to animal health products; and £34 million by the seed industry, animal breeders (pigs and poultry), and animal feed producers.[7]

The food industry in the United Kingdom is dominated by a few large companies with R&D expenditures of between 1 and 2 percent of sales. The major firms supporting in-house R&D include Unilever, Associated British Foods, and Rank Hovis McDougall. As might be expected, the activities of these companies are targeted in part toward international markets. Smaller companies tend to have lower research intensities, and the BBSRC (1995) estimated expenditures at some £250–300 million. These probably include expenditures for product development, and so the lower value is reported in Figure 6.1 as an estimate of food industry research efforts.

Public Agricultural Science Policy in the Post–World War II Period

Major political changes in the governing of the United Kingdom since World War II have had important ramifications for science policy, particularly because of the integrated nature of legislative and administrative functions within the British parliamentary system of government (in which the prime minister and the cabinet head both the legislative assembly and all central government agencies). This section therefore examines the history of public agricultural science policy in the United Kingdom in the context of the major changes that have occurred in general science policy and economic policy during the post–World War II period.[8]

---

7. These estimates were obtained in several ways. Expenditures on animal health R&D in 1993/94 (estimated to be about £40 million) were obtained by assuming that R&D in this sector was 21 percent of industry sales in 1994, which is the R&D intensity for the rest of the pharmaceutical industry (OST 1995b:Vol. 3, Table 11). The other estimates were all provided by industry insiders, with the help of the trade associations, and are more reliable.

8. This section relies heavily on material presented in Thirtle et al. (1991) for the history of U.K. agricultural research policy prior to the mid-1990s and on Cunningham and Nicholson (1991) and Gummett (1991) for insights into general science and technology policy.

The postwar history of U.K. agricultural research policy can be divided into four phases.

- 1945/46–1955/56: A period of emphasis on a centralized agricultural research system under the control of the Ministry of Agriculture.
- 1956/57–1970/71: A period in which publicly funded agricultural research clearly operated under a dual funding system in which public agricultural research was funded by the central government through both the Ministry of Agriculture (in its various incarnations) and an Agricultural Research Council (ARC; also in various incarnations).
- 1971/72–1978/79: A period in which three important changes took place. First, as a result of a governmentwide review of general science policy, called the Rothschild Report, successive Conservative and Labour governments reallocated resources toward the Ministry of Agriculture and away from the ARC. Second, partly spurred by U.K. entry into the European Economic Community (EEC), environmental interest groups began to question the benefits from agricultural research. Third, successive budget crises led to slowdowns in the growth of public spending on agricultural research.
- 1979/80–1994/95: A period in which, within the context of a dual funding system, a more radically Conservative government cut total funding for agricultural research (as well as other scientific and technological research) in real terms between 1982 and 1987 and then between 1987 and 1994 increased aggregate funding in real terms above its 1979 levels by expanding support for basic research. The government also to some degree privatized the research functions of the agricultural ministry, supported the creation of statutory bodies to carry out near-market research funded by levies on agricultural producers, and introduced more competition among universities and independent research institutes for public agricultural research funds through expanded competitive-grants programs and by linking funding for academic departments in higher education institutions to periodic evaluations of their research productivity.

We begin with a brief description of the U.K. agricultural research system in 1945 and then examine the changes that occurred in each of the four phases outlined above.

Agricultural Research Policy, 1945

In 1945, immediately at the end of World War II, four organizations were involved in the administration of agricultural research in England and Wales: the Ministry for Agriculture and Fisheries, the Ministry of Food, the Development Commission, and the ARC. Prior to the 1930s, the Ministry of Agriculture dispensed research funds under the direction of the Development Commission, which was established in 1909 to administer a national agricultural research system and which also administered its own budget. Snelling (1976) has noted that the division of tasks between the ministry and the commission was ambiguous partly because, in contrast to other research areas, it was harder to draw a line between basic and applied research in agriculture. Thus, in 1931 the ARC was established to coordinate the activities of the two bodies. However, the ARC was not successful in accomplishing that goal, mainly because it was not given oversight authority over ministry or commission agricultural R&D expenditures.

The independence of the Ministry of Agriculture and the Development Commission was largely consistent with the recommendation of a 1918 Parliamentary Committee of Enquiry chaired by General Richard Haldane, which set up a broad framework within which all U.K. governments have subsequently implemented general R&D policy. The Haldane Committee recommended that a dual funding system be established for all U.K. publicly funded science and technology research. The committee argued that research required by a government department should be managed by that department, but that research relevant to several departments should be funded by more autonomous research councils. The Haldane recommendations were accepted and became the basis for successive governments' general science R&D policies. As a result, by 1949 four separate research councils had been established, including the ARC.

Agricultural Research Policy, 1945/46–1955/56

The postwar Labour government led by Clement Atlee viewed the crossed lines of responsibility for agricultural research between the Development Commission and the Ministry of Agriculture as inappropriate. Atlee's government placed all administrative authority over agricultural research funding in the ministry's hands. In addition, in 1946, in response to the criticism that too little attention had been paid to linking agricultural research and advisory or extension functions within U.K. agricultural research, the Atlee government also created the National Agricultural Advisory Service (NAAS).[9] NAAS was to supply agricultural advice and information through a system of regional field advisors, who are roughly comparable to extension agents in the United States. However, these advisors were to be supported by researchers located in a system of 13 experimental husbandry farms and horticultural stations.

In Scotland and Northern Ireland arrangements were slightly different. In Scotland all publicly funded agricultural research became the responsibility of the Department of Agriculture and Fisheries, Scotland (DAFS). In Northern

9. In 1971 NAAS was renamed the Agricultural Development and Advisory Service (ADAS). In 1992 ADAS became an autonomous agency with a mandate to charge fees for services to private-sector clients.

Ireland DANI similarly was solely responsible for public agricultural research funds. These institutional arrangements were maintained by successive governments until 1956, when the agricultural research system in England and Wales experienced a substantial organizational change. However, the structure for funding public agricultural research in Scotland and Northern Ireland has not changed much since 1946.

One other important administrative change took place in this period. In 1955, as soon as food rationing in Britain ended, the Ministry of Food was closed and its functions integrated within the Ministry of Agriculture and Fisheries, which was then renamed the Ministry of Agriculture, Fisheries and Food (MAFF). This name change carried few short-term implications for MAFF's agricultural research agenda, but when concerns about food safety increased in the 1970s and 1980s, MAFF was identified as the agency responsible for food-safety research.

Agricultural Research Policy, 1956/57–1970/71

In 1956 a major change occurred in the public agricultural research system for England and Wales. Control over research funding was transferred to the ARC and the Development Commission disappeared. It is unclear why this change came about. However, Shattock (1991) has noted that during the 1950s a close relationship developed between the minister responsible for science research and all of the research councils and the universities. It was in this general environment that the ARC was given independent responsibility for agricultural research and MAFF's influence declined correspondingly. However, MAFF remained in charge of substantial research funds through its mandate to provide applied research support for the NAAS. Thus the public agricultural research system still remained a dual system.

Public-sector agricultural research funding expanded rapidly between 1956 and 1971, although some administrative changes did occur. As part of the Wilson Labour government's attempts to channel the "white heat of science" into effective improvements in economic productivity, funding for all of the research councils was reorganized under the 1965 Science and Technology Act.[10] This act transferred responsibility for these councils to the newly created Department of Education and Science (which later became the OST). As a

10. In the 1964 election campaign, Wilson (then leader of the opposition Labour party) emphasized the need for an aggressive science and technology policy to prevent a "brain drain" of talented researchers to the United States and to enhance economic productivity. In response the Conservative government established a committee of enquiry into the organization and management of research, which was chaired by Sir Burke Trend. The Trend Committee made several recommendations for reorganization, including the creation of a science research council and a development authority to handle R&D applications. The Wilson Labour government, which was elected in 1964, accepted most of these proposals, but placed responsibility for industrial R&D in a new Ministry of Technology.

result, the dichotomy between the two components of Britain's agricultural research system became institutionalized; each became the clear responsibility of a separate government agency.

Agricultural Research Policy, 1971/72–1978/79

In 1971 a new parliamentary review, *The Organization and Management of Government R&D,* led to major changes in agricultural research funding. This report, which was largely written by Lord Rothschild, was part of a comprehensive review of government operations initiated by a newly elected Conservative administration in 1970. The Rothschild Report advocated that a customer-contractor principle be implemented for all science R&D. Under this principle, research would be oriented to the needs of the agencies commissioning the research on a customer-contractor basis.[11] In relation to agriculture, Rothschild argued that the ARC had been too divorced from the needs of its clients—the private agricultural community (farmers, agricultural input suppliers, and food processors) and MAFF—in its research agenda. This had important ramifications for the independent agricultural research institutes funded by ARC; apparently they also were not doing their job adequately, at least as that job was perceived by influential U.K. agricultural lobby groups.[12]

The Rothschild Report recommended that MAFF, which acted as a customer on its own behalf and a more effective principal agent for the U.K. agricultural community, should be given a much larger direct say over the disposition of agricultural research funds for applied research. This implied that resources should be reallocated from the ARC to MAFF. As a result, within two years the ARC had lost a considerable proportion of its budget to MAFF and MAFF then became an important client for the agricultural research institutes. In addition funding was increased for MAFF's agricultural advisory service, which (with expanded responsibilities that included veterinary research and services) was reborn as the ADAS in 1971.

The Rothschild Report also argued for institutional changes to improve communications between the ARC and MAFF. Thus, a Joint Consultative Organization was established to create links between the ARC, MAFF, and the Departments of Agriculture for Scotland and Northern Ireland, all of which had

11. The Rothschild Report created a furor among many leading scientists, who claimed it was a threat to the flexibility and academic freedom of researchers. In fact the report did not invoke any new principles. In 1918 the Haldane Report had argued that research councils should exist to fund research that spanned the interests of several departments but not that they should be entirely autonomous entities.

12. Part of the ARC's public-relations problem was that it had to allocate substantial funds to support basic research and defense-related research rather than applied research (Thirtle et al. 1991). In addition the division of responsibilities for research between ARC and MAFF had led ARC-funded agricultural research institutes to assume that MAFF's National Agricultural Advisory Service would be responsible for translating their laboratory results into useful inputs for agricultural producers (see Winifirth 1962; Ruttan 1982).

R&D responsibilities. Five specialist advisory boards were created within the Joint Consultative Organization to provide advice on research policy, and board members were drawn from user groups (farmers, agricultural input suppliers, and food processors), civil servants, and scientists.[13]

This structure for the administration of research funds was maintained throughout the 1970s and into the 1980s. For most of the 1970s agricultural research funding continued to expand, although successive government budget crises caused funding for all research, including agriculture, to oscillate from year to year (Table 6.2). Nevertheless, there were clear signs that public agricultural research funding would be less likely to expand in the 1980s, no matter which political party was in charge of Westminster. The change in the climate for public agricultural research funding was closely linked to widespread concerns about the need for reduced government intervention in the economy and lower levels of government spending in general (Thirtle et al. 1991). However, there also was increasing skepticism about the potential benefits from additional scientific research. The United Kingdom's entry into the EU is relevant in this context.

The United Kingdom achieved full membership in the EU in the mid-1970s (after a three-year transition period beginning in 1973) and by 1976 was a full participant in the Common Agricultural Policy (CAP). The CAP generally guaranteed EEC crop and livestock producers prices for their products that were well in excess of world market levels. Prior to U.K. entry into the EU, U.K. producers received lower prices than EU producers, especially for cereals. Under the CAP, however, U.K. crop producers immediately enjoyed sharp increases in prices and responded by adopting more intensive and extensive cropping practices. In response, fertilizer, herbicide, and pesticide expanded rapidly and more land came into production, often through the removal of hedges and walls. As a result water pollution increased and the diversity and size of bird populations declined (Woods et al. 1988). In addition, heavier grazing patterns were adopted for livestock that changed the rural landscape. These changes led influential environmental and conservation groups like the Royal Society for the Protection of Birds to question both the value of the CAP and the appropriateness of agricultural research that encouraged chemical-intensive agricultural production practices.

Agricultural Research Policy, 1979/80–1994/95

The election of a Conservative government under Margaret Thatcher in 1979 had radical implications for many aspects of government policy. With a perceived mandate to reduce the scale of government, cut taxes, and lower the

13. Separate advisory boards were created for animals, arable and forage crops, food science and technology, and engineering and buildings.

TABLE 6.2 Expenditures of the public agricultural research system, 1973–1994

Year	AFRC[a]	MAFF[b]	DAFS[c]	HEIs[d]	DANI[e]	Other[f]	Total R&D	Food[g]	Total Agriculture
				(millions 1993 pounds sterling)					
1972/73	146.6	32.0	42.6	6.5	9.0	2.9	240.0	19.7	220.0
1973/74	148.2	33.2	42.8	6.7	9.0	2.9	243.0	24.6	218.0
1974/75	159.9	35.2	45.9	5.9	9.0	5.3	261.0	26.7	234.0
1975/76	150.6	38.1	57.4	5.9	9.0	5.4	266.0	29.3	237.0
1976/77	156.6	52.5	48.8	6.7	9.0	5.1	279.0	22.6	256.0
1977/78	151.3	55.3	44.8	7.3	9.0	4.5	272.0	15.3	257.0
1978/79	156.5	52.8	46.2	7.2	9.0	4.9	277.0	16.6	260.0
1979/80	153.6	53.9	45.4	7.2	9.0	7.5	277.0	19.3	258.0
1980/81	157.1	61.9	46.8	8.0	9.0	8.4	291.0	20.4	271.0
1981/82	162.9	70.2	47.4	8.9	9.0	12.6	311.0	25.4	286.0
1982/83	160.2	79.3	47.6	11.1	8.6	13.0	320.0	28.2	292.0
1983/84	156.5	74.6	47.7	11.9	9.8	11.0	311.5	29.4	282.0
1984/85	146.9	69.3	55.0	14.4	10.3	13.2	309.2	30.1	279.2
1985/86	144.4	68.9	53.9	15.2	11.3	16.9	310.5	34.6	275.9
1986/87	145.2	72.6	52.2	15.2	10.5	18.2	313.8	37.3	276.5
1987/88	131.6	65.7	49.6	16.2	11.6	23.6	298.4	36.7	261.7
1988/89	132.1	68.8	50.5	18.0	11.5	18.6	299.4	44.0	255.5
1989/90	134.1	65.0	49.3	26.0	12.8	14.3	301.6	38.6	263.0
1990/91	132.5	65.1	49.2	27.4	11.3	13.2	298.7	40.4	258.3
1991/92	123.0	74.9	49.5	29.6	11.2	17.0	305.3	40.7	264.6
1992/93	122.1	88.3	46.5	41.2	8.6	13.6	320.4	43.1	277.2
1993/94	112.6	84.4	45.2	45.8	8.3	14.4	310.7	37.8	272.9

SOURCE: AFRC annual reports; MAFF (various years) *Report on Research and Development,* from 1972/73 to 1981/82; Cabinet Office *Annual Review of Government R&D* (various years), from 1982/83 to 1990/91; OST (1995a), from 1991/92 to 1993/94.

[a]AFRC, Agricultural and Food Research Council. AFRC income, plus income to the institutes, minus pensions.

[b]MAFF, Ministry of Agriculture, Fisheries and Food. MAFF in-house, including Kew Gardens, minus fisheries research.

[c]DAFS, Department of Agriculture and Fisheries, Scotland, SOAFD, Scottish Office of the Agricultural and Fisheries Department. Core funding from DAFS/SOAFD only, minus fisheries research.

[d]HEIs, higher education institutions. Contracts from the AFRC and MAFF but not other institutions or the Higher Education Funding Council.

[e]DANI, Department of Agriculture for Northern Ireland. DANI in-house, minus fisheries research: prior to 1981/82, estimates from Thirtle (1989).

[f]All MAFF commissions, except for fisheries research and with the AFRC and the HEIs. In later years, the AFRC commissions are subtracted from the totals that the councils show in *Forward Look.*

[g]Food research including all commissions up to 1985/86. Then AFRC and MAFF only, plus estimates for DAFS and DANI.

government budget deficit, the Thatcher administration was willing to scrutinize carefully all aspects of public research funding. In 1982 budgets for all research councils were reduced and the ARC budget continued to decline throughout most of the 1980s. By 1987/88 public funding for all food and agricultural research had fallen in real terms by about 9.3 percent to £298.4 million from its peak 1982 level of £320.0 million and the ARC's expenditures had declined even more rapidly (by 19.3 percent, from £162.9 million in 1981/82 to £131.6 million in 1987/88).[14]

The decline in public agricultural research funding was not simply due to a general decline in public funds for scientific research. The Conservative administrations of the 1980s and 1990s continually argued that the government should not do what the private sector can accomplish through the market. This view led to policy innovations under which some government departments were privatized or closed down during the late 1980s and 1990s. Many agricultural research institutions were particularly susceptible on these grounds because of the applied or near-market nature of their research, and cuts in funding during the first part of the 1980s resulted in the amalgamation and consolidation of the research institutions funded by ARC from 18 separate facilities to eight. In this process, some research institutions were privatized. In 1988 the government sold the National Seed Development Organization (which was previously operated by MAFF) and part of ARC's plant breeding institute to Unilever. The Liscombe Experimental Husbandry Farm was also transferred to the private sector in 1989. Between 1989 and 1990 several horticultural experiment stations operated by MAFF were closed and the ARC's Institute of Horticultural Research was transferred to the British Horticultural Society, which then had to obtain support for its work on a competitive, client-oriented basis from ARC, MAFF, and the private sector.

Some important institutional changes in the management of public agricultural research were also made. The Joint Consultative Organization, which was created in the 1960s to coordinate research by MAFF and the ARC, had been criticized as cumbersome and overly bureaucratic. In 1984 this organization was replaced by a Priorities Board. In its first report, which was released in 1985, the Priorities Board (whose membership explicitly included expanded representation for the food industry) identified the expansion of food research as a major priority for agricultural research. In response, in 1984 the ARC was renamed the Agricultural and Food Research Council (AFRC); funding for food research was increased sharply until such research accounted for 14.6

14. Thirtle et al. (1991) argue that there is some confusion about the actual decline in real funding levels over this period because it is hard to identify what constitutes public agricultural research expenditures and published data also include expenditures on retirement buyouts and pensions.

percent of the AFRC's 1988/89 budget. In part this increase in funding reflected increased concerns for food safety among the general population—in the late 1980s, concerns had been raised in the United Kingdom about the use of bovine somatotropin in milk production and bovine spongiform encephalopathy (BSE; mad cow disease) in meat, as well as salmonella in eggs. The increase in funding also reflected an increase in the political influence of the food-processing and distribution lobby. In 1989 the Priorities Board's membership was changed to include food retailing and consumer interests.

In the 1990s John Major's Conservative government continued to emphasize privatization and reduced government involvement in market-related activities, and these broad policy concerns continued to have important implications for agricultural research. In 1990 the Priorities Board determined that all MAFF funding for near-market research should cease by 1991/92, which resulted in a £30 million reduction in MAFF's applied research budget. Some of these funds were reallocated to the Department of Education and Science to fund higher levels of basic research at the universities and, through the AFRC, the agricultural research institutes.[15]

In 1990 the Priorities Board also recommended that Project LINK be created. Under this project, AFRC was to use some of its resources to fund joint research projects with industry. Thus some of the funds to be transferred from MAFF to the AFRC would be allocated to develop technology-transfer projects. In 1995 570 projects, with a total budget of about £300 million, were being jointly funded under Project LINK, but most of the support for these projects was provided by the private sector; the government's contribution was only about £10 million.

Also in 1990, as part of a general initiative to reorganize the funding and management of government, the Conservative administration announced that some government departments would become executive agencies. In the past these departments' agencies had dual roles of providing services to government departments and to private firms. Henceforth, instead of providing service to the public for free, these executive agencies would have to sell their services to the private sector and only receive part of their budgets from general government revenues. Between 1990 and 1992, four research-related components of MAFF were given executive agency status: the Veterinary Medical Directive (1990), the Central Veterinary Laboratory (1990), the Central Scientific Laboratory (1992), and the ADAS (1992). On being granted

15. The 1990 Priorities Board recommendation that funding for near-market research should end was based on the 1987 Barnes Report, an internal study commissioned by MAFF (Read 1989). This report had recommended that MAFF funding for near-market research be cut by about £24 million. An earlier report by ADAS (MAFF 1984b) had also recommended that ADAS charge private-sector users for services.

agency status, ADAS (the largest of these agencies) also experienced a 55 percent cut in MAFF funding for its research and technology-transfer services. ADAS's commercial income rose by 40 percent between 1990 and 1994, but this did not compensate for the loss of MAFF funds and thus its workforce declined from about 2,900 people in 1991 to about 2,500 in 1994.

The private sector was expected to increase its applied research to offset the reduction in direct government support. In part this was supposed to occur through increased research funding by the statutory bodies. These agencies derive their resources from legislated mandatory levies on the output of specific commodities. Priorities for the use of these funds are established by each statutory body's governing board, which largely consists of representatives of commodity groups or industry.

In the United Kingdom legislatively sponsored commodity and industry groups have had a long history of providing research support on commodity-specific issues. However, between 1980 and 1995 many of these groups were reorganized, largely because of EU policies. In the 1960s most of them operated as marketing boards with, in some cases, explicit mandates to maximize producer revenues through price discrimination. These groups also provided research funds to independent research laboratories and universities. Most of these boards' operations were incompatible with EU CAP regulations, and, by 1994, even the Milk Marketing Board had been disestablished. This left an organizational vacuum with respect to research funding by commodity groups. Beginning in 1986 the Conservative government moved to fill that vacuum by creating a series of new research and market development institutions. These include the Horticultural Development Council, the Apple and Pear Research Council, and the Oil Seed Council (all founded in the 1980s) and, more recently, the Milk Development Council (founded in 1995), the Potato Research Council, and the Egg Research Council (both established in 1997). In addition the government required some existing commodity groups such as the Meat and Livestock Commission (MLC) to increase, or at least maintain, their research budgets. Thus, for example, the Meat and Livestock Commission research expenditures increased by more than 300 percent between 1986 and 1994 (from about £500 thousand to about £2 million). Total research funding from these bodies was expected to be about £22 million in 1994/95, a substantial increase over such funding in the mid-1980s.

The above changes in government agricultural research policy represent a significant and interesting shift in emphasis. To some degree between 1990 and 1995, in relation to agricultural research, the U.K. government abandoned the customer-contractor principle for funding agricultural research that was emphasized in the Rothschild Report (that is, the client for the research should have oversight of the research). This change was made not because the government did not believe in the principle in relation to applied research, but because it did not believe that the government should be carrying out near-market research at all.[16] While cutting MAFF budgets for applied research, the government largely maintained funding for the AFRC, which received a more substantial mandate to support basic research rather than applied research.

The government also implemented several organizational changes to increase the independence of the agricultural research councils. In 1994 the Department of Education and Science was reorganized as the OST and the research councils were reorganized into six new entities. In this process, the AFRC was subsumed within the more general BBSRC and the AFRC institutes were redesignated as Biotechnology and Biological Science Research Institutes.

Conservative administrations of the 1980s and 1990s have strongly argued that market competition improves economic efficiency and productivity. This view is consistent with and at least partly responsible for some important innovations in the administration of public R&D funds. Since the mid-1980s three important changes have been made. First, in 1987 the research outputs of all university academic departments became subject to periodic peer reviews and the outcomes of these reviews are now used to determine five-year, block-grant research funding levels. Second, a larger share of total R&D funds has been allocated through competitive-grants programs for which university, government, and independent research institute scientists are eligible. Third, in higher education institutions, independent research institutes, and government laboratories, scientists are increasingly employed on the basis of short-term contracts rather than long-term contracts. All of these trends have been reflected within agricultural research departments within universities, the government, and independent research institutes (Thirtle et al. 1991; Scottish Crop Research Institute 1993).

There are some indications that the pace of change in U.K. public agricultural research policy and organization has slowed (Table 6.3). In its 1995 annual review of public research organization and funding, *Forward Look,* the government stated that most of the recommendations for change in science policy developed in the early 1990s had been put into effect. In particular, no new initiatives have been put forward to privatize or in other ways devolve parts of the public agricultural research system into the private sector. Nor are there any major proposals to reduce funding levels.

Postwar Trends in Public Agricultural R&D Expenditures and Funding

This section presents data on agricultural R&D expenditures by public institutions over the period 1947/48 to 1993/94 and public funding for agricultural R&D over the period 1972/73 to 1993/94. Public expenditures and public

16. In 1990 the Priorities Board (1990:2) stated that "all government funding of near-market R&D . . . will have been withdrawn by 1991–92."

TABLE 6.3 Chronology of major changes in agricultural and food research

Year/Period	Event/Change	Comment/Description
1971	Rothschild Report	Customer/contractor principle: Critical of ARC[a]
	ADAS formed[b]	Strengthen "relevant" applied R&D and diffusion
	JCO formed[c]	Link ARC, MAFF, Scottish and Irish departments[d]
1976/80	Funding stops growing	General cuts on public sector to control budget
1979	Thatcher elected	Government intent on reforming the public sector
1983–1984	Budget cuts begin	Growth ceases: agricultural research targeted
1984	ARC becomes AFRC[e]	Emphasis on food research due to value added
	Priorities Board formed	Replaces JCO to streamline the system
1985–1988	AFRC restructuring	Closure and amalgamation of institutes: from 18 to 8
1987	ADAS extension charges	Services no longer free: increase in private advice
	Barnes Report	Industry to be the customer for near-market R&D
	Privatization	AFRC's PBI and NSDO sold to Unilever[f]
	LINK Project	To encourage collaborative research
1988	BSE and salmonella strike[g]	BSE made a notifiable disease; minister resigns over salmonella
1989–1990	ADAS horticulture cuts	Three centers closed, one transferred to industry funding

1990	AFRC horticulture cuts	AFRC Institute now Horticulture International (mixed funding)
1990–1992	Next steps agencies	Four MAFF and one SOAFD executive agencies formed[h]
1991–1992	Near-market funds cut	MAFF near-market R&D funding of £30m supposed to end
1992	OST formed[i]	Headed by cabinet minister to indicate importance
1993	Realizing our potential	First full review of science and technology since the Rothschild Report
1994	AFRC incorporated in BBSRC[j]	Broader research council to capture biotechnology synergies
1995	OST transferred to DTI[k]	From cabinet office: technology seen as wealth creation
1995–1996	New statutory bodies	Three new bodies formed to raise levies on producers

[a]ARC, Agricultural Research Council.
[b]ADAS, Agricultural Development Advisory Service.
[c]JCO, Joint Consultative Organisation.
[d]MAFF, Ministry of Agriculture Fisheries, and Food.
[e]AFRC, Agricultural and Food Research Council.
[f]PBI, Plant Breeding Institute; NSDO, National Seed Development Organisation.
[g]BSE, bovine spongiform encephalopathy.
[h]SOAFD, Scottish Office of the Agricultural and Fisheries Department.
[i]OST, Office of Science and Technology.
[j]BBSRC, Biotechnology and Biological Sciences Research Council.
[k]DTI, Department of Trade and Industry.

funding differ because some research at public institutions is funded by the private sector and some public funds are used to support research in private institutions.

Public expenditures and public funding increased rapidly over the period 1947/48 to 1970/71 at an annual growth rate of 8 percent, more slowly and more sporadically in the 1970s at an annual growth rate of about 3 percent, and then peaked in 1982/83. Public expenditures and public funding declined in the mid-1980s, but expenditures then recovered between 1987/88 and 1993/94 to levels that were only slightly below those of the early 1980s. However, public funding did not increase, although it did not continue to decline. In addition, between 1972/73 and 1987/88 public funds were shifted away from more-basic research funded and executed by the ARC toward more-applied research funded by MAFF. However, in the late 1980s and 1990s public funds were reallocated back toward basic research funded by the ARC, reflecting the Conservative government's desire to allocate public monies to more-basic research. In addition, during the 1980s public expenditures on food research increased at the expense of more traditional agricultural research, reflecting the increasing influence of the food-processing sector on agricultural R&D policy.

Expenditures

Consistent annual data are not available for total expenditures by the public agricultural research system described in Figure 6.1. However, such data are available for expenditures by the AFRC, MAFF, DAFS, DANI, and the universities (Thirtle 1989), which jointly accounted for 99 percent of all expenditures between 1947/48 and 1972/73 and more than 92 percent between 1972/73 and 1993/94. In addition, data on other expenditures by the system are available for the latter period. These data, which are presented in Figure 6.2, show that between 1947/48 and 1976/77 real total public expenditures (in 1993 prices) increased from about £25 million to about £282 million at an 8 percent annual growth rate. However, periods of economic instability in the 1970s associated with successive oil crises and industrial strife led to a series of budget crises, cuts in public spending, and much lower and more variable annual average growth rates in public agricultural research expenditures. During this decade the average annual growth rate in public agricultural R&D declined to about 3 percent. Expenditures continued to increase in the very early 1980s; peaked in 1982/83; and then, during a period of Conservative government retrenchment, declined until 1987/88. After that year expenditures recovered, and by 1993/94 they were close to the peak levels attained in 1982/83.

More complete data on public agricultural R&D expenditures are available for the period 1972/73 to 1993/94. These data, which include the category "other expenditures" shown in Figure 6.2, are presented in Table 6.2.[17] Al-

17. Other expenditures include the Department of Education and Science direct grant to,

FIGURE 6.2 Major public R&D expenditures in the United Kingdom, 1947–1993

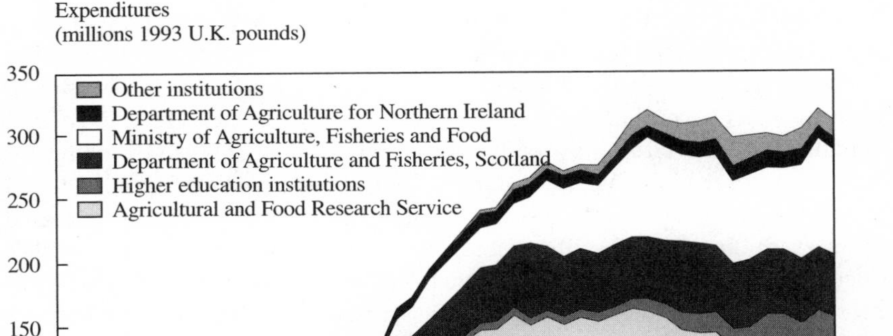

SOURCE: Prior to 1973 from Thirtle (1989), thereafter as for Table 6.5.

though other expenditures have been relatively small, during the period 1982/83 to 1987/88, when deep cuts were being made in the MAFF and ARC R&D budgets, they increased sufficiently rapidly (by about £10 million) to offset about 20 percent of the cuts in traditional agricultural R&D programs. In addition public expenditures on food safety also rose sharply during this period, from £28 million in 1982/83 to £39 million in 1987/88. These increases also mitigated the adverse effects of other cuts on total public R&D expenditures for food and agriculture, which, therefore, declined only by 7 percent, from £320 million to £298 million, over the same period.[18]

Total expenditures on agricultural R&D (total expenditures less food expenditures) are shown in the last column of Table 6.2. Between 1983 and 1989 these expenditures declined by 12.7 percent from £292 million to £255 million at almost twice the rate of decline in total public food and agricultural R&D outlays. Between 1988 and 1994, as noted, public expenditures on total food and agricultural recovered to close to 1982/83 levels. Public expenditures

---

and MAFF commissions with, the Natural and Environmental Resource Council; the food research associations and other contractors, including other government departments; the Home Grown Cereals Authority; National Institute of Agricultural Botany; and the Eggs Authority.

18. Note that data on R&D expenditures by the Scottish and Irish departments, DAFS and DANI, do not identify food and agricultural research separately. It is assumed here that these funds were allocated to agricultural R&D rather than food R&D.

on agricultural R&D also increased, but by 1993/94 were still 3.5 percent lower than their 1982/83 peak levels.[19]

Figure 6.2 and Table 6.2 also provide insights about the changing roles of public research institutions. Expenditures by AFRC institutes (largely for basic research) peaked at £163 million in 1981/82, but by 1993/94 had fallen by 30 percent to £112 million. However, at the same time the AFRC increased funding for research at universities by substantial amounts, and funding for university research by MAFF also rose. Thus, between 1982/83 and 1993/94 expenditures on university research (also largely basic research) increased from £9 million to £46 million, thus almost offsetting the decline in AFRC spending. Expenditures by the Scottish system (again largely for basic research) remained relatively stable over the same period. Thus between 1982/83 and 1993/94 real public expenditures for basic scientific research appeared to decline by about £16 million, or 7 percent.

The research portfolio of the AFRC institutes changed substantially during the 1980s and 1990s. In 1981/82 about 47 percent of institute expenditures were categorized as expenditures on improvements in technology; in 1993/94 98 percent were listed as expenditures in the most-basic research category (ARC 1984; OST 1995b). If these categorizations are accurate, public expenditures on more-basic research actually increased by about £20 million, or 9 percent, between 1981/82 and 1993/94, although care must be taken in interpreting the data because definitions of different research categories changed over the period. However, it does seem reasonable to conclude that public expenditures on basic scientific research did increase. In addition most of the increase in basic research expenditures took place over the last six years of the period, between 1987/88 and 1993/94.

MAFF's mandate for R&D led it to focus on policy-related, public-interest, and applied research. Intramural expenditures by MAFF (including Kew Gardens) increased 260 percent, from £32 million in 1972/73 to about £84 million in 1993/94. Most of this increase occurred between 1973 and 1983, when MAFF's intramural R&D expenditures peaked at £79 million. The considerable increase in MAFF's expenditures in the 1970s can in large part be attributed to the Rothschild Report recommendations that government research be driven by the customer-contractor principle. In the mid-1980s, as was the case with the AFRC, MAFF's expenditures declined by 17 percent to less than £66 million in 1987/88, when public agricultural R&D expenditures reached their lowest point under the Thatcher government.

---

19. These data do not include MAFF's technology-transfer budget, which has only been published since 1987/88. This budget, shown in Table 6.7, declined from £37.7 million in 1987/88 to £19.4 million in 1993/94. If this budget is included, total public R&D expenditures in 1993/94 were 2 percent lower than in 1987/88. This issue is discussed in more detail in the section on recent changes in the agricultural and food research system.

The removal of government funding for MAFF's near-market research, which was recommended by the Barnes Report in 1988 and the Priorities Board in 1990 (see the section on public agricultural science policy in the post–World War II period), was expected to reduce MAFF's budget by a further £30 million in the early 1990s (Webster 1989). Although MAFF did reduce expenditures on such research in the early 1990s, it did not withdraw completely from near-market research activities because the statutory bodies, which were supposed to take on this responsibility, had raised only an extra £6 million by 1993/94. In fact, incomplete withdrawal from near-market research coupled with a rapid increase in public-interest research (on food safety, diet, pollutants, and preserving the rural environment) resulted in a 28 percent increase in all MAFF R&D expenditures between 1990/91 and 1993/94.

The regional systems for Scotland and Northern Ireland experienced cuts in R&D expenditures in the later 1980s and 1990s rather than the early 1980s. Joint peak expenditures by the two systems occurred in 1986/87 at £62.7 million, but by 1993/94 expenditures had fallen by about 20 percent to £53.5 million.

Data on the total numbers of researchers in the public agricultural system also provide insights about research activity levels. These data, which are available only for the ARC, are presented in Figure 6.3. Between 1948 and 1982 the ARC's complement of research staff rose by more than 600 percent, from 500 to 3,085. By 1982 the total science staff, including assistant research officers, was more than 400 and the total ARC staff was 6,796. From 1982 to 1994 expenditures fell by 30 percent, but total and scientific staff were reduced by 45 percent, to 2,190 and 3,716, respectively. Also, 38 percent (825) of these had become short-term appointments rather than permanent posts, reflecting a shift in government hiring policy. Because the AFRC tried to protect science staff, cuts in support staff were even more drastic. Employment cuts were more severe than expenditure cuts largely because of the adoption of more capital-intensive biotechnological and other sophisticated scientific techniques.

Funding

Public funding of R&D and public expenditures on R&D differ because the private sector provides some funding for publicly executed R&D expenditures. Data on core public funding of agricultural R&D are presented in Figure 6.4 and Table 6.4 for the period 1972/73 to 1993/94. During this period total core public funding (Table 6.4, column 8) followed a similar pattern to total public expenditures. Core public funding increased fairly steadily from £229.2 million in 1972/73 to £300.5 million in 1982/83, but fell between 1982/83 and 1989/90 by 13.5 percent, from £300.5 million to £259.7 million (almost twice the rate of decline in public expenditures, which were bolstered by substantial increases in contract funding from new sources and direct funding of the universities).

FIGURE 6.3 Scientific and other staff of the Agricultural and Food Research Council (AFRC), 1981–1994

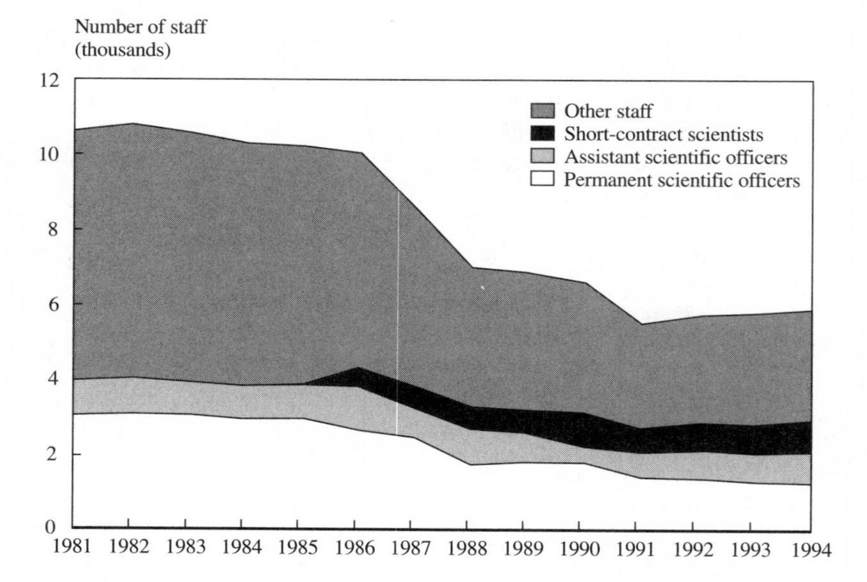

SOURCE: AFRC annual reports.

NOTES: The figures for 1981–1984 are actual numbers employed; the 1985–1994 numbers include vacant posts.

The data in Table 6.4 also provide insights about shifts in funding policies with respect to the ARC and MAFF. In the 1970s and early 1980s total funding for the ARC from its direct grant (Table 6.4, column 2) and MAFF commissions (Table 6.4, column 3) increased slowly, and direct funding for the ARC through its science grant dropped during this period. At the same time MAFF funding rose sharply both for commissioned research from the ARC (which was zero prior to 1973/74) and for in-house and other (non-ARC) contract research. Much of the shift in emphasis in public funding toward research directed by MAFF was a direct consequence of the Rothschild Report recommendations that the customer-contractor principle be implemented in relation to agricultural research and that autonomous funding for the ARC be reduced. Thus between 1975/76 and 1988/89 more than 50 percent of the ARC's total annual budget was controlled by MAFF.

In the 1980s, as has been discussed above, funding cuts were implemented for both the ARC and MAFF. However, in the late 1980s MAFF was judged to have performed poorly with respect to the customer-contractor principle and, in addition, the Conservative government chose to allocate public

FIGURE 6.4 Core public R&D funding and new contract funding, 1972–1992

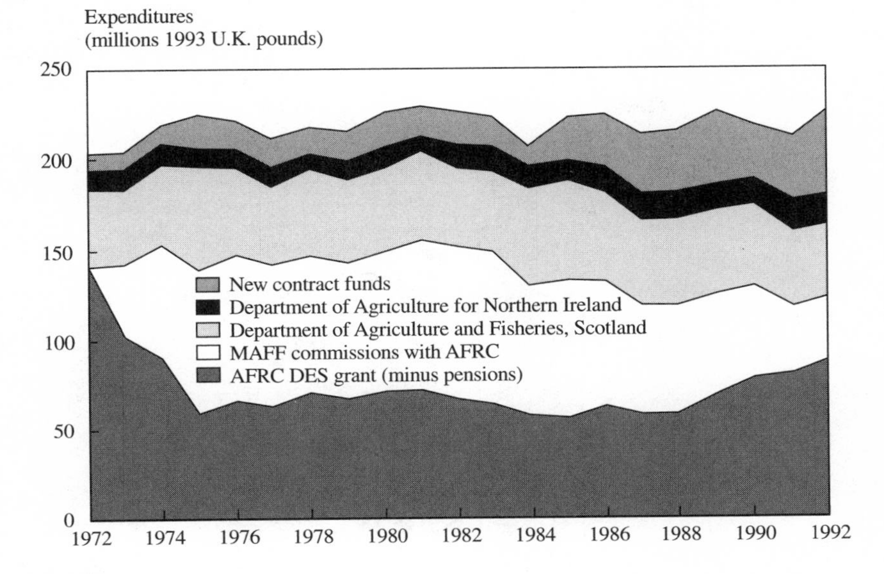

SOURCE: Table 6.4.

NOTE: Variables as defined in notes to Table 6.2.

funds for basic research rather than applied, near-market research. Thus after 1988/89 the ARC was given direct control over an increasingly large share of its total budget, and by 1993/94 the AFRC's direct grant made up more than 66 percent of its total budget. In contrast, MAFF not only lost control of funds for commissioned research from the AFRC in the 1990s, but plans were made to cut MAFF's funds for intramural and other research by £30.0 million. However, the BSE (mad cow disease) crisis, outbreaks of salmonella, and increasing concern over pollution and the rural environment led to rapid increases in public-interest research, which fell into MAFF's domain. Therefore, instead of falling, funding for MAFF intramural and other research increased by about 21 percent, although in 1993/94 it was still only 89 percent of its 1984/85 peak level of £101.4 million.

Total public core funding and total public expenditures for agricultural research are presented in columns 8 and 9 of Table 6.4. The difference between them (column 10) is new contract funding for specific projects from both public and private sources (including statutory bodies). The importance of these new sources of funds has approximately doubled since the beginning of the 1980s, reflecting the pressure that the government has put on all public

TABLE 6.4 Core public funding of agricultural research and new contract funding, 1973–1994

Year	AFRC Science Grant Minus Pensions[a]	MAFF Commissions with the AFRC[b]	AFRC Total	MAFF Internal and External (not AFRC)	DAFS[c]	DANI[d]	Total Core Funding	Total Expenditure	New Contract Funding
				(millions 1993 pounds sterling)					
1972/73	142.2	0.0	142.2	35.4	42.6	9.0	229.2	240.0	10.8
1973/74	102.9	40.5	143.5	38.7	42.8	9.0	233.9	243.5	9.7
1974/75	91.1	62.5	153.6	42.7	45.9	9.0	251.2	263.2	12.0
1975/76	60.0	80.0	140.0	44.4	57.4	9.0	250.8	270.4	19.6
1976/77	66.7	81.3	147.9	59.0	48.8	9.0	264.8	281.9	17.1
1977/78	63.2	79.0	142.1	61.4	44.8	9.0	257.3	274.1	16.8
1978/79	71.1	76.9	148.0	59.1	46.2	9.0	262.4	278.1	15.7
1979/80	68.0	76.2	144.2	62.4	45.4	9.0	261.0	279.6	18.7
1980/81	71.2	78.5	149.7	68.2	47.0	9.0	273.9	295.9	22.0
1981/82	72.3	83.4	155.7	80.4	48.6	9.0	293.6	311.0	17.4
1982/83	67.4	85.1	152.5	91.9	47.6	8.6	300.5	319.9	19.3
1983/84	64.5	85.2	149.7	87.1	47.7	9.8	294.3	311.5	17.2
1984/85	57.7	73.0	130.7	101.4	55.0	10.3	297.5	309.2	11.7

1985/86	56.1	77.6	133.7	86.2	53.9	11.3	285.1	310.5	25.4
1986/87	63.1	69.8	132.9	87.4	51.0	11.9	283.2	313.8	30.6
1987/88	58.5	61.3	119.8	83.0	49.6	11.6	264.2	298.4	34.3
1988/89	59.3	60.7	120.0	81.9	50.5	11.5	263.8	299.4	35.6
1989/90	69.8	56.2	126.0	73.6	47.3	12.8	259.7	301.6	41.9
1990/91	78.7	51.8	130.5	78.3	47.3	11.3	267.3	298.7	31.4
1991/92	81.5	37.2	118.7	90.8	47.5	11.2	268.2	305.3	37.1
1992/93	88.2	36.5	124.7	91.4	46.5	8.7	271.3	320.4	49.1
1993/94	89.4	36.0	125.4	89.3	45.2	8.4	268.3	310.7	42.4

SOURCE: Sources are as for Table 6.2, except that MAFF includes direct grants until 1980/81.

NOTE: Up to 1980/81, the new contract funding column includes new Agricultural Research Council funds and expenditures, which are almost constant. It is not really possible to record core funding, but it is almost synonymous with expenditures, except in the case of the AFRC. From 1981/82 the calculation really is AFRC, MAFF, DAFS, and DANI's public funding from the Cabinet Office Review of Government (R&D) and from *Forward Look* (OST 1995a). This part of the series covers the important changes.

[a]AFRC, Agricultural and Food Research Council.
[b]MAFF, Ministry of Agriculture, Fisheries and Food.
[c]DAFS, Department of Agriculture and Fisheries, Scotland.
[d]DANI, Department of Agriculture for Northern Ireland.

institutions to diversify their funding sources and compete for research contracts. These new sources also reflect the growth in funding from statutory bodies through commodity-based checkoffs.

Recent Changes in the Agricultural and Food Research System

The history of agricultural research policy and public expenditures and funding has been examined above. Clearly, major structural changes took place in the U.K. public agricultural research system in the 1980s and 1990s. However, lack of annual data precluded a similar assessment of changes in private-sector R&D during this period. Insights about relatively recent changes in both public and private U.K. agricultural research between 1987/88 and 1993/94 can be gleaned by comparing both public and private agricultural R&D expenditures and funding in those two years. The data on public and private expenditures for 1993/94 presented in Figure 6.1 were discussed above. The data for 1987/88 are based on the work of Thirtle et al. (1991), but are revised in light of more recent information provided to the authors by industry sources. Throughout this section, institutions are described by their current names rather than their names in 1987/88.

Recent Changes in Public-Sector Expenditures and Funding

Between 1987/88 and 1993/94, public expenditures and funding for agricultural R&D and employment of scientific researchers within the BBSRC (previously the AFRC) institutes all stabilized. Two countervailing policy forces were at work during this period. Conservative government initiatives to devolve near-market R&D activities to the private sector tended to reduce funding, but the government's renewed emphasis on the economic importance of scientific and technological research led to some increases in basic research funding. However, most of the recovery in public agricultural R&D expenditures during this period was caused by rapid growth in public-interest research and some expansion in private funding for public research institutions.

Expenditures by public-sector research agencies for 1987/88 and 1993/94 are presented in Table 6.5. Over this period, real public expenditures on agricultural and food research increased by 5 percent, from £296 million to £311 million.[20] However, public funding for technology transfer was almost halved so that total science and technology expenditures fell by 1 percent. If the intramural R&D programs carried out by MAFF, SOAFD, and DANI are viewed as applied research and all other public R&D expenditures are viewed

20. The data presented in Figure 6.2 and Table 6.2 are slightly different from the data in Table 6.5 because the time series in Table 6.2 omitted some items in the interests of consistency. For example, privately funded public research in the Scottish system is excluded as is MAFF's technology-transfer budget. Conversely, in Table 6.5 MAFF's expenditures with the food research associations were included as public research, whereas here they are treated as private research.

as basic research, then public expenditures on basic research increased by about £7 million, or 3 percent. However, in 1987/88 32 percent of BBSRC institute expenditures were classified as improvement of technology, whereas in 1993/94 98 percent were listed in the most basic category (HMSO 1984; OST 1995a). This seems to imply that basic research increased much more substantially, although this conclusion is mitigated by the fact that, as noted above, category definitions changed between 1987/88 and 1993/94.

As noted in the section on postwar trends in public agricultural R&D expenditures and funding, the organization of basic research changed during this period. Expenditures by BBSRC institutes fell by £20.0 million, but the BBSRC compensated by increasing its funding of university research from £12.5 million to £35.5 million, bringing it into line with the other research councils. MAFF also increased its funding for university research from £6.0 million to £10.3 million and, as a result, university research funding from these sources rose by 280 percent from £16.3 million to £45.8 million. Total university R&D expenditures in 1993/94 amounted to £54.8 million, including £9.0 million provided directly to the universities by the HEFC.

Despite deep cuts in public funding for near-market research, total public applied research expenditures increased by more than 8 percent between 1987/88 and 1993/94 because of increased support for public-interest research, particularly on the environment and food safety. MAFF is responsible for such research and, because of increased demands for public-interest research, its intramural R&D expenditures increased by £17 million (Table 6.5). In 1983/84, for example, public expenditures on food-safety research were only £5 million, but by 1993/94 had risen to almost £30 million (if R&D expenditures on BSE are included). In fact, in 1993/94 more than 60 percent of MAFF's total public R&D funds were allocated to research to protect the public (£55 million) and farm animals (£4 million) and to enhance the rural environment (£20 million). Only £49 million was allocated for research to improve economic performance (MAFF 1994).

The total income and expenditures of the BBSRC in 1987/88 and 1993/94 are presented in Table 6.6. Funding via the direct grant from OST to the BBSRC increased substantially from £76.9 million to £109.4 million, but this increase was largely offset by the reduction in MAFF commissions from £61.9 million to £36.0 million. Funds from other sources increased very slightly and, therefore, total BBSRC funding rose by almost 5 percent. This shift in the mix of public funding for publicly executed research was, as noted above, the outcome of a deliberate policy decision to give the BBSRC more control over its budget with the objective of increasing public funding for basic research. However, expenditures for research at the BBSRC institutes declined, partly because of increased funding by the BBSRC for university research (discussed above) and partly because expenditures on pensions increased from £17.8 million to £20.0 million.

TABLE 6.5 Public agricultural research expenditures

	1988[a]		1994		1994 as Share of 1988
	Expenditure	Share	Expenditure	Share	
R&D by	(millions 1993 pounds sterling)	(percentages)	(millions 1993 pounds sterling)	(percentages)	(percentages)
Biotechnology and Biological Sciences Research Council[b]	133.00	44.99	113.00	36.39	84.96
Ministry of Agriculture, Fisheries and Food[c]	58.50	19.79	75.10	24.19	128.38
Kew Gardens	7.60	2.57	9.40	3.03	123.68
Scottish Agricultural and Biological Research Institutes	39.50	13.36	34.00	10.95	86.08
Scottish Agricultural College	11.60	3.92	13.30	4.28	114.66
Scottish Office Agriculture and Fisheries Department	2.80	0.95	5.00	1.61	178.57
Department of Agriculture for Northern Ireland	11.80	3.99	8.40	2.71	71.19
Universities	16.30	5.51	45.80 (54.80)[d]	14.75	280.98
Other[e]	14.50	4.91	6.50	2.09	44.83
Total R&D	*295.60*	*100.00*	*310.50*	*100.00*	*105.04*
MAFF technology transfer	*37.70*		*19.40*		*51.46*
Total R&D and technology transfer	*333.30*		*329.90*		*98.98*
Of which basic science	208.00		215.50		103.61
Of which applied R&D	87.60		95.00		108.48

SOURCE: AFRC annual reports (1987/88, 1993/94); Cabinet Office *Annual Review of Government R&D* (1990); OST *Forward Look* (1995a).

[a]1988 pounds were converted to 1993 pounds, using an inflation factor of 1.4.

[b]The Biotechnical and Biological Science Research Council was called the Agricultural and Food Research Council in 1987.

[c]Represents intramural expenditures, minus fisheries research, minus Kew Gardens.

[d]This includes funding from Higher Education Funding Council; no equivalent figure is available for 1988.

[e]For these items, see Figure 6.1.

TABLE 6.6 The Biotechnology and Biological Science Research Council

	1988		1994		1994 as Share of 1988
	Expenditure	Share	Expenditure	Share	
	(millions pounds sterling)	(percentages)	(millions pounds sterling)	(percentages)	(percentages)
Funding					
OST science grant[a]	76.90	47.18	109.40	64.05	142.26
MAFF commissions[b]	61.90	37.98	36.00	21.08	58.16
Other council funds	9.50	5.83	10.80	6.32	113.68
Other institute funds	14.70	9.02	14.60	8.55	99.32
Total funding	*163.00*	*100.01*	*170.80*	*100.00*	*104.79*
Disbursements					
To HEIs[c]	12.00	40.27	35.50	61.42	295.83
Pensions	17.80	59.73	20.00	34.60	112.36
Other councils and departments	0.00	0.00	2.30	3.98	na
Total disbursements	*29.80*	*100.00*	*57.80*	*100.00*	*193.96*
Intramural R&D expenditures	*133.20*		*113.00*		*84.83*

SOURCE: AFRC annual reports (1987/88 and 1993/94).

NOTE: na indicates not available.

[a]OST, Office of Science and Technology.
[b]MAFF, Ministry of Agriculture, Fisheries and Food.
[c]HEIs, higher education institutions.

Table 6.7 presents a similar decomposition of sources of funding and expenditures for MAFF. Between 1987/88 and 1993/94 the decreases in MAFF's core funding and its technology-transfer budget were only partly offset by increases in receipts from fees and contracts for services, and total funding declined by 15 percent. Moreover, substantial changes were made in how these funds were spent. MAFF expenditures on commissioned research from the BBRSC were cut from £61.9 million to £36.0 million. Conversely, support for university research increased from £4.3 million to £10.3 million. In contrast, MAFF funding of contracts with private industry, public corporations, and other agencies fell largely because 1987/88 was an abnormal year (see Table 6.2). Finally, support for Kew Gardens was increased. In all these disbursements decreased by about 28 percent between 1987/88 and 1993/94, almost completely offsetting the decrease in total funding for MAFF R&D. Thus MAFF intramural science and technology expenditures fell by only 2 percent. In fact, if the technology-transfer budget is excluded, MAFF intramural R&D expenditures actually rose by 28 percent.

The redirection of resources toward the universities and away from the agricultural research institutes represents the outcome of three policy initiatives introduced by the Conservative government in the 1990s to open the BBSRC funds to competition. The first was the drive toward an increased allocation of public funds for basic research, the second was a movement away from public support for research into traditional production agriculture, and the third was an attempt to increase competition among public agencies for public research funds. Another reason for this reallocation may have been that the nature of agricultural R&D also was changing over the period as genetic and other laboratory-based microbiological research activities became more important.

Similar trends were exhibited in the allocation of R&D resources by the SOAFD. Table 6.8 shows that public funding for the SABRIs declined substantially from about £39 million to about £34 million, despite the fact that contract research by those institutes doubled. In contrast, intramural expenditures by SOAFD increased by more than £2 million, but only because SOAFD's newly formed executive agency (the Scottish Agricultural Science Agency) ran over budget. However, public funding for the Scottish Agricultural College, which was reorganized, increased. Finally, funding for DANI was also reduced by about £3 million, or more than 20 percent (although a detailed breakdown of changes in the distribution of that budget is not available).

Recent Changes in Private-Sector Expenditures and Funding

Consistent time-series data are not available on private-sector agricultural and food R&D in the United Kingdom. However, aggregated and disaggregated data on estimated private R&D expenditures in 1987/88 and 1993/94 have been constructed by the authors. These estimates are presented Figure 6.1 and Table 6.9. The estimates for 1993/94, although the best available, were gener-

TABLE 6.7 Ministry of Agriculture, Fisheries and Food (MAFF)

	1988		1994		1994 as Share of 1988
	Expenditure	Share	Expenditure	Share	
	(millions pounds sterling)	(percentages)	(millions pounds sterling)	(percentages)	(percentages)
Funding					
From the exchequer	145.70	75.18	125.30	76.12	86.00
MAFF receipts	10.40	5.37	19.90	12.09	191.35
Technology-transfer budget	37.70	19.45	19.40	11.79	51.46
Total funding	*193.80*	*100.00*	*164.60*	*100.00*	*84.93*
Disbursements					
To BBSRC[a]	61.90	63.42	36.00	51.36	58.16
To HEIs[b]	4.30	4.41	10.30	14.69	239.53
Kew Gardens	7.60	7.79	9.40	13.41	123.68
Other councils and departments	6.70	6.86	7.00	9.99	104.48
Private industry, public corporations, and other	17.10	17.52	7.40	10.56	43.27
Total disbursements	*97.60*	*100.00*	*70.10*	*100.01*	*71.82*
MAFF and agencies S&T expenditures[c]	96.20		94.50		98.23
MAFF and agencies R&D expenditures	58.50		75.10		128.38

SOURCE: Cabinet Office, *Annual Review of Government R&D* (1990); OST *Forward Look* (1995a).

[a]BBSRC, Biotechnology and Biological Sciences Research Council.
[b]HEIs, higher education institutions.
[c]S&T, science and technology.

TABLE 6.8 The Scottish Office Agricultural and Fisheries Department (SOAFD) and the Department of Agriculture for Northern Ireland (DANI)

	1987/88		1993/94		1993/94 as Share of 1987/88
	Expenditure	Share	Expenditure	Share	
	(millions 1993 pounds sterling)	(percentages)	(millions 1993 pounds sterling)	(percentages)	(percentages)
Expenditures by					
Scottish Biological and Agricultural Research Institutes	39.50	73.28	34.00	65.01	86.08
In-house research and Royal Botanical Gardens	2.80	5.19	5.00	9.56	178.57
Scottish Agricultural College	11.60	21.52	13.30	25.43	114.66
Total expenditures	*53.90*	*99.99*	*52.30*	*100.00*	*97.03*
Core funding from SOAFD	50.40		45.20		89.68
Funds from new contracts	3.50		7.10		202.86
DANI: in-house and Queens University	11.80		8.40		87.29

SOURCE: Cabinet Office, *Annual Review of Government R&D* (1990); OST *Forward Look* (1995a); annual reports of the institutes and the Scottish Agricultural College.

TABLE 6.9 Private agricultural research expenditures

Input Industries	1988		1994		1994 as Share of 1988
	Expenditure	Share	Expenditure	Share	
	(millions 1993 pounds sterling)	(percentages)	(millions 1993 pounds sterling)	(percentages)	(percentages)
Agrochemicals	140.00	46.51	150.00	52.45	107.14
Animal breeding	15.00	4.98	10.00	3.50	66.67
Animal feedstuffs	7.00	2.33	4.00	1.40	57.14
Animal health inputs	30.00	9.97	40.00	13.99	133.33
Farm machinery and equipment	78.00	25.91	60.00	20.98	76.92
Seed industry	30.00	9.97	20.00	6.99	66.67
Input industries total	*300.00*	*99.67*	*284.00*	*99.30*	*94.67*
Statutory bodies in-house	1.00	0.33	2.00	0.70	200.00
Total agricultural research	*301.00*	*100.00*	*286.00*	*100.00*	*95.02*
Food industry laboratories	140.00	91.74	250.00	91.91	178.57
Food research associations	12.60	8.26	22.00	8.09	174.60
Food industry total	152.60	100.00	272.00	100.00	178.24
Total agriculture and food	*453.60*		*558.00*		*123.02*

SOURCE: A revised version of Thirtle (1991), updated to include data for 1994. All figures are based on unpublished estimates, except for the expenditures of the food research associations, which are from annual reports.

ally obtained from industry sources and are much less reliable than the data presented above on public R&D outlays.

Between 1987/88 and 1993/94 total estimated food and agricultural R&D expenditures by the private sector increased by about 28 percent, from £454 million to £558 million. However, total R&D expenditures by agricultural input industries declined by about 6 percent, from £300 million to £284 million. Because the only other source of private expenditures involved trivial amounts of in-house research by statutory bodies, a similar decline occurred in total private expenditures on production-agriculture research. In contrast, estimated total private expenditures on food R&D increased by 78 percent from £153 million to £272 million. However, the 1993/94 estimate for food industry R&D should be viewed with some skepticism because the food industry often treats market development expenditures as research investments and research even includes such things as the invention of new recipes.[21] The estimate for 1987/88 food industry expenditures was developed by MAFF (1990, *Agriculture in the United Kingdom*) and may be more realistic. Thus the estimate that food R&D increased by 78 percent over this period should also be viewed with caution.

Almost all private-sector agricultural R&D has been carried out by the agricultural input industries in private laboratories, although a small amount of contract research was purchased from BBSRC institutes (about 2 percent of the total in 1993/94). In Table 6.9 disaggregated data on estimated R&D expenditures are reported for agricultural chemicals, farm equipment and machinery, animal health, seed production, pig and poultry breeding, and feed production.

Expenditures on R&D by the agricultural chemical industry (fertilizers, herbicides, and pesticides) increased by about 7 percent, from £140 million to £150 million.[22] However, these estimates overstate agricultural chemical R&D relevant to U.K. agriculture because they include all U.K. R&D expenditures by several multinational corporations such as the Industry Commission that operate worldwide research programs in their U.K. facilities. Expenditures on R&D by farm machinery and equipment producers declined by 30 percent, from £78 million to £60 million over this period.[23] The industry attributes

21. The estimate for food research was obtained as follows. The Biotechnology and Biological Sciences Research Council (BBSRC 1995:paragraph 16) recently stated that "about 0.5 percent of sales revenue is spent on R&D, amounting to some £250–300 million per annum (based on Food and Drink Federation Data) . . . by U.K. food companies." The lower bound is reported here as a "best guess" about current food industry R&D and should be viewed as an extremely optimistic estimate because of the food industry's broad definition of R&D.

22. These estimates are based on industry sources. Lower estimates reported by Thirtle et al. (1991) were derived from evidence that the agricultural chemical industry employed about 2,300 R&D staff.

23. The 1987/88 estimate assumes that producers of farm machinery and equipment spent 4 percent of the value of sales revenues (estimated at £1.5 billion) on R&D. Rothwell (1978) previously reported that equipment producers allocated 1–3 percent of sales revenues for R&D

these cutbacks to poor sales associated with the depressed state of U.K. agriculture over the past few years. Expenditures on animal health R&D by the pharmaceutical industry were estimated to have increased by about 33 percent, from £30 million to £40 million, largely because of increased research intensities in the pharmaceutical industry.[24] Expenditures on R&D by seed companies are estimated to have declined by about 33 percent, from £30 million to £20 million, apparently because of a decline in industry profits.[25] Expenditures by animal breeding companies (mainly pig and poultry producers) were estimated to have declined by 33 percent, from £15 million to £10 million, also probably because of a decline in these sectors of U.K. agriculture.[26] Finally, research expenditures by animal feed producers also declined by more than 40 percent, from £7 million to £4 million, again because of depressed conditions in the livestock sector.

Thus in an era when the government implemented cuts in public funding for agricultural research while claiming that the private sector would pick up the slack, R&D expenditures by the private sector declined by about 3.5 percent, partly because of adverse economic conditions within the livestock sector and a decline in demand for agricultural machinery and equipment. This outcome was somewhat ironic and, to the extent that it was due simply to a cyclical decline in agricultural activity, government intervention might have been warranted to provide incentives to sustain private R&D programs.[27]

Food industry R&D is estimated to have increased dramatically by almost 80 percent between 1987/88 and 1993/94, from £140 million to £250 million, although (as noted above) the estimated amount is probably too large. Most of these expenditures were for services provided by privately owned laboratories. However, about 10 percent of food industry R&D expenditures funded R&D by industry food research associations. These expenditures, which are more

---

and tractor manufacturers spent 5–10 percent of revenues on R&D. The industry estimates that it now spends less than 5 percent of revenues on R&D.

24. In 1993/94 the pharmaceutical industry allocated approximately 21 percent of total sales revenues to R&D (OST 1995a). The estimate for animal health R&D in 1993/94 assumes that veterinary pharmaceutical companies spend a similar proportion of their revenue on R&D. In 1987/88 pharmaceutical companies allocated 13 percent of sales revenues to R&D (OST 1995a), and thus the 1987/88 estimate for animal health R&D reported here is lower.

25. Thirtle et al. (1991) report an estimate for seed R&D in 1987/88 based on the assumption that 15 percent of industry revenues were allocated to R&D. Industry sources indicate that this was an overestimate, and thus the estimate for 1987/88 presented here is lower. In 1993/94, seed breeding for cereals, pulses, and potatoes accounted for £18.6 million (according to the British Society of Plant Breeders). Research related to vegetable production was about £0.5 million and involved only two companies. Most vegetable seed is imported from Denmark, Japan, the Netherlands, and the United States.

26. This estimate is based on the R&D expenditures of major firms and their market shares.

27. The possibility that private-sector R&D would decline has long been part of the debate over whether public R&D for near-market research should be cut (ABRC 1986). Nelson (1982) has argued that one appropriate role for government R&D policy is to ensure that private R&D is not the first casualty of poor business conditions.

reliably estimated from the annual reports of the food research associations, increased from about £13 million in 1987/88 to about £22 million in 1993/94.[28] Funding sources for food research have also become more diverse as MAFF has provided less direct support, but has encouraged joint EU projects (Leatherhead Food Research Agency 1993).[29]

In addition to its own in-house expenditures, the private sector also funds a substantial amount of research conducted by public institutions. The data on new funding sources for public R&D in Table 6.3 include funding of public research by the private sector. However, more-detailed information is available for 1987/88 and 1993/94 from the annual reports of public R&D institutions. These data, which are presented in Table 6.10, show that between 1987/88 and 1993/94 private funding of the public system increased from £40 million to £48 million. Thus total private funding for all agricultural research increased by 23 percent, from about £494 million to £606 million over this period (assuming that the estimates of food industry R&D are valid). However, even if all of the increase in private funding of public research between 1987/88 and 1993/94 was for traditional agricultural research, it would still be the case that total private funding for traditional agricultural research (private in-house and privately funded public research) declined between 1987/88 and 1993/94 by about £7 million, or about 2 percent. Therefore there is compelling evidence that while private expenditures on food research increased, private expenditures on production-agriculture research decreased and public expenditures on near-market agricultural research were also being cut.

In this context, it is also important to note that while MAFF almost doubled receipts from fees for service to the private sector (in line with government policy), industry, trust, and foundation funding fell by 52 percent (although the latter provided only modest funding to MAFF even in 1987/88). Thus, although funding from the statutory bodies also increased by 30 percent, the net increases in MAFF income from these private sources were not sufficient to replace the reductions in MAFF's public funding caused by the loss of its near-market research budget. Clearly the private sector has not yet chosen to step in to take over the public sector's previous role as a provider of near-market research.

Agricultural R&D Policy Reforms, 1982–1995

In 1971, the Rothschild Report raised the issue of value-for-money in scientific research and led to changes intended to improve responsibility and accounta-

---

28. The Campden Food and Drink Research Association had a turnover of £7.1 million by 1993/94 and BIBRA had an income of £3.8 million. The Leatherhead Food Research Association had a 1993/94 turnover of £8.7 million and the Flour Milling and Baking Research Association, which is now merged with the Camden Food and Drink Research Association, had a turnover of £2.6 million.

29. For example, the Campden Food and Drink Research Association (1994) reported that funding from MAFF had fallen from 26 to 12 percent of total turnover over a three-year period.

TABLE 6.10 Private funding of publicly performed agricultural research

Source	1988		1994		1994 as Share of 1988
	Expenditure	Share	Expenditure	Share	
	(millions 1993 pounds sterling)	(percentages)	(millions 1993 pounds sterling)	(percentages)	(percentages)
Statutory bodies	15.80	39.50	20.40	42.24	129.11
Input and food industries to AFRS[a]	9.60	24.00	6.00	12.42	62.50
Trusts and foundations to AFRS	4.20	10.50	2.00	4.14	47.62
MAFF receipts[b]	10.40	26.00	19.90	41.20	191.35
Total private funding of public agricultural R&D	40.00	100.00	48.30	100.00	120.75
Total private funding of all agricultural R&D	493.60		606.30		122.83

SOURCE: AFRC annual reports (1987/88, 1993/94); Cabinet Office *Annual Review of Government R&D* (1990): OST *Forward Look* (1995a), plus various annual reports and unpublished estimates.

[a]AFRS, Agricultural and Food Research Service (AFRC plus Scottish institutes).
[b]MAFF, Ministry of Agriculture, Fisheries and Food.

bility through the implementation of the customer-contractor principle. Ruttan (1982) raised the possibility that by the late 1970s "a cozy bilateral monopoly relationship between customer (MAFF) and supplier (ARC)" had developed. This suggests that in the early 1980s even independent commentators believed it was time for a new review of the U.K. public agricultural research system. In fact, a review of the entire science budget had already been initiated by the Thatcher government (ABRC 1982), which led to the substantial changes in agricultural R&D policy that took place between 1982 and 1995. This section provides a brief assessment of those reforms.

To ensure that R&D makes a maximum contribution to wealth creation, the public and private research portfolios should produce an optimal quantity of economically appropriate technologies in the most efficient manner possible. The recent changes in U.K. agricultural R&D policy were in part driven by government initiatives founded on the belief that competitive markets are the key to achieving both of these goals, except in special cases where the market fails and public intervention is required. Skeptics have noted that neither "agriculture" nor "food" appear in the title of the new agricultural research council and that, with only a modest proportion of MAFF's budget available for productivity enhancing research, support for agricultural R&D is now even lower than when Spedding (1984) complained that there really was no agricultural research as such in the United Kingdom.

Although the administrative reforms of the 1980s and 1990s make some economic sense, it should also be recognized that creating a competitive marketplace for agricultural R&D, at least on the supply side, may be counterproductive. In terms of productivity gains, the adverse effects of increasing transaction costs in that market may exceed any benefits from increased competition among researchers and accountability. There is also a danger of short-termism, in that markets do not necessarily handle projects with long gestation periods well and much scientific activity is long term. In addition, partly because of short-termism, more competition among researchers in markets in which wages are regulated may lead to a socially suboptimal allocation of intellectual resources as gifted potential researchers eschew science for other fields.[30]

Conclusion

The funding, governance, and institutional structure of agricultural research in the United Kingdom have all changed substantially since World War II. To

30. At least one historian's version of endogenous growth theory (Crafts 1996) suggests that France failed where Britain succeeded in the eighteenth century because the most able Frenchmen chose to be bishops, generals, lawyers, and civil servants rather than entrepreneurs and scientists. This element of the incentive system has been overlooked in the recent reforms, but was noted by John R. Hillman (Scottish Crop Research Institute 1993:8).

understand the reasons for these changes in research policy, we have placed them in a broader policy context, making it possible to distinguish changes in agricultural research policy that were specific to that sector from those that reflected broader trends. Some regional differences between Scotland, Northern Ireland, and England and Wales were also noted.

In the immediate postwar period (1945–1956) agricultural research was controlled by the agriculture ministry. However, in 1956 a dual funding system was introduced in which near-market agricultural R&D was the responsibility of MAFF, whereas more-basic research was the domain of the ARC. Funding for both types of research grew substantially during this period, and the dual system has survived to the present. However, there have been major changes in emphasis between the two lines of research and the ways in which they are managed, financed, and executed.

Two important changes took place in the 1970s. The Rothschild Report of 1970 continued to uphold the principle of a dual funding system, but also advocated that the customer-contractor principle be adopted. The report argued that the ARC had become too distant from its client groups in the agricultural industry and even from MAFF, which was viewed as a more effective principal agent for the agricultural sector. Therefore control of over one-quarter of the ARC's funding was given to MAFF, whose own expenditures also increased because a higher priority was placed on applied, near-market research funding (both in-house and contracted).

The most recent major shift in research policy began in the early 1980s, with the Thatcher government's initiatives to reduce the size of the public sector and allow market forces to play a greater role in the U.K. economy, including the allocation of agricultural research resources. Six major developments have been identified. First, the government has redefined the roles of the public and private sectors, with productivity-enhancing research, near-market research, and technology-transfer activities becoming the responsibility of the private sector. Public R&D is now concentrated on broadly based, basic research (culminating in the restructuring of the research councils, in which the AFRC was incorporated in the BBSRC) and public-interest research. Second, this led to reductions in the expenditures of the public system of 7 percent and cuts in core funding from taxation of 13 percent. Third, where possible, fees for services to the private sector have been introduced; statutory bodies have been created to fund commodity-specific R&D with levies on producers; and some public research institutions have been privatized, amalgamated, or closed. Fourth, to create a competitive market in research, a greater role has been given to the universities and other institutions, and competitive bidding for research projects has, to an extent, replaced automatic program funding. Fifth, to extend market incentives to within the organizations, accountability has been increased by the formation of executive agencies, monitoring and evaluation procedures have been strengthened, and increasing numbers of staff have been hired on short-term contracts. Sixth, the government has committed itself to providing clear policy leadership and improving information flows by means of the regular and organized provision of data on research expenditures, more frequent white pages outlining policy directions, improved management structures, and increased collaborative research. In particular, the establishment of the OST under a cabinet minister reflects the government's commitment. The relocation of the OST from the Cabinet Office to the Department of Trade and Industry is an indication that the government now views the central role of science as wealth creation. Agricultural science policy has stabilized and so have public funding and expenditures on R&D, at 1987/88 levels, but private agricultural R&D has declined.

7 Financing Agricultural R&D in the Netherlands: The Changing Role of Government

JOHANNES ROSEBOOM AND HANS RUTTEN

The Netherlands is a densely populated country with a highly specialized and productive agricultural sector.[1] In value terms, the Netherlands is the world's third-largest exporter of agricultural products. In 1994, for example, total agricultural exports (including processed food) were valued at Dfl 67.6 billion (US$37.1 billion), imports amounted to Dfl 40.6 billion (US$22.3 billion), and the agricultural sector's net trade surplus was Dfl 27.0 billion (US$14.8 billion). The competitive position of Dutch agriculture is based on a combination of specialization in production (vegetables, flowers, dairy, and intensive livestock), marketing, and transport. Trade and transportation are important characteristics of the entire Dutch economy. In the Netherlands export earnings account for about 60 percent of GDP, and Dutch transport companies enjoy a disproportionately large share in Europe's commodity trade.

Dutch agriculture is also highly productive. Despite the country's high population density of 451 persons per square kilometer (compared with 28 persons per square kilometer in the United States), 64 percent of the available land is used for agriculture. Agricultural production is intensive, and output per hectare as well as per unit of agricultural labor is among the highest in the world. Most Dutch farmers operate at the forefront of agricultural technology, supported by a sizable and productive agricultural research system.

The beginnings of the Dutch agricultural research system can be traced to the late nineteenth century. From the start the government played an important and often dominant role in the organization and funding of agricultural research. However, rather than growing according to any agreed-upon plan, the Dutch agricultural research system developed through a large number of ad hoc initiatives. Over time a patchwork of agricultural research institutes and stations emerged, each focusing on a certain discipline, problem, commodity,

1. This chapter represents the personal views of the authors and not those of the agencies for which they work. The authors thank Julian Alston, Philip Pardey, Vincent Smith, and A. P. Verkaik for comments on earlier drafts of this chapter.

region, or a combination of these. The institutes and stations also varied with respect to financing and legal structures.

The need to improve coordination has been a recurrent theme throughout the more recent history of Dutch agricultural research. The establishment of the National Council for Agricultural Research (Dutch acronym, NRLO) in 1957 and the creation of a Directorate of Agricultural Research within the Ministry of Agriculture (MOA) in 1962, resulted in a period of consolidation, rationalization, and more coordination of agricultural research.[2] A central research project administration was introduced in 1966, and the first five-year plan for agricultural research was presented to Parliament in 1972.

In the early 1980s questions about the role of government began to dominate Dutch politics. The credo was "Less government, more market," and a political consensus to reduce government intervention emerged.[3] In 1986 the government placed agricultural R&D on a list of government activities targeted for privatization. What followed was a protracted period during which the MOA's agricultural research activities were restructured and consolidated in preparation for the privatization, which was completed in 1999.

This chapter provides an overview of the development of the Dutch agricultural research system and the way it is governed and financed. We give special emphasis to the effects of the recent reforms on the total investment in public agricultural research and the recent institutional changes in funding arrangements.

Agricultural Research in the Netherlands

This section provides an overview of the current structure of the Dutch agricultural research system and its historical development, highlighting the changes that have taken place since the mid-1980s.

Present Structure and Linkages

A brief overview of the present structure of the Dutch agricultural research system is presented in Table 7.1. The term "system" is used here to refer to the various agencies that deal with agricultural research in the Netherlands. The agricultural sector is taken to include primary agricultural production, as well as agricultural input and processing industries.

---

2. The complete name of the ministry is at present the Ministry of Agriculture, Nature Management, and Fisheries. For ease of exposition, we will use the name Ministry of Agriculture and the acronym MOA throughout the text.

3. Dutch governments are usually formed by a coalition of two or three political parties. Therefore, policy changes are generally more gradual than in Australia or the United Kingdom, both of which have one-party governments.

The Dutch MOA is responsible for agricultural research and extension, as well as agricultural education. In 1995 policy formulation for these areas was entrusted to a single directorate within the ministry, the Directorate of Science and Knowledge Transfer (DSKT). At the same time the National Council for Agricultural Research (NRLO) lost its coordination and planning functions, becoming an advisory board with responsibilities for conducting science and technology foresight studies. Insofar as central planning and coordination are deemed necessary, these functions have been assumed by the DSKT, the single most important funding agency for agricultural research in the Netherlands. DSKT funds or purchases research conducted by the Agricultural Research Department (Dutch acronym: DLO); the Organization for Applied Research in Agriculture (a temporary organization that clusters a group of nine experiment stations conducting more-applied research); Wageningen Agricultural University (WAU); and, to a substantially lesser extent, the Nutrition and Food Research Institute of the Netherlands Organization for Applied Scientific Research (TNO-Food) and the Faculty of Veterinary Sciences of the University of Utrecht.

In 1996 the minister of agriculture decided to merge the DLO and WAU into a single organization: Wageningen University Research Centre (Wageningen UR). In preparation for this merger, the boards of the DLO and WAU were merged and a new chairman was appointed in 1997. Further details of the merger are currently being worked out. It is expected that parts of the experiment stations will also be merged into Wageningen UR in the near future. Those not included in the merger will be handed over to various farmer organizations or closed down. The only two entities not directly affected by this recent reorganization are TNO-Food and the Utrecht's Faculty of Veterinary Sciences.

The private sector, which had an estimated intramural research budget of Dfl 745 million in 1995, is an important component of the Dutch agricultural research system (CBS 1997). Private R&D expenditures include intramural research outlays by the agriculture, forestry, and fisheries industries (Dfl 89 million); the food and drinks industries (Dfl 478 million); and an estimated 10 percent of the research expenditures of chemical and pharmaceutical companies (Dfl 178 million). In addition, in 1995 the public sector (especially TNO-Food) executed about Dfl 126 million of agricultural and agriculture-related research funded by the private sector. In the Netherlands, private agricultural research expenditures approximately match public agricultural research expenditures. This chapter focuses mainly on the research system's public components.

Historical Development

Public agricultural research in the Netherlands began in 1877, when the first state agricultural experiment station was established in Wageningen, a small town near the Rhine, where a year earlier, an agricultural school had been

TABLE 7.1 Present structure of the Dutch agricultural research system

Organization	Description	Research Budget 1995
Directorate of Science and Knowledge Transfer (DSKT)	Created in 1995 and responsible for the agricultural research, extension, and education policies of the Ministry of Agriculture (MOA).	—
National Council for Agricultural Research (NRLO)	Originally a coordinating and planning body, but reestablished in January 1995 as a ministerial advisory board in charge of conducting science and technology foresight studies for the agricultural sector.	—
Wageningen University Research Centre (Wageningen UR)	Merger of the Agricultural Research Department and Wageningen Agricultural University. Parts of the experiments are expected to be integrated as well.	—
Agricultural Research Department (DLO)	Executing branch of DSKT. Detached from MOA in 1999. Comprises the following entities: (1) Institute for Forestry and Nature Management (IBN-DLO); (2) Winand Staring Centre for Integrated Land, Soil and Water Research (SC-DLO); (3) Agrotechnological Research Institute (ATO-DLO); (4) Centre for Plant Breeding and Reproduction Research (CPRO-DLO; including the Centre for Genetic Resources); (5) Research Institute for Plant Protection (IPO-DLO); (6) Research Institute for Agrobiology and Soil Fertility (AB-DLO); (7) Institute for Animal Science and Health (ID-DLO); (8) Netherlands Institute for Fisheries Research (RIVO-DLO); (9) Agricultural Economics Research Institute (LEI-DLO); (10) Institute of Agricultural and Environmental Engineering (IMAG-DLO); (11) State Institute for Quality Control of Agricultural Products (RIKILT-DLO); and (12) Centre for Agricultural Publishing and Documentation (PUDOC-DLO). DLO institutes focus primarily on strategic and basic research.	Dfl 370 million[a]

Organization for Applied Research in Agriculture	Executing branch of DSKT. Comprises nine experiment stations and their affiliated regional research centers. They conduct applied agricultural research for the following subsectors: (1) arable crops and field vegetables; (2) fruits; (3) bulbs; (4) arboriculture; (5) floriculture and glasshouse vegetables; (6) mushrooms; (7) poultry; (8) pigs; and (9) cattle, sheep, and horses. Traditionally funding of the stations is shared between the MOA and the farmers.	Dfl 107 million
Nutrition and Food-Processing Division of the Netherlands Organization for Applied Research (TNO-Food)	TNO-Food is a private, nonprofit research organization that conducts research for both the public and the private sector. MOA has a client relationship with TNO-Food. TNO itself resorts under the Ministry of Economic Affairs.	Dfl 104 million
Wageningen Agricultural University (WAU)	Resorts under the responsibility of DSKT-MOA. Has a primary role to play with regard to basic research, but also conducts some strategic and applied research.	Dfl 319 million
Faculty of Veterinary Sciences (FVS), University of Utrecht	Resorts under the responsibility of the Ministry of Science, Education, and Culture. Conducts primarily basic research.	Dfl 39 million
Private sector	Includes research departments of private companies as well as a few research institutes fully financed and managed by the industry, such as the Institute for Efficient Sugar Production and Netherlands Institute for Dairy Research.	Dfl 745 million

NOTE: — indicates not applicable.

[a] In 1995 the Dutch guilder equaled 0.62 US dollars.

established.[4] Over time, Wageningen developed into the main center for agricultural research in the Netherlands. Many agricultural research institutes were established around Wageningen, and the school developed into what is still the only agricultural university in the country.

Most of the early agricultural experiment stations in Europe were modeled after the agricultural experiment station in Möckern (Saxony, Germany). The Möckern station, which was established in 1852, was the first to receive a government grant. Similarly, the first station at Wageningen was financed by a government grant and income from services provided. The first director of the Wageningen station was Adolf Mayer, a German professor. Like most early experiment stations in Europe, the Wageningen station functioned mainly as a chemical laboratory for testing fertilizers and soils. In 1887 the Dutch experiment station system was substantially expanded through the creation of stations at Groningen, Hoorn, and Breda (relocated to Goes in 1893). In 1898 an additional station was established at Maastricht, thereby completing the development of a network of stations with comprehensive geographical coverage. In 1899 a separate seed-testing station was established in Wageningen.

During the first 30 years of their existence, state agricultural experiment stations in the Netherlands primarily performed regulatory, testing, and monitoring functions. Although research was part of their mandate, relatively little research was actually carried out. Thus, to encourage more research, the regulatory and research functions of the experiment stations were split in 1907. This reorganization failed, and so around 1913–1916 a more drastic approach was introduced. The stations at Goes, Maastricht, and Wageningen and the seed-testing station in Wageningen were given exclusive regulatory mandates and the stations at Groningen and Hoorn were given exclusive research mandates. In this new setup, Hoorn focused on dairy research and Groningen on crop and grassland research. Similar "research-only" state experiment stations were established for forestry research at Wageningen in 1919 and poultry research at Beekbergen in 1921.

The state agricultural school in Wageningen became a state agricultural college in 1904. Several research institutes emerged in conjunction with the college, including the Sugar Laboratory (1903), the Institute of Agricultural Machinery (1905), the Phytopathological Institute (1906), the Institute of Plant Breeding (1910), and the Laboratory of Bulb Research (1917). Most of these institutes and laboratories were directed by professors of the college. The college gained university status in 1918.

---

4. This section draws on Broekema (1938), Verkaik (1972), Maltha (1976), an article in *Landbouwkundig Tijdschrift* (Anonymous 1950), and Hissink (1916). Although the history starts here, there were earlier, private initiatives to establish experiment stations. In Deventer, for example, an experimental garden was opened in 1860. However, the first recorded state intervention dates from 1877.

During the 1920s and 1930s the number of agricultural research institutes continued to grow. At the same time, the private sector became more explicitly involved in agricultural research. Several agricultural research institutes, such as the Laboratory for Soil Research, the Institute for Sugarbeets, and the Institute for Animal Feed, were established by private interests and managed and financed by users.

Another institutional innovation in the 1930s was the introduction of public research foundations. These foundations enjoyed some structural advantages over the older state experiment stations. The governance of a foundation was vested in a board that included representatives of farmer organizations, agricultural industry, the MOA, and other interested parties. This arrangement substantially improved links with client groups and helped the institutions to secure and manage financial support from the private sector. In the 1930s several research institutes specific to agriculture were also established as foundations. After World War II the foundation became the dominant organizational structure for public agricultural research institutes and experiment stations.

By the end of the 1930s, the uncoordinated growth of agricultural research had resulted in a multitude of different agricultural research entities. In a first attempt to improve coordination, the government established a Directorate of Agricultural Research in 1938, followed by a Central Agricultural Research Institute in 1939. In a parallel development, the Netherlands Organization for Applied Scientific Research (Dutch acronym: TNO) was created in 1932. This organization was mandated by law to promote, coordinate, and manage all applied scientific research in the Netherlands. The TNO gradually assumed its responsibilities and eventually took charge of coordinating agricultural research in 1943. The plan was ultimately to integrate all agricultural research entities operating under the MOA into the TNO. In the meantime, the TNO established several new agricultural research institutes during the late 1940s and early 1950s.

However, the planned transfer of the MOA's research institutes and experiment stations to the TNO was kept on hold because of resistance within the agricultural sector (including the ministry). To solve this problem the government introduced a rather awkward division of responsibilities in the mid-1950s. The MOA assumed control of all agricultural research agencies managed by the TNO (except for the food and nutrition research agencies), while the TNO took control of the NRLO. The NRLO—established in 1957—was given the mandate to promote and coordinate agricultural research and to provide policy advice on agricultural research matters.

NRLO's initial board of management had eight members: two representatives of the MOA, three representatives of the scientific community (two appointed by WAU and one by the Netherlands Royal Academy of Sciences), and three representatives of farmers' organizations appointed by the Agricultural Board.[5] The NRLO created five divisions (crops and pastures, horticulture, livestock, dairy, and animal health), which corresponded to the divisions of the Agricultural Directorate of the MOA at that time. Initially the NRLO's mandate only included the 37 research entities of the MOA. In 1970, its mandate was broadened to encompass the coordination of the agricultural research activities of WAU, the Faculty of Veterinary Sciences of the University of Utrecht, TNO-Food, and some scattered agricultural research entities.

Whereas the NRLO mainly had a coordinating and advisory role, the MOA administered and financed most of the agricultural research entities that had operated more or less independently. In an attempt to consolidate, the MOA established a Directorate of Agricultural Extension and Research in 1962, which was subsequently split into a directorate for research and one for extension in 1968. Initially, only the more upstream research institutes dealing with crop and livestock issues were transferred to this new directorate, while forestry, fisheries, veterinary research institutes, and the experiment stations and farms remained within their subject-matter directorates. The creation of this directorate (along with the introduction of a central research project administration in 1966) led to greater administrative oversight and facilitated a centralized planning process for agricultural research (a consolidated five-year plan for agricultural research was presented to Parliament for the first time in 1972; Ministerie van Landbouw en Visserij 1972, 1977).

The number of agricultural research entities had expanded rapidly during the years immediately following World War II. In 1962, however, a long period of consolidation and rationalization began. At the instigation of the Directorate of Agricultural Research, several research entities were merged, some had their mandates modified, and others were closed. To save costs, a central technical service and a central publication and documentation service were established.

Further consolidation of agricultural research within the MOA occurred in 1981. The number of directorates with administrative responsibility for one or more agricultural research entities was reduced from nine to three: (1) the Directorate for Agricultural Research, (2) the Directorate for Arable Farming and Horticulture, and (3) the Directorate for Animal Husbandry and Dairy. The research institutes under the Directorate for Agricultural Research focused primarily on strategic and basic research. Nearly all of the institutes were organized as semipublic entities governed by a board and received most of their funding from the government. The Directorates for Arable Farming and Horticulture and for Animal Husbandry and Dairy were given administrative responsibility for about 10 agricultural experiment stations and some 50 re-

5. The Agricultural Board (Landbouwschap) was a semipublic body with administrative and regulatory powers within the agricultural sector. Both the farmer organizations and the labor unions were represented on the board.

gional experimental farms and gardens as well as for the MOA's extension service.

Each experiment station dealt with a specific branch of agriculture and was governed by a board dominated by representatives of the Agricultural Board and commodity boards. The experiment stations conducted applied research and collaborated closely with a dense network of regional experimental farms and gardens where new technologies were tested and demonstrated under normal farm conditions.[6] The MOA and agricultural producers shared responsibility for funding the experiment stations, which served to locally adapt and fine-tune technologies in the process of extending them.

Recent Changes, 1985 to Present

This section highlights some of the recent major institutional changes within the NRLO, the MOA (including the DSKT, DLO, and experiment stations), TNO-Food, and WAU, as well as the recent establishment of WURC.[7]

NATIONAL COUNCIL FOR AGRICULTURAL RESEARCH. The NRLO was transferred from the TNO to the MOA in 1986. For many years it had been disadvantageous for the council to be linked to the TNO instead of the MOA. The division between the financing of public agricultural research, which was channeled through the MOA, and the policy formulation and oversight responsibilities of the NRLO frequently caused coordination problems. After the transfer, the council was significantly restructured, partly because the industry and the government were increasingly dissatisfied with the NRLO's role. The council was considered too bureaucratic and burdened with too many subcommittees. Moreover, the council had no effective influence over the implementation of its recommendations. Participation in the NRLO's decisionmaking apparatus was on a voluntary basis, and the council had no budgetary powers. The new structure sought to streamline consultations among the different participants but left the voluntary character of participation unaltered (Enzing 1989; Enzing and Smits 1989).

The NRLO brought together three groups of participants:

- government entities that fund or cofund agricultural research and therefore have claims on public policies regarding agricultural R&D (in addition to the MOA, five other ministries direct some of their funds to agricultural research);

---

6. In recent years the number of regional experimental farms and gardens has been cut quite substantially to about 20 at present.

7. The description of recent changes draws upon a wide range of published and unpublished materials. Of particular interest are NRLO (1981, 1988b, 1990a,b, 1995); Ministerie van Landbouw en Visserij (1985, 1987); Ministerie van Landbouw, Natuurbeheer, en Visserij (1992, 1995b, 1996); Ministry of Agriculture, Nature Management, and Fisheries (1991, 1992); and LUW (various years a and b).

- nongovernmental organizations representing users (for example, the Agricultural Board and commodity boards), specific interest groups (for example, the Netherlands Society for the Protection of Animals), and representatives of private firms (for example, seed companies); and
- research-executing agencies.

In addition to the general council, the NRLO had five sector clusters, or "chambers" (plant production; animal production, health, and welfare; processing and marketing; land use and nature and landscape conservation; and agriculture and society). Each of the sector chambers consisted of a cluster of committees and subcommittees. Despite the efforts to reorganize and streamline the NRLO, it still comprised about 100 committees and subcommittees in 1988. The NRLO directory for 1988 (NRLO 1988a) identified more than 1,000 persons involved in the NRLO's committee work, with many participating in more than one committee.

With the establishment of the Directorate of Science and Technology (DST) in 1989 (see next section), the secretariat of the NRLO was integrated into the DST. This gave the ministry a dominant position in the council vis-à-vis other participants. The same staff formulated policies for the MOA as well as the NRLO. However, in January 1995 the NRLO ceased to be a coordinating and planning agency and was transformed into an advisory board to conduct science and technology foresight studies for the agricultural sector and rural areas. All the committees and subcommittees were abolished, and the five sector chambers were restructured into less-formal sounding boards. The NRLO now reports directly to the minister of agriculture.

The NRLO structure became increasingly obsolete in the course of the 1990s for several reasons. Perhaps the most important reason was that the research requirements of the various participants drifted too far apart. For example, the research demands expressed by representatives of the floriculture branches became so specific that it became increasingly difficult to reach agreements with representatives of other specialized branches, such as the sugarbeet production branch. Another reason was the change from block-grant funding to research institutes to funding earmarked for specific research outputs, combined with the increasing adherence to a "user-pays" principle, which led research institutes to consult directly with their own clients and develop their own strategic research plans rather than work through a consensus-building process mediated by the NRLO. Finally, insofar as coordination was still appropriate and required, the MOA directorate responsible for agricultural research policy (DST, now DSKT) began to take a more direct interest in this task.

Thus the functions of the NRLO have changed substantially over time. Initially the council focused primarily on the promotion and coordination of agricultural research (1957–1970). After 1970, medium-term planning became one of the NRLO's more important functions in addition to coordination. After

1985 the NRLO's coordination role gradually declined and its primary focus shifted toward medium-term planning and technology foresight studies. Since 1995 the council has only conducted foresight studies.

MINISTRY OF AGRICULTURE. In the 1980s, a political consensus emerged that ministries should concentrate on policy formulation and delegate the implementation of policies as much as possible to semipublic entities or the private sector. In 1986 the cabinet placed agricultural research on a list of activities to be devolved from the MOA and placed under the auspices of an autonomous, semipublic agency.[8] As a first step toward such privatization, the MOA split the Directorate of Agricultural Research into the Directorate of Science and Technology (DST) and the Department of Agricultural Research (DLO) in May 1989. Under this new arrangement the DST dealt primarily with policy and was to remain within the ministry, whereas the DLO was to oversee and manage the research institutes. During the following 10 years the DLO developed a corporate identity, adopted a more decentralized and private-sector-like management style, and completely restructured itself. In early 1999 the DLO was detached from the ministry: DLO staff are no longer government employees, and the relevant physical infrastructure was taken from the ministry and placed under the direct control of the department.

After many years of growth, agricultural research funding by the MOA began to decline or stagnate in real terms after 1978. The economic crises of the late 1970s and early 1980s forced the Dutch government to cut government expenditures quite substantially. Despite attempts to improve operational efficiencies (for example, by merging some institutes), by 1985 most institutes were operating with tight budgets and with little flexibility to shift resources. It was also clear that few additional resources could be expected from the government to address these problems. As a consequence some drastic changes were proposed in the MOA's *Development Plan for the Agricultural Research Institutes and Experiment Stations 1987–90.* The total number of research and support staff positions was to be reduced by 11 percent over four years. The money freed up in this way, as well as a one-off contribution from the MOA, was to be used for

- a major program of capital investment in the construction of new buildings and renovation of old buildings, as well as new research equipment;[9]
- an increase in operating budgets per researcher;
- an increase in maintenance budgets; and
- an increase in salary budgets to contract better-qualified research staff.

---

8. The decision to privatize agricultural extension had been taken earlier by the cabinet.

9. Many research buildings were built in the early 1950s, and after 40 years they needed major renovations or had to be replaced by new buildings. In addition the mergers and reorganizations of institutes often required modification of existing office and laboratory space or the construction of new buildings.

In addition the plan proposed a further consolidation of the research institutes from 22 to 17 entities and an increase of external financial support (for example, industry-funded, contract research) up to 20 percent of the total funds for agricultural R&D. By 1991 most of the proposed changes had been implemented.

In 1990 the minister of agriculture presented his *Agricultural Structure Memorandum: Policy on Agriculture in the Netherlands in the 1990s.* The memorandum included a program of additional expenditures of nearly Dfl 2 billion for the period 1990–1994 to restructure the agricultural sector. Nearly half of these funds were to be spent on environmental issues (now a major concern for Dutch agriculture) and about Dfl 175 million was earmarked for agricultural education, extension, and research. For the first time in many years, a real increase in the MOA's support to agricultural research was anticipated. However, much of the proposed increase did not materialize because of new budget cuts.

In 1991 the DST presented its *Agricultural Research Policy Plan 1991–94.* The plan proposed another round of consolidation of the DLO research institutes from 17 to 12, as well as a geographical concentration of institutes in Wageningen and to a lesser extent in Lelystad. Sharing resources such as equipment, laboratories, and expertise among the institutes—and also with the departments of WAU—was explicitly identified as a way of saving costs. In this restructuring plan the DLO faced increasing stepwise cuts in its budget to total Dfl 12 million per year in 1994. In addition, by 1992, the DLO's income had declined by about Dfl 6 million per year because the DLO had stopped producing hoof-and-mouth disease vaccine. On the expenditure side, additional monies had to be set aside to meet reorganization costs and tax payments due to the change in the organization's tax status. In total an annual reduction of Dfl 25 million (about 8 percent of the DLO's budget) had to be accommodated. It was expected that most of these cuts could be offset by increasing contributions from other sources to 30 percent of total expenditures.

The ink on this plan had barely dried when, in 1992, the government decided to cut another 150 positions at the DLO institutes as part of a broader "efficiency operation," a euphemism for cutting government expenditures to reduce the budget deficit. Under the 1991 plan, the budgets of the experiment stations and farms remained largely untouched and no major structural changes were proposed. The 1991 policy plan also called for a major change in the programming and financing of the DLO research institutes. Instead of the traditional system under which government provided the institutes with a lump-sum grant (so-called input financing), the new system was to be based on output financing, which meant that funding would be targeted more specifically to agreed programs of work.

In May 1993 the MOA presented an agricultural knowledge policy memorandum (Ministerie van Landbouw, Natuurbeheer en Visserij 1993). Instead of separate research, extension, and education policies, the MOA presented one "knowledge policy" to operate for and within an agricultural knowledge system. One reason for adopting this new approach was that the rapid changes within each of the system's components meant that links between those components were becoming increasingly problematic. Moreover, the traditional cohesion and division of labor within the knowledge system had disappeared.

The new knowledge-system approach formed the basis for the 1995 merger of the agricultural research, extension, and education directorates within the MOA into a single directorate, the Directorate of Science and Knowledge Transfer (DSKT). Additionally, the nine experiment stations (including the 20-odd regional experimental farms and gardens attached to them), which previously came under the Directorate of Arable Farming and Horticulture and the Directorate of Animal Husbandry and Dairy, were also transferred to the DSKT. For the first time in the history of the MOA, all the components of the agricultural knowledge system were integrated into one directorate. An agricultural knowledge policy plan for the period 1996–1999 was presented to Parliament in 1996 (Ministerie van Landbouw, Natuurbeheer, en Visserij 1996).

Since their transfer to the DSKT in 1995, the agricultural experiment stations have had a rather insecure position for several reasons. These include:

- the MOA's policy of reducing its contribution to near-market research;
- the collapse of the Agricultural Board in 1995; together with the commodity boards this semipublic entity collected the levies that partially finance the experiment stations; the Agricultural Board also represented the agricultural producers on the boards of the experiment stations; and
- a historical carryover, whereby the ministry maintained two separate approaches for executing agricultural research—one for the more-basic and strategic research (DLO) and one for more-applied research (conducted by the experiment stations).

By 1995, the nine experiment stations were clustered in a temporary organization (the Organization for Applied Research in Agriculture). Since then, the DSKT has tried to bring the management of the experiment stations on a par with the DLO research institutes and to provide funding on the basis of agreed-upon research programs rather than an undirected lump-sum grant (Ministerie van Landbouw, Natuurbeheer, en Visserij 1995a). As a consequence, the DSKT ended its one-to-one funding arrangement with the experiment stations in 1998. At present, discussions are taking place about which parts of the experiment stations will be integrated into the WURC, which will be handed over to regional farmer organizations, and which will be closed.

TNO-FOOD. TNO-Food is part of the TNO, which conducts research in industrial technology, defense, health, environment and energy, building technology, transport and infrastructure, technology policy, and food and nutrition. TNO includes 18 research institutes and laboratories, which operate relatively independently from each other (TNO 1991). TNO-Food is one of the larger TNO institutes. It consists of several laboratories and centers engaged in food processing, nutrition, and health research. These are located throughout the country and are sometimes linked to universities. Since the early 1980s TNO-Food has substantially strengthened its corporate identity and rationalized its research capacity through mergers and relocation of research facilities. TNO-Food has a strong market orientation and in 1995 derived about 70 percent of its budget from research contracts, up from 64 percent in 1985. Interestingly, 22 percent of these contracts are with international clients, reflecting TNO-Food's strong orientation toward foreign markets (TNO-Food various years).

WAGENINGEN AGRICULTURAL UNIVERSITY. Since the mid-1980s, WAU has undergone major change. First, because of demographic developments as well as a nationwide shortening of study programs, the number of students enrolled in regular programs dropped by about 20 percent, and this decline is expected to continue for some years. Second, during the 1980s WAU introduced a series of international master of science programs in order to attract more foreign students. Although successful, the number of foreign students enrolled in these programs is severely limited by the number of study grants available. Third, despite (or perhaps because of) the reduction in its teaching program, the university has substantially expanded its research endeavors. The number of full-time-equivalent research positions increased from 503 in 1985 to 833 in 1994. Eighty-eight percent of this expansion was realized by attracting funding from sources other than the MOA (for example, the European Union (EU), Netherlands Organization for Scientific Research, Ministry of Development Cooperation, and private industries).

Although WAU traditionally conducts basic research, it also carries out a considerable amount of applied research. A strong culture within the university encourages the production of "practical" (not only theoretical) outputs that have an immediate and direct application.

In the 1990s the Ministry of Education, Science, and Culture began developing so-called research schools within the Dutch university system. These thematic research schools bring together research expertise across departments, faculties, and universities. Nonuniversity research organizations can also participate in each research school. WAU hosts or participates in seven schools in the fields of experimental plant sciences, production ecology, animal sciences, environment and climate, toxicology, food technology, and socioeconomics. The Research School on Experimental Plant Sciences, for example, draws together researchers from 12 departments of WAU as well as scientists from the University of Nijmegen, the University of Utrecht, and three DLO institutes. About 100 research schools currently operate within the Dutch university system.

WAGENINGEN UNIVERSITY RESEARCH CENTRE. A study of the future of the Dutch agricultural knowledge system commissioned by the minister of agriculture in 1995 concluded that the system suffered from considerable overlap and unnecessary competition (Peper 1996). One of the report's important recommendations was to merge the DLO and WAU. Although this was a controversial idea at the time, the minister adopted the recommendation immediately and has made the integration of the DLO and WAU one of his major policy objectives.

The new organization, Wageningen UR, will be one of the world's largest agricultural science conglomerates. As a first step, the boards of the DLO and WAU were merged, and a new chairman was appointed in 1997. In 1998 further details of the merger were still being worked out. It was expected that parts of the nine experiment stations would merge with Wageningen UR and be reorganized into an applied livestock research cluster and a cluster for applied crop research. However, the 20 experimental farms and gardens currently attached to the experiment stations will most likely be transferred to regional farmers' organizations and linked more closely to extension. This is in line with the MOA's present policy of devolving its responsibility for near-market research.

Financing Agricultural Research

In this section we review the policy objectives of the Dutch government with regard to the financing of agricultural research and examine how these objectives have changed over time. We also review trends in Dutch agricultural research investments during the past 25 years and offer a more detailed description of the financing practices of each component of the agricultural research system.

Policy Objectives: Past and Present

Over the past 20 years Dutch government' involvement in the financing and provision of agricultural research has been influenced significantly by three major policy developments: (1) the changing role of government in society, (2) public administration reform, and (3) the reorientation of agricultural policy toward broader societal concerns.[10]

The rapid expansion of the Dutch government sector during the 1960s and 1970s was followed by a period in which further growth in the public sector was considered undesirable and, above all, infeasible because of rapidly growing government deficits and debts. In addition, weak economic growth throughout much of the 1980s and early 1990s forced all Dutch governments (regardless of their political composition) to scrutinize and prioritize their

10. See Roseboom and Rutten (1998) for a more detailed overview of the policy changes in the Netherlands.

expenditures, including those on agricultural research. As a result, the Dutch government adopted a more critical and restrictive attitude toward its role in society in the 1980s. At the same time it looked for ways to increase the effectiveness and efficiency of government and adopted many of the new public administration ideas that emerged in the early 1980s.

One of the more prominent aspects of the new public administration ideas adopted by the Dutch government in the mid-1980s was the concept of a stricter separation between policy formulation and policy implementation. Ministries sought to restrict themselves to policy formulation and, to the extent possible, contract the implementation of policies to semipublic entities or the private sector. For the MOA, this shift meant that several of its executing agencies, including its research and extension agencies, were selected as candidates for privatization as autonomous executing agencies.[11] The privatization process took place in three steps:

- a separation of policy formulation and implementation activities within the MOA (second half of the 1980s);
- a period during which the executing agencies were restructured and reorganized in preparation for privatization, while the policymaking entities developed new ways of overseeing the executing agencies; and
- the formal privatization of the executing agencies (the extension service in 1994 and the DLO in 1999).

Privatization has changed the way in which these new, autonomous agencies are financed by the MOA. Instead of traditional lump-sum input transfers to finance R&D, the MOA now provides its funding through contracts that specify the expected research and extension outputs.

Increased public concerns about the environment, the sustainability of agricultural production systems, food safety, animal welfare, and land use have also had important effects on Dutch agricultural policies in the 1980s and 1990s. The traditional focus of the MOA on increasing agricultural production and productivity (which largely coincided with the interests of farmers) yielded ground to these more general societal concerns. In the mid-1980s the MOA changed its mission statement to "creating or improving the conditions for a competitive, safe and sustainable agriculture."[12] Despite this new mission

---

11. For more details on the privatization of the extension service see Bos (1989).

12. For many years the principal objective of the MOA was to raise the general level of welfare of the agricultural population and assure food security for the population at large through an increase in agricultural production and lower food prices. In this context the research and technology policy was straightforward and focused almost exclusively on increasing production and productivity. Measured in these terms, this policy has been most successful (Meer, Rutten, and Dijkveld Stol 1991; Rutten 1992).

statement, the MOA often finds itself caught between an agricultural sector that is reluctant to accept (or to carry the financial burden of) stricter regulations and other sections of society that call for such regulations. Nonetheless, the MOA has moved away from identifying itself exclusively with the production interests of the agricultural sector toward a position that represents broader societal interests such as a clean environment and food safety.

In keeping with these new objectives, the MOA has sought to reorient its research and technology policy toward broader societal interests. This change is reflected most explicitly in how the MOA now distinguishes between two types of research: (1) basic research and research on environmental issues, food safety, and nature conservation, for which the MOA acts as the primary client; and (2) production-oriented research whose benefits are more appropriable and that the agricultural producers and industries should thus finance or cofinance (Ministerie van Landbouw, Natuurbeheer en Visserij 1996).

The series of piecemeal policy changes since 1970 has collectively brought about a major shift in the way agricultural research is financed in the Netherlands. The MOA's share in financing public agricultural research declined from 76 percent in 1970 to about 52 percent in 1995. The year 1978 was a turning point in the financing of agricultural research because, after a long period of growth, total expenditures began to stagnate and the amount of funding provided by the MOA began to decline. This development did not represent a deliberate policy change on the part of the MOA, but resulted from a series of crises in government finances. A rapidly increasing government deficit forced the cabinet to curtail expenditures across all government activities. More recently, however, the government has explicitly reconsidered its role in the provision of agricultural research services, sought to broaden the base of funding for agricultural research, and encouraged other parties (including agricultural producers) to contribute more. To quote an MOA policy document: "users should take more responsibility themselves, in particular with regard to extension, applied research, and training" (Ministerie van Landbouw, Natuurbeheer, en Visserij 1995b:3).

Investment Trends in Agricultural Research, 1970–1995

Measuring investments in agricultural research is a complex task. Different approaches may result in very different time series. One approach is to use the science-and-technology statistics collected and published by the Central Bureau of Statistics of the Netherlands. The bureau reports research expenditures by type of organization (public, university, and business) and by objective/area of science. For our purposes, the objective "agricultural productivity and technology" is closest to what we seek when, say, estimating agricultural research intensity ratios. These ratios express public agricultural research expenditures as a percentage of agricultural GDP.

Another approach is to construct expenditure time series for the various executing research agencies within the public component of the research system on the basis of annual budgets and reports. For the university and private-sector components more detailed time-series data were not available. Instead we used the more aggregated science-and-technology statistics for these two components. The disadvantage of using these data is that they overestimate the relevant public research effort. A substantial and increasing part of public agricultural research is on topics of no direct relevance to primary agricultural production, such as research on nature and the environment, outdoor recreation, and food safety and processing. Moreover not all expenditures by research institutes are necessarily research related; many institutes also perform some service and advisory functions. However, using more detailed expenditure data, where available, provides much richer insight into the financing issues.

Figure 7.1 plots the agricultural research intensity ratios based on these two data series. The two ratios follow roughly the same trend, although they diverge over time. By 1995 the science-and-technology estimate of the intensity ratio was 29 percent lower than the intensity ratio based on an aggregation of agency budget data (not including TNO-Food).

In the rest of this chapter, we will refer to the budget time-series data only. This expenditure time series includes outlays by the DLO research institutes, the experiment stations, WAU, the Faculty of Veterinary Sciences, and TNO-Food (Table 7.2). In addition expenditures for agricultural and agriculture-related research that is both funded and performed by the private-business sector are reported for the years 1981–1995. They include the R&D expenditures by the agriculture, forestry, and fishing industries; the food-processing industries; and 10 percent of the R&D expenditures by the chemical and pharmaceutical industries.[13] Together agricultural and agriculture-related R&D expenditures by the private-business sector are presently on a par with the expenditures by the public agricultural research agencies. In terms of growth, the private sector grew faster than the public component of the national agricultural research system.

Total expenditures by public agencies grew rapidly between 1970 and 1978, stalled or shrank during most of the 1980s, increased a little in the late 1980s and early 1990s, and contracted again during the last two years for which data were available (1994 and 1995). There are indications that this decline continued after 1995. However, trends differ among the various public compo-

13. This figure of 10 percent is a very rough estimate based on some partial data from the United Kingdom and the United States. A more in-depth study on the structure of the Dutch agricultural chemical and pharmaceutical industries is needed to improve the precision of this estimate.

FIGURE 7.1 Agricultural research intensity ratios based on budget data and science and technology statistics, 1970–1995

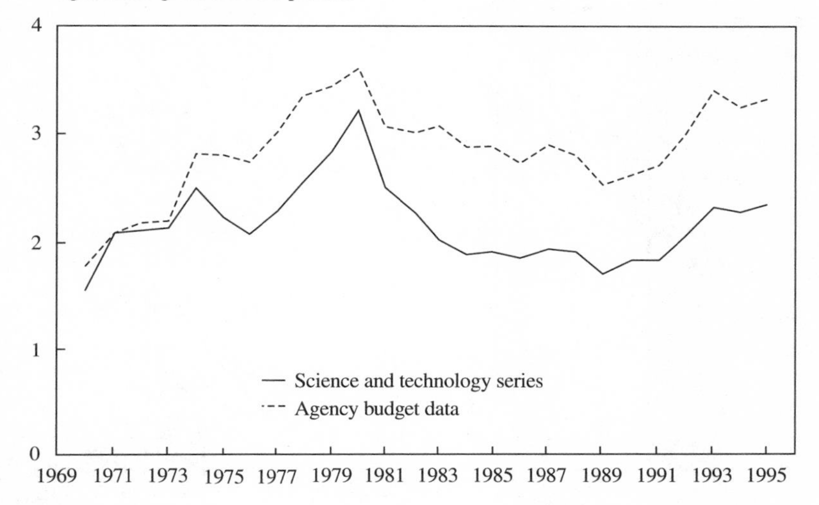

SOURCE: See Table 7.2.
NOTE: These intensity ratios cover only public agricultural research expenditures.

nents of the national agricultural research system. The DLO research institutes experienced major budget cuts after 1978, whereas TNO-Food and WAU expanded their research budgets. Over time this has led to a significant change in the structure of the system (Table 7.3).

Traditionally, the MOA has been the principal source of funding for public agricultural research. However, Figure 7.2 shows that the MOA's share of public agricultural research funding has declined from 76 percent in 1970 to 51 percent in 1995. In particular, since 1978 other sources of funding such as other ministries, the Netherlands Organization for Scientific Research, the EU, the Agricultural Board, commodity boards, and private firms contracting public research institutes have become more important. The MOA's contribution to public agricultural research peaked in 1978, but declined by some 17 percent in real terms between 1978 and 1995. This decline was more than offset by contributions from various other sources of funding. Together these other sources of funding increased by more than 90 percent between 1978 and 1995. As a result the sources of funding for public agricultural research has become more complex.

TABLE 7.2 Agricultural research expenditures by executing agency, 1970–1995

	Public Sector								
	DLO Research Institutes[a]	Experiment Stations	WAU[b]	FVS[c]	Subtotal Agriculture	TNO-Food[d]	Subtotal Agriculture and Food	Private Sector	Total
	(million 1995 Dutch guilders)								
1970	182.6	61.3	31.2	8.5	283.7	38.4	322.1	na	na
1971	247.0	61.1	44.7	13.1	365.9	44.7	410.5	na	na
1972	262.6	71.0	55.3	16.8	405.7	52.0	457.6	na	na
1973	271.7	68.1	63.9	19.8	423.5	59.3	482.8	na	na
1974	290.7	73.2	70.7	20.2	454.8	66.6	521.4	na	na
1975	311.1	81.2	75.1	22.0	489.4	67.0	556.4	na	na
1976	320.1	86.6	79.0	23.5	509.3	67.4	576.7	na	na
1977	342.7	74.7	83.5	25.2	526.2	67.8	594.0	na	na
1978	357.5	77.4	103.3	34.4	572.6	64.1	636.6	na	na
1979	342.6	67.5	93.4	37.4	540.9	60.4	601.2	na	na
1980	346.8	69.5	103.5	35.4	555.2	58.6	613.9	na	na
1981	342.1	88.4	97.0	32.3	559.8	56.9	616.7	535.3	1,151.9
1982	338.0	88.3	106.7	38.0	571.0	62.6	633.6	517.2	1,150.8
1983	337.7	83.2	108.2	36.1	565.1	61.3	626.4	530.2	1,156.6

1984	339.8	82.6	107.9	39.1	569.4	65.2	634.6	534.4	1,169.0
1985	315.1	78.4	121.1	39.6	554.3	68.7	623.0	603.0	1,226.0
1986	308.9	72.8	130.3	47.7	559.7	76.8	636.5	614.1	1,250.6
1987	308.3	85.5	134.8	44.5	573.1	80.9	654.0	664.8	1,318.8
1988	322.2	83.3	122.8	39.4	567.7	82.2	649.9	672.6	1,322.5
1989	335.5	84.8	129.3	45.8	595.4	87.0	682.5	758.0	1,440.5
1990	327.7	85.5	153.3	42.5	609.0	91.9	700.9	721.4	1,422.3
1991	336.6	89.3	158.0	43.6	627.4	96.7	724.1	704.8	1,428.9
1992	344.1	93.4	169.4	41.6	648.5	96.8	745.3	693.1	1,438.5
1993	351.4	94.2	170.3	40.7	656.7	96.8	753.5	819.8	1,573.3
1994	336.6	94.7	170.4	39.8	641.5	111.2	752.8	682.0	1,434.7
1995	327.4	91.7	171.0	39.0	629.1	104.3	733.3	745.4	1,478.7

SOURCE: Ministerie van Landbouw, Natuurbeheer en Visserij (various years); CBS (various years, 1996, 1997); EUROSTAT (various years); TNO-Food (various years); and World Bank (1997).

NOTE: na indicates not available. The 1995 figures reported here do not correspond with the totals reported in Table 7.1 and Figure 7.3 because the expenditure data for the DLO research institutes and the experiment stations exclude the expenditures for land and buildings, and the expenditure time series for WAU only cover the agricultural sciences component of WAU's research expenditures. In 1995, one Dutch guilder equaled 0.62 U.S. dollars.

[a]DLO, Agricultural Research Department.

[b]WAU, Wageningen Agricultural University.

[c]FVS, Faculty of Veterinary Sciences, University of Utrecht.

[d]TNO-Food, Nutrition and Food-Processing Division of the Netherlands Organization for Applied Research.

TABLE 7.3 Expenditure shares of executing agencies in Dutch public agricultural research

	1970	1980	1990	1995
		(percentage)		
DLO research institutes[a]	56.7	56.5	46.8	44.6
Experiment stations	19.0	11.3	12.2	12.5
TNO-Food[b]	11.9	9.6	13.1	14.2
Wageningen Agricultural University	9.7	16.9	21.9	23.3
Faculty of Veterinary Sciences	2.6	5.8	6.1	5.3
Total	100	100	100	100

SOURCE: See Table 7.2.

[a]DLO, Agricultural Research Department.

[b]TNO-Food, Nutrition and Food-Processing Division of the Netherlands Organization for Applied Research.

FIGURE 7.2 Ministry of Agriculture (MOA) funding as a component of total public agricultural research, 1970–1995

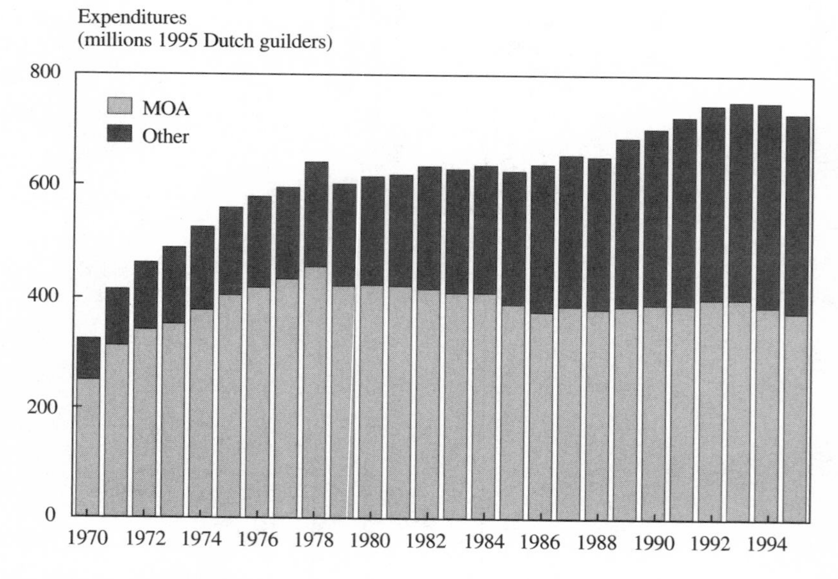

SOURCE: See Table 7.2.
NOTE: See Table 7.2.

Current Funding Practices

Figure 7.3 provides a sense of the flow of funds among funding and executing agencies within the Dutch agricultural research system in 1995. It depicts the complex way in which agricultural research is funded. The total public and private budget for the system is estimated at Dfl 1,679 million (US$1,041 million) in 1995. The various executing agencies have distinctive funding profiles as well as particular relationships with the funding agencies. We discuss these in more detail in the rest of this section.

FUNDING AGENCIES. The DSKT of the MOA is still the most important funding agency for agricultural research. However, as shown in Figure 7.2, the MOA's relative importance has declined over time. The DSKT supports the DLO research institutes; the experiment stations; WAU; and, to a much lesser extent, TNO-Food and the Faculty of Veterinary Sciences. Each entity has its own financial and organizational arrangements with the DSKT and various third parties. Most of the DSKT's funds are now disbursed on the basis of agreed-upon research programs. In addition to the regular budget for the entities, the DSKT keeps about 10 percent of its research budget in reserve to finance immediate research needs and new initiatives and agricultural research executed by other agencies. This provides the DSKT with an opportunity to mobilize expertise outside the traditional group of agricultural research entities. It also introduces an element of competition for the DLO research institutes.

Other agencies that fund agricultural and agriculture-related research include the Ministry of Education, Science, and Culture (which pays for the core research budget of the Faculty of Veterinary Sciences); the Ministry of Economic Affairs (which pays for the core budget of TNO-Food); the Ministry of Development Cooperation (which funds tropical agricultural research projects, many of which are executed by WAU); the Ministry of Environment; and the Ministry of Health. In addition the Ministry of Education, Science, and Culture funds the Netherlands Organization for Scientific Research, which operates a competitive-grant system (total 1995 budget, Dfl 363 million) for basic research at the universities (NWO 1995). In recent years the Ministry of Education, Science, and Culture has increased the Netherlands Organization for Scientific Research's budget for university research at the expense of the ministry's direct contribution to the university research budgets. In this way the ministry is trying to tie research at universities more closely to national priorities.

The EU, through its multiannual Framework Programs coordinated by the Commission's DG XIII, also makes a modest contribution to the financing of agricultural research in the Netherlands. Every four years the EU identifies thematic fields for which funding will be made available through the Framework Programs. Competition for EU funding is rigorous—on average only one

FIGURE 7.3 Flows of funds in the Dutch agricultural research system, 1995

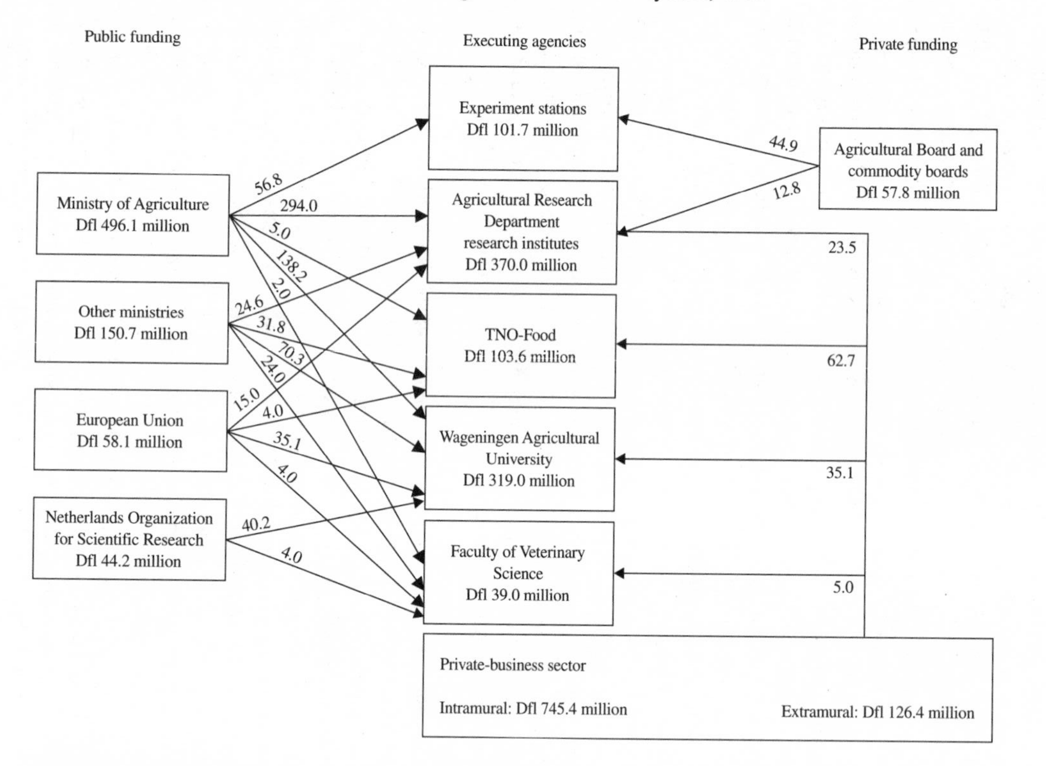

SOURCE: See Table 7.2.

NOTE: The total sum of funding is higher than reported in Table 7.2. These data include (1) expenditures on land and buildings for the Dutch Agricultural Research Department (DLO) and the experiment stations (Dfl 52 million) and (2) all research expenditures by Wageningen Agricultural University (WAU), not simply those classified as agricultural science research (Dfl 148 million).

in five project proposals gets funded. At the same time, however, the allocation of EU research budgets is kept roughly in line with each country's contribution. A basic requirement for projects funded by the EU is that they are conducted collaboratively with at least one agricultural research organization in another EU country.

Two private sources of agricultural research financing are important: (1) agricultural producers who fund research through collective funding arrangements, and (2) private industries that contract out some of their research to public agricultural research agencies (although most of their research budgets are spent on intramural efforts). Until 1998 contributions by agricultural producers were collected in two ways: (1) about half were collected by the now defunct Agricultural Board through a levy based on farm size that each farmer is legally obliged to pay, and (2) the other half were collected by commodity boards, usually at the point of entry of a commodity into the market (generally, the processing industry or auction).[14]

In late 1995 the Agricultural Board, which was created in 1954, finally collapsed because labor unions unilaterally ended their participation in the board. This decision by the labor unions was triggered by a dispute over a general salary agreement between the federation of farmer organizations and the labor unions. Horticulture farmers voted against the agreement their own organization had reached with the labor unions after many months of negotiations. The Agricultural Board was phased out over a period of approximately two years. Starting in 1998, the levy formerly collected by the Agricultural Board was collected by the commodity boards.

To boost private-sector R&D investments, the Dutch government introduced tax breaks for private R&D in 1994.[15] These tax breaks are given to subsidize research personnel costs. An absolute limit on the total amount of tax reduction per firm ensures that the tax break provides more support for small and medium-sized businesses than for large ones. The indirect subsidy provided by these tax breaks constitutes about 5–10 percent of total R&D costs. The subsidy also applies to publicly performed research under contract with the private sector. In 1995 the Ministry of Economic Affairs budgeted Dfl 350

---

14. Research levies collected by the Agricultural Board are based on the size of the farms in terms of "standard farm units" and the agricultural products produced. The level of the research levy depends on the funds needed in each branch and upon the number of farms in that branch (and the collective volume of their standard farm units). The levy per farm differs significantly between branches. For example, the research levy for the average arable farmer is less than Dfl 100 per year, whereas the average mushroom farmer pays about Dfl 3,000 per year. In recent years the number of commodity boards has been reduced quite substantially through mergers. Currently there are four remaining commodity boards for livestock, horticulture, arable crops, and fisheries. Commodity boards bring together all interested parties in the production, marketing, and processing chain of a particular commodity group.

15. Bureau Bartels B.V. and Bakkenist Management Consultants (1996) made a first evaluation of this new tax facility.

million for R&D tax breaks, of which 10–15 percent was spent on private agricultural and agriculture-related R&D.[16]

EXECUTING AGENCIES. In 1995 the DLO research institutes received, on average, about 75 percent of their operating budget from the DSKT.[17] The remaining 25 percent was provided by other ministries, the EU, private companies, the Agricultural Board, and commodity boards. About half of the other sources of funding were private. The balance between ministerial funds and other sources has shifted over time from 90:10 in the early 1970s, to 85:15 in the mid-1980s and to 75:25 in 1995 (Figure 7.4). Other funding sources may become even more important in the future.[18]

In the case of the DLO research institutes, other funding sources have only partially compensated for reductions in ministerial funding. In real terms the DLO institutes experienced nearly a 20 percent reduction in MOA funding between 1978 and 1987. In recent years the MOA's financial support to the DLO stabilized but is expected to decline again in the coming years.

Broadening of the funding base of the DLO research institutes has taken place only gradually. It reflects the fact that new sources of funding do not emerge overnight. Also, at least initially, new sources of support tend to be small and fragmented. As a consequence, DLO institutes must put considerable effort into developing and maintaining relationships with more clients than just the MOA. Although there are some advantages to being more responsive to the needs of clients, the associated transaction costs are considerable.

Contracts with private industries have been a bone of contention in policy debates in the past. In the mid-1980s the ministry ruled that contracts with the private sector should not constitute more than 10 percent of an institute's research portfolio and research results under these contracts cannot remain secret for more than three years. With the recent privatization of the DLO research institutes, this rule seems to have been abandoned.

Current ministerial support for the DLO research institutes consists of three components: (1) core funds for maintaining the institute's knowledge base (about 15 percent), (2) funding for services DLO institutes provide to the ministry or for services they are required to provide by law (for example, certain control functions), and (3) program funds. Program funds constitute the largest part of the ministerial support (70–75 percent). As the responsible financing and policy entity, the DSKT sets out the broad research priorities for the MOA's financial contribution to the DLO. The DLO research institutes

16. Tax rebate claims are lodged with the Ministry of Economic Affairs. If the total claims within a year exceed the budgeted amounts, the rate of tax relief is adjusted down accordingly.

17. Other sources report somewhat lower percentages, but these usually only look at the research programs of the DLO institutes.

18. The ratios presented here are averages. Depending on the research focus of a particular research institute, other sources of funding may represent a considerably smaller or larger part of the funding.

FIGURE 7.4 Dutch Agricultural Research Department (DLO) research expenditures by source of funding

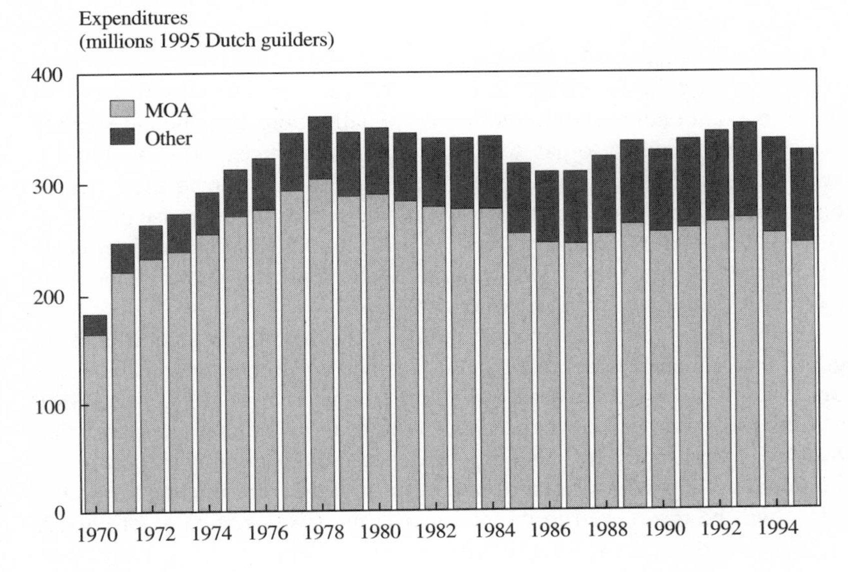

SOURCE: See Table 7.2.
NOTE: Not included here are the expenditures on land and buildings, which are fully paid by the Ministry of Agriculture (MOA).

collectively submit program proposals that the DSKT screens and comments upon. Under this rubric, most funding is tied to agreed programs of work on the premise that this will enhance performance evaluation and increase accountability. However, institutes now spend more time and energy on securing their funding from the MOA than under the previous lump-sum financing arrangements.

A longstanding tradition in Dutch agriculture has been that the MOA and agricultural producers should jointly share responsibility for the deficits of the regional experimental farms and gardens. The tradition stems from the times when many of these stations and regional farms were established by farmer organizations, which subsequently sought financial assistance from the government. After World War II the collection of farmers' dues was delegated to the Agricultural Board and the commodity boards. Although the general rule was that the costs should be split between the MOA and agricultural producers on a 50:50 basis, in practice this rule was not applied very strictly. By the early 1970s, for example, the ratio was closer to 75:25. In some of the weaker subsectors (for example, fruit), the MOA picked up a considerably larger share.

However, since the mid-1970s the MOA has insisted on a stricter adherence to the 50:50 rule and on that basis has made significant cuts in its contribution to the experiment stations.[19] However, the costs for land and buildings are fully paid by the MOA.

For equity (or, perhaps, political) reasons, the ministry strives to maintain a certain balance in its grants to the experiment stations. This means that subsectors do not automatically get proportionally more government subsidy when they are prepared to pay for more research at the experiment stations. However, subsectors can fund research projects in addition to the research program agreed upon between the MOA and the subsector. Likewise the MOA can fund additional research projects. At present additional research projects fully financed by either the MOA or the subsector represent a small fraction of the research funding of the experiment stations.

As mentioned, the position and funding of the experiment stations and the regional experimental farms and gardens has become rather tenuous in recent years. The MOA ended the shared-funding arrangement in 1998. The DSKT now limits its funding to the experiment stations to include only research that reflects MOA priorities. It is also expected that in the upcoming merger of the experiment stations with Wageningen UR, most of the experimental farms and gardens will be transferred to the regional farmers' organizations or closed.

TNO-Food is noteworthy because contracts with the private sector currently account for about two-thirds of its total budget. This reflects TNO-Food's close contacts with Dutch food-processing and agrochemical industries and its rapidly expanding portfolio of research contracts with foreign (both public and private) clients (approximately Dfl 12 million in 1995). TNO-Food has an active policy of operating in foreign research markets and in recent years has opened offices in Prague and Tokyo together with other TNO institutes. After the completion of the privatization of the DLO in 1999, it is expected that some of the DLO institutes will also orient themselves more actively toward foreign markets.

WAU has a special status within the Dutch university system because it is the only university that does not come under the responsibility of the Ministry of Education, Science, and Culture. Instead it falls within the MOA. However, the financing rules and regulations for WAU are largely the same as for the other universities. Although the universities have academic freedom, their operations are closely tied to government regulations (for example, salaries of professors are bound to government salary scales). Research at WAU receives three types of funding: (1) a core contribution from the MOA, (2) the Netherlands Organization for Scientific Research, and (3) contract research. In university jargon, these are the first, second, and third funding sources.

---

19. In real terms, the MOA's 1995 contribution to the experiment stations was only 81 percent of the amount provided in 1976.

With regard to the first funding source (the MOA's core contribution), controls have become much tighter over the years. Until the mid-1970s WAU did not have a coordinated program of research. University departments were free to determine their own research priorities, with core funding for research linked to departmental teaching loads. Beginning in the late 1970s WAU gradually developed a more integrated program of research, which involved a more structured and explicit priority-setting process. Initially this process involved little more than reporting on the research activities undertaken by the departments. Now, however, once every four years each department is required to submit a research program that is reviewed by a sectoral research committee of the university. These committees report to the Permanent Science Committee of the University Council. In the past such a centralized coordination, review, and priority-setting mechanism did not exist.

In principle, however, WAU was still free to determine its own research policies and priorities. Even though the MOA was the major funding agency, it was not supposed to interfere with such matters. In practice, however, a good deal of coordination took place between WAU, the MOA, and other interested parties. The NRLO played an important role in this regard. In the mid-1980s the MOA introduced tighter controls on its core contribution to WAU. This was in line with procedures developed by the Ministry of Education, Science, and Culture, which were introduced in the early 1980s at the other universities. Instead of providing an undirected lump sum on the basis of a general research plan, the MOA now requests that the university submit a comprehensive four-year plan for education and research every two years. This plan is subsequently reviewed and commented upon by the MOA. Together with the internal reviews by the sectoral research committees, this new system of financing is called conditional financing.

Figure 7.5 shows that the number of full-time-equivalent research positions at WAU almost tripled between 1976 and 1994. Most of this expansion was due to a significant increase in the number of research positions funded by research contracts (the third source of funding). However, the first and second sources of funding have also contributed to the expansion of the research capacity at WAU, although more modestly. The relative shares of the different funding sources have changed from 70:10:20 for the first, second, and third funding sources, respectively, in 1976 to 43:13:44 in 1994.

The increase in research positions funded by the MOA's core contribution contrasts with the development of the MOA's overall contribution (covering both research and education) to the university, which in 1995 was 13 percent lower in real terms than in the record year 1978. This decline affected the education budget of the university, as the number of students has fallen quite dramatically (by 20 percent between 1985 and 1995). Therefore, the MOA's contribution to the research budget of the university may have increased rather than decreased.

FIGURE 7.5 Research positions at Wageningen Agricultural University by source of funding

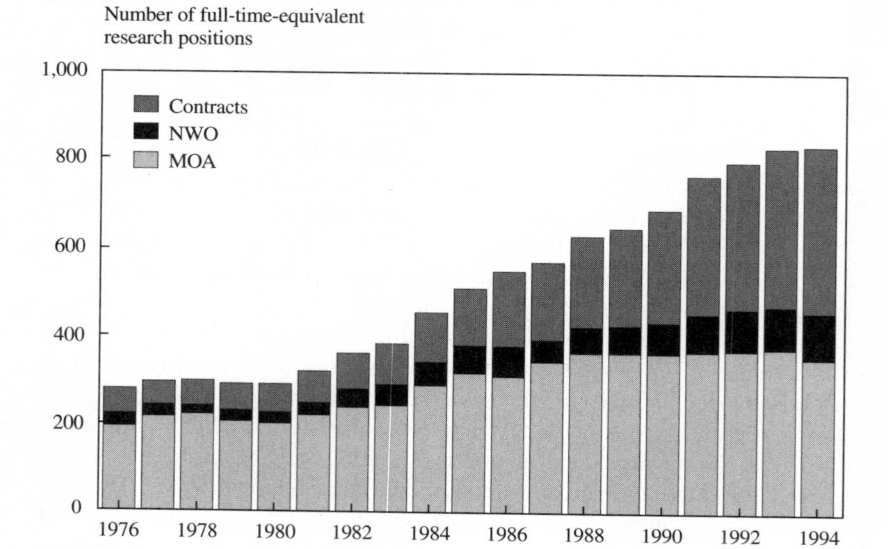

SOURCE: Landbouwhogeschool (various years) and Landbouwuniversiteit Wageningen (various years).

NOTE: MOA, Ministry of Agriculture; NWO, Nederlandse Organisatie voor Wetenschappelijk Onderzoek.

Like WAU, the Faculty of Veterinary Sciences receives research funding from three different sources: (1) a core contribution by the Ministry of Education, Science, and Culture; (2) the Netherlands Organization for Scientific Research; and (3) contract research. The last funding source currently accounts for close to 30 percent of the faculty's research portfolio.

Conclusion

The Dutch agricultural knowledge system is in the midst of a major transformation. Important components of the system have been divested from the MOA and reestablished as autonomous, semipublic entities. At the same time the relationship between the MOA and the executing agencies has become more businesslike, with an emerging provider-client relationship. Since the early 1980s streamlining of the agricultural research system has resulted in substantially fewer agricultural institutes and a clustering of agricultural research agencies in and around Wageningen.

The MOA increasingly acts as a client in the market for research. This has major implications not only for the executing agencies, but also for the ultimate clients of the knowledge system and the linkages between the different components of the system. Much uncertainty still exists among all participants concerning how these changes will settle down over time and whether they will lead to a more effective and efficient agricultural knowledge system.

Along with these changes in institutional roles, there have been some important changes in the nature of funding arrangements and funding patterns for agricultural research in the Netherlands and in the relationships between the private and public sectors. Important among these are the following:

- In real terms, total funding for public agricultural research grew rapidly during the 1970s, stagnated during the 1980s, and grew modestly during the first half of the 1990s.
- In 1995 total agricultural research spending (including the private sector) amounted to an estimated Dfl 1,679 million (or about US$1,041 million).
- Of this total, 44.4 percent was spent by private businesses, 34.3 percent by government research agencies, and 21.3 percent by universities.
- Of the funding for the public component of the national agricultural research system (government agencies and universities), 53.2 percent came from the MOA, 16.2 percent from other ministries, 13.5 percent from the private-business sector, 6.2 percent from the EU, 6.2 percent from the agricultural producers (collected by the Agricultural Board and the commodity boards), and 4.7 percent from the Netherlands Organization for Scientific Research.
- The MOA's dominant role in financing public agricultural research has declined substantially as its share declined from about 76 percent in 1970 to about 52 percent in 1995.
- The Dutch agricultural research intensity ratio (excluding TNO-Food and the private sector) increased rapidly throughout the 1970s to a peak of 3.6 percent in 1980, declining to 2.5 percent in 1989, but more recently trending upward again to reach 3.3 percent in 1995.
- The MOA's research priorities have shifted to encompass environmental concerns, food safety, nature management, and outdoor recreation in addition to the more traditional productivity or competitiveness objectives.
- In 1995, 42 percent of the research at WAU was contract research, up from 20 percent in 1976.

Many similar changes have also taken place in the other countries examined in this study, although with important differences in the timing of events and the detail of the changes. Most countries have experienced in common the shift in the research agenda to encompass the new issues, the rapid rise in funding followed by the recent decline in real terms, and the growing importance of private-sector agricultural R&D. In the Netherlands these changes have been perhaps more pronounced than in the United States. In contrast to

Australia, where checkoff funding has expanded rapidly in recent years, at the moment the long-standing tradition of contributions by the agricultural producers to the funding of the agricultural experiment stations in the Netherlands is in disarray owing to changes in the organizational structure of the agricultural sector as well as the agricultural research system.

8 Agricultural R&D Policy in New Zealand

VERONICA JACOBSEN AND GRANT M. SCOBIE

The funding of agricultural R&D in New Zealand has changed radically since 1984. Public spending on R&D has been centralized and allocated on a contestable basis among competing science providers according to government research priorities. Government research agencies have been reorganized along commercial lines and have become responsible for obtaining their own funding to carry out their mandates.

The system of agricultural research in New Zealand before 1984 was characterized by high levels of government intervention. The government funded and conducted research through its own research agencies; relatively little research was funded through universities and private research organizations. Although total private and public funding has been relatively low by international standards, the government's share has been high.

The practice of government funding by parliamentary appropriation to departments that then carried out the research and provided advice to the minister of agriculture began in the 1920s. It continued, largely unchanged, until 1984. At this time the economy was heavily regulated and economic performance highly unsatisfactory. Following the 1984 elections, however, a new Labour government began an unprecedented process of governmentwide structural reform and liberalization. In that process funding for research was reduced, with government agencies expected to make up the shortfall with contracts from the private sector. Although the funding cuts were intended both to increase the efficiency of the science agencies and to improve equity by increasing private investment, in fact the government had no clearly articulated science policy. A period of confusion followed.

In 1989 the government announced that it would establish a Ministry of Research, Science and Technology (MRST), which would be responsible for policy advice and priority setting, together with a Foundation for Research, Science and Technology (FRST), which would manage a contestable public fund (known as the Public-Good Science Fund [PGSF]) to buy research for national priorities and government's research needs. Access to this fund on a contestable basis was to be open to a wider range of potential public and private suppliers of research services who could bid competitively for research contracts. FRST was to act as the government's agent in buying research that would provide widespread public benefits and that would not otherwise be funded by the private sector.

The government's science agencies were restructured into 10 Crown research institutes (CRIs) in 1992. They remained under the ownership of the Crown but were modeled on private companies with ministers as shareholders. Unlike other state-owned organizations, to date the CRIs have not been required to generate a commercial return to the state as the shareholder. CRIs continue to win most of the research contracts awarded by FRST, an outcome that raises questions about the degree to which the public fund controlled by FRST is actually contestable.

There have been two priority-setting rounds since this funding system was established. These determine the outcomes that the government wishes to achieve with the science outputs purchased from science providers through FRST. The priorities not only provide guidelines about the type of research that will be funded from the PGSF, but also specify the total amount available for each outcome, given the government's research budget. The first round was carried out for the five fiscal years beginning in 1992/93. Although spending for social science was increased in accordance with the public-good principle, total funding for agriculture was unchanged. However, agricultural research spending did show some shifts from production to processing research in line with the government's strategy to fund research into "value-added" activities to contribute to growth and employment. A second priority-setting exercise was concluded in 1995 for the five fiscal years from 1996/97. Overall, funding for agriculture was scheduled to fall from 47 percent to 38 percent of the PGSF.

Private-sector investment in research has started to grow, albeit from a very low base. The Commodity Levies Act 1992 has enabled groups of agricultural producers to levy themselves to fund research. Significant features of the act are that (1) it requires proponents of a levy to show that voluntary funding is not feasible; (2) producers decide by means of a referendum to levy themselves; and (3) the mandate to impose a compulsory levy should lapse after five years unless renewed by a further referendum. Statutory marketing boards, which have funded research and other activities from compulsory levies imposed on producers, are increasingly under threat as the monopoly, government-granted powers they enjoy have become questioned in a new trade and domestic environment. The Commodity Levies Act provides an alternative mechanism for funding research in industries with marketing boards, such as meat and wool.

The guiding economic principle for public funding of agricultural research is that it increases net social benefits. However, judgment must be reserved on whether the reforms have increased the return to the taxpayers' investment. Despite evidence that private investment in agricultural research has increased, a significant proportion of the PGSF is still directed at agriculture. Government-owned CRIs continue to undertake most of the research. To date the new systems of priority setting and resource allocation have not yet explicitly addressed the difficult trade-offs that must be made to maximize total returns to the public investment. Importantly, the system still provides ample opportunity for rent seeking by all stakeholders, especially in the context of processes for setting the research agenda.

This chapter starts by describing the historical patterns of funding agricultural research. Although the reforms have addressed all areas of R&D, this chapter specifically focuses on agricultural R&D. In the next section we describe the economic conditions prevailing in 1984 and the philosophical and theoretical basis of the reform process that resulted first in reductions in research funding and later in the tripartite restructuring into MRST, FRST, and CRIs. In the final section we present an assessment of the reforms and conclude that the new framework for funding has involved new initiatives. As in any process of reform, there is still some unfinished business.

Research Funding before 1984

New Zealand Agriculture

New Zealand agriculture is characterized by two features that affect the funding and provision of R&D. First, the contribution of agriculture to GDP and export receipts is large by high-income-country standards so that investment in R&D in this sector is likely to be high. Second, before 1984 New Zealand was characterized by high levels of government intervention in the overall economy and especially in agriculture. In this climate government undertook both the funding and provision of agricultural research. The highly interventionist approach ultimately proved to be unsustainable. Policy changes in 1984 not only removed a distorted structure of incentives for agriculture and improved its competitiveness, but also involved a significant change in the provision and funding of government services, including research.

Before World War II the comparative advantage of the agricultural sector was so strong that the industry grew with little government support. By the 1950s, following a period of strong demand for food and high prices for primary products, New Zealand was relatively rich, with the world's third-highest per capita gross national product (GNP; Johnston and Frengley 1991). However, agricultural commodity prices began to fall in the 1960s. In response and to compensate for the invisible costs imposed by import protection policies for manufacturing, the government provided the agricultural sector with grants, subsidies, export incentives, and tax concessions. A wide range of assistance to agriculture was introduced in response to Britain's entry into the Common Market and the Organization of Petroleum Exporting Countries

(OPEC) oil shocks in the early 1970s. Price and capital subsidies were provided to offset rising costs. Funding for research, extension, and quality control was increased. Tax incentives were provided to increase stock numbers, and price stabilization schemes were introduced that involved heavily subsidized loans to producer boards.

By 1984 assistance to agriculture amounted to about 34 percent of the value of output and 10 percent of total public expenditure and was recognized as contributing to the overall poor economic performance of the economy (Valdés 1993). The economy was highly regulated, with controls on interest rates, prices, wages, and the exchange rate; import licensing; capital and investment regulation; and producer board monopolies on trade in the most important agricultural commodities. The government was running a significant budget deficit and public debt had reached 51 percent of GDP (Bollard and Mayes 1993). Falling terms of trade, continued balance of payments problems, the declining value of the New Zealand dollar, low productivity, and growth rates below those of trading partners culminated in a major currency crisis, which marked the start of the reform process.

Public Funding and Provision of Agricultural R&D

Agricultural R&D in New Zealand traditionally involved a high level of government intervention in both the funding and provision of R&D; the two were seldom considered separately. By the mid-1980s the structure of science policy had remained virtually unchanged for over half a century. The major share of total investment was made by the government. Most public science funding went to finance the state's own science agencies, with a particular emphasis on agriculture. The balance went to universities, industry research associations (many of which involved the agricultural sector), and some specific areas, including medicine and social science.

The state chose to intervene in the market for R&D for a number of reasons, none of which were clearly articulated (Bollard, Harper, and Theron 1987). The standard justification for such a large government role was that, if left to itself, the market would underinvest in research. This market failure was variously ascribed to the riskiness of R&D, its long lead times, the inability of private investors to appropriate sufficient amounts of the benefits to repay their investment, and the indivisibility, or "lumpiness," of the capital requirements of the science industry. These arguments were seen as particularly strong in the agricultural sector, which was characterized by large numbers of owner-operated farms. A further argument was that the government should support R&D to build up strategic competitiveness. The number of agricultural producers, their importance to the national economy, and the consequent political power of the rural lobby may explain the historical pattern of agricultural R&D (Jardine 1986).

The bulk of science funded by the government was undertaken through its own science organizations, principally the Department of Scientific and Industrial Research (DSIR) and the Ministry of Agriculture and Fisheries (MAF). Relatively little research funding went to universities. Until the reforms, public-sector spending on agricultural research was focused largely on production R&D, but some processing R&D was publicly funded through contributions to private-sector research associations. Real agricultural research expenditures grew significantly from the 1920s until 1984 (Table 8.1).

In contrast to the average for countries of the Organization for Economic Cooperation and Development (OECD), where 16 percent of overall R&D expenditure was publicly funded, the New Zealand government financed 46 percent of all science. Despite the relatively high public contribution, overall research expenditures as a proportion of GDP remained low at less than 1 percent, compared with the OECD average of 2 percent. The largest amounts were spent on R&D in the agricultural, fishing, and forestry sectors, together with the environment (including soil and water conservation), meteorology, and geophysics. The DSIR estimated that 43 percent of government science work involved agricultural R&D and another 40 percent concerned other aspects of primary production and natural sciences.

Parliamentary appropriations were spent by science agencies on research activities as they saw fit. Traditionally, public-sector research managers had little incentive to evaluate the returns to R&D, identify high-payoff areas, and allocate budgets between competing projects. Nevertheless, the evidence on rates of return to agricultural R&D in New Zealand indicates that the return to public investment was in keeping with the returns estimated for other countries. Using the best available estimate for the period over which research benefits accrue (23 years), the average real rate of return for the period 1926–1984 was estimated as 30 percent per year (Scobie and Eveleens 1987).

Private Funding and Provision of Agricultural R&D

Although a limited number of agricultural industries invested in research without government subsidies, in others research associations were funded by levies on producers, processing firms, and matching government grants. The research was focused largely on processing and was carried out within each industry's research organization. The three principal research organizations were the Dairy Research Institute (DRI), the Meat Industry Research Institute of New Zealand (MIRINZ), and the Wool Research Organization of New Zealand (WRONZ). Levies on producers made up a large part of the income of MIRINZ and WRONZ. However, this investment by producer boards constituted only a small part of their overall levy income. In contrast to public funding, which was concentrated in the production sector, two-thirds of private research investment was in processing.

TABLE 8.1 Agricultural research funding, 1927–1984

Year Ending	Government	Research Associations	Universities	Total	
				Nominal	Real
		(New Zealand dollars)			(1984 New Zealand dollars)
1927	71,060	na	1,930	72,990	1,100,086
1928	73,444	na	3,682	77,126	1,155,919
1929	70,242	21,816	1,832	93,890	1,411,115
1930	86,464	27,304	7,608	121,376	1,866,091
1931	77,502	22,548	9,044	109,094	1,818,120
1932	53,596	28,936	6,474	89,006	1,603,006
1933	55,948	25,896	23,700	105,544	2,005,336
1934	50,542	20,576	23,550	94,668	1,770,440
1935	60,404	26,148	24,582	111,134	2,004,902
1936	61,824	27,804	34,412	124,040	2,164,841
1937	95,752	39,976	34,238	169,966	2,780,696
1938	140,640	49,352	65,252	255,244	4,052,140
1939	301,120	77,368	946	379,434	5,783,805
1940	280,954	74,124	1,890	356,968	5,204,448
1941	na	na	na	556,218	7,832,368
1942	na	na	na	636,097	8,661,265
1943	na	na	na	406,748	5,421,283
1944	na	na	na	472,780	6,185,971
1945	na	na	na	548,512	7,081,753
1946	589,676	44,252	20,578	654,506	8,379,554
1947	846,864	73,212	21,302	941,378	11,689,231

1948	1,126,082	107,224	23,450	1,256,756	14,465,518
1949	1,257,304	113,112	36,370	1,406,786	15,918,816
1950	1,555,522	133,460	41,990	1,730,972	18,549,096
1951	1,621,880	107,732	38,958	1,768,570	17,073,870
1952	1,868,772	130,468	42,498	2,041,738	18,293,699
1953	1,823,084	144,904	36,924	2,004,912	17,187,710
1954	1,856,136	166,868	43,532	2,066,536	16,930,428
1955	2,042,126	190,288	51,136	2,283,550	18,261,583
1956	2,206,998	223,280	54,728	2,485,006	19,199,224
1957	2,522,124	300,216	53,992	2,876,332	21,767,496
1958	2,690,654	258,564	50,752	2,999,970	21,736,091
1959	2,847,356	291,644	54,752	3,193,752	22,295,926
1960	2,977,802	306,400	60,100	3,344,302	23,180,815
1961	3,423,262	319,044	65,000	3,807,306	25,920,642
1962	3,699,232	537,800	58,598	4,295,630	28,502,768
1963	3,827,448	616,880	67,940	4,512,268	29,322,901
1964	4,072,746	736,708	61,794	4,871,248	30,616,008
1965	4,881,298	852,560	138,620	5,872,478	35,694,540
1966	5,613,848	1,177,336	101,230	6,892,414	40,761,097
1967	6,335,357	1,367,904	60,000	7,763,261	43,306,145
1968	7,016,308	1,097,624	70,200	8,184,132	43,763,053
1969	7,558,786	1,234,138	102,000	8,894,924	45,324,777
1970	7,776,029	1,592,444	104,950	9,473,423	45,299,956
1971	9,554,436	1,726,168	110,950	11,391,554	49,341,913
1972	11,818,780	2,424,626	95,650	14,339,056	58,093,506
1973	13,372,000	2,390,400	126,300	15,888,700	59,491,023

(*continued*)

TABLE 8.1 *Continued*

				Total	
Year Ending	Government	Research Associations	Universities	Nominal	Real
1974	16,160,000	2,497,000	179,040	18,836,040	63,533,838
1975	21,631,150	3,045,600	229,800	24,906,550	73,223,207
1976	24,418,450	3,351,000	270,300	28,039,750	70,418,083
1977	26,577,710	3,672,400	390,300	30,640,410	67,269,542
1978	31,056,470	4,322,000	434,500	35,812,970	70,236,418
1979	37,225,460	5,114,000	549,200	42,888,660	74,032,681
1980	40,795,290	5,527,000	714,500	47,036,790	69,284,707
1981	51,052,511	7,052,000	846,100	58,950,611	80,606,705
1982	65,265,000	7,590,000	848,950	73,703,950	91,604,213
1983	71,366,077	8,438,000	1,044,950	80,849,027	84,360,095
1984	75,282,729	8,988,000	953,000	85,223,729	85,223,729

SOURCE: Scobie and Eveleens (1987).

NOTE: na indicates not available.

Research Policy and Funding since 1984

Macroeconomic Conditions in 1984

The reforms that began in 1984 had their origins in unsustainable policy interventions and poor economic performance. They fundamentally transformed the manner in which agricultural research was financed and conducted. The institutional arrangements determine how much is invested in research, what research is undertaken, by whom and for whom it is conducted, what the returns can be, and who benefits. In other words, they influence both the efficiency and equity of the research process. They also shape incentives and thus the behavior of all the parties involved: politicians, scientists, industry groups, and bureaucrats. A full appreciation of the causes and effects of the reform of agricultural research therefore requires an examination of the macroeconomic conditions that made change inevitable and the theoretical framework that underpinned it. These are discussed here, together with an assessment of the effects of each of the major stages in the reform process.

By 1984 it was generally recognized that structural economic reform was required. Since the mid-1970s overall economic performance had been poor and agricultural growth low despite a raft of subsidies, regulation, and stabilization policy interventions. In June 1984 a new Labour government was elected and began a rapid program of economic liberalization and structural reform that has continued sporadically, under different governments, until the present.

Theoretical Framework of Reform

The preferential policies that had supposedly favored agriculture reflected a widely accepted role of state intervention. However, by 1984 a new conventional wisdom was questioning the efficacy of public-sector performance and regulation. The Western world was adopting more liberal economic policies. Ideas originating in the writings of Milton Friedman and F. A. Hayek, who argued that a free market is the cornerstone of political freedom and economic prosperity, and the analysis of new microeconomic models provided a conceptual framework for the reforms.

Bollard and Mayes (1993) identified four new theories as being of fundamental importance in informing the reform process.

- The theory of public funding based on public-choice theory. This approach recognizes the importance of transaction costs, bureaucratic and government failure, and the structure of incentives facing individuals. It suggests that government should be smaller and rejects most of market failure as grounds for government intervention. It indicates that provision should not be an automatic solution for the problems of market failure, but that public funding of services provided by the private sector could overcome the problems of "government failure." This theory highlights the need to separate policy advice from regulation and for accountability and transparency in all forms of government.

- The theory of ownership based on principal-agent theory. This suggests that the incentive structures facing firms make them typically more efficient at providing services than government bureaus. It implies that the government should not own trading activities that could be provided by the private sector and looks to corporatization and privatization to improve efficiency.
- The theory of public provision. This suggests that government provision of goods and services crowds out the private sector. It indicates that with the withdrawal of public provision and a reduction of the tax and regulatory burden on firms, the private sector can provide goods and services more efficiently.
- The theory of regulation. This suggests that existing or potential competition in the market can regulate firm behavior more effectively than government control.

These microeconomic theories provided the basis for the deregulation of factor and product markets, as well as the reform of the public sector. In 1985 the government embarked on a program to reorganize state-trading activities into state-owned enterprises based on the following general principles (Duncan and Bollard 1992).

- Noncommercial functions should be separated from trading functions.
- The state should specifically contract with the enterprise to provide any noncommercial activities.
- Regulation and policy advice should be separated from trading functions.
- Managers should run trading organizations as business enterprises.
- The enterprise should operate without competitive advantages or disadvantages.

Policymakers relying on these principles sought to improve the efficiency of state-trading activities by requiring state funding enterprises to imitate private firms. They provided a framework for the establishment of state enterprises in a wide range of activities, including the provision of agricultural research. However, the corporatization of R&D into CRIs only took place in 1992. The path to corporatization was gradual, beginning with reductions in funding in 1984 and cost recovery for public R&D in 1985. By 1990, however, the government had developed a coherent policy framework that was consistent with the political and philosophical basis of the reforms, with the establishment of a contestable pool for public funding of research through the FRST and the separation of policy advice with the establishment of the MRST.

Funding Reductions

By 1984 the case for reform of science policy was strong. Although evidence on rates of return to R&D in New Zealand was scant, it would have been difficult to argue that too many resources had been allocated to R&D, thus driving rates of return to unacceptably low levels. The evidence of more than adequate rates of return for agricultural R&D suggested that there were no gross shortcomings in the management and execution of science programs. However, there was potential for improving both the efficiency and equity of publicly funded research. Bulk funding was granted to the government's own research organizations through parliamentary appropriations on their advice. The organizations' accountability for those funds was through the minister in charge. The state was the owner of science organizations that provided research services to both the private and public sector. There was little, if any, attempt to set research priorities among research projects or research organizations in a way that maximized net social benefits. There were few incentives for scientists to cater to the market demand for research. The bulk of agricultural research that benefited the farming sector was funded by taxpayers. In comparison with other OECD countries, there appeared to be significant underinvestment by the private sector.

The first steps toward a major restructuring of public research were taken in 1984 and 1985 in concert with the structural adjustment and liberalization of the economy and the reorganization of the public sector. However, the initial stages did not involve any clearly articulated policy analysis. Rather the government introduced a series of funding cuts, with graduated cost-recovery beginning in 1985. By 1988/89 the overall amount allocated via MAF and DSIR had been reduced by 19 percent and the principal recipient of public funding for agricultural research, MAFTech, was expected to obtain more than 14 percent of its funds from external sources. This period of funding reductions was associated with a disruption in the collection of data related to research. The best estimates of public research funding of agricultural research to MAF and DSIR are provided in Table 8.2. No figures are available for agricultural R&D funded through research associations or universities. These estimates show a significant decline in funding in real terms over the period.

Although the reasons were never clearly articulated, the government probably instituted these funding reductions for reasons of both equity and efficiency. Policymakers perceived an overinvestment of public funds that crowded out the private sector, as well as inefficient management in government departments. They assumed that the funding cuts would force managers to weigh more carefully the nature and extent of the costs and benefits of research. Constraints on funding would sharpen the focus of research and increase productivity. Payments from users would orient research toward the needs of the market and improve accountability and transparency. Further-

TABLE 8.2 Agricultural research funding, 1985–1989

Year Ending	Government	Research Associations	Universities	Total Nominal	Total Real
	(1984 New Zealand dollars)				(1984 New Zealand dollars)
1985	81,014,000[a]	na	na	81,014,000	72,524,598
1986	81,064,000[a]	na	na	81,064,000	63,272,519
1987	94,904,000[a]	na	na	94,904,000	62,649,046
1988	101,625,000[a]	na	na	101,625,000	61,535,522
1989	85,733,000[a]	na	na	85,733,000	49,936,753

SOURCE: Sandrey (1990).

NOTE: na indicates not available.

[a]Estimate.

more, the size of the public sector and its deficit were increasingly seen as necessitating the search for greater public efficiency. Cutting government funding to encourage private investment would allow the savings to be used to reduce the deficit without reducing the overall research effort. In addition, the cuts were aimed at increasing the equity of public funding. Taxpayers were funding a very large share of the total R&D investment, while specific groups, industries, and sectors were enjoying the benefits. The agricultural sector in particular reaped the benefits of public funding of research. A reduction in public funding would reduce the burden on taxpayers and induce greater private research investment.

At the same time, reforms provided little overall direction to science policy. These reductions, and a focus on the "user-pays" principle became the sole, rather simplistic instrument for achieving reforms. The institutional framework was left largely untouched; the legal and commercial rules that had been designed for the operation of government departments continued to apply, despite new commercial demands on the agencies to obtain contributions from the beneficiaries of research. Also, although the private sector was assumed to be underinvesting in research, no mechanisms were established to encourage individual firms or industry groups to make up the shortfall.

The results of this policy were predictable. Merely cutting public funding was a blunt instrument for achieving the multiple needs of reform. Total funding fell, and new capital investment and recruitment dried up. There was a "brain drain" of competent scientists overseas; scientists shifted their attention from research to finding "clients" who would pay and were redeployed as managers and marketers. Public funds were used to cross-subsidize commercial activities, with one state agency often competing with another. The sale of research services was largely driven by supply rather than market demand.

Despite the requirement to act commercially, research agencies remained limited in their ability to behave like private firms. They could not take risks that compromised the Crown, raise equity or capital, or enter into joint ventures. Their flexibility was hampered by their traditional bureaucratic and hierarchical nature. At the same time, they were able to compete unfairly with private suppliers of R&D, thus reducing the overall efficiency of the national research effort. Furthermore, nothing was done to ensure that public funding was directed to areas of high social payoff. Accountability remained limited to that of any minister responsible for a parliamentary appropriation.

The effect of the funding reduction on private-sector investment is unclear. One of the casualties of the reform process was data collection: there is scant information on the private funding of agricultural R&D during this period. It is probable that in the past, some private-sector investment had been crowded out by generous public funding. Yet the political and policy uncertainty that accompanied the reform process probably discouraged all private-sector investment, including R&D investment. In R&D in particular, lobbying by the scientific community—who exhorted the government to implement the recommendations of the Beattie Report (1986), which called for increasing spending on research—may have added to the uncertainty and further limited private investment. In addition, the funding reductions coincided with a period of high interest rates and the share market crash of 1987, which affected the New Zealand economy particularly badly. At the same time, the agricultural sector was undergoing substantial reform that involved the removal of subsidies, which left farmers in a weak position to invest in research. The emerging venture capital market was also not fully developed, thus making it difficult for risky investments such as R&D. In this climate, private-sector funds for R&D investment were limited.

The real costs of the reforms were probably significant. The loss and inefficient use of human capital, coupled with commercial, bureaucratic, and labor constraints and the likely shortfall in private-sector investment, almost certainly led to a reduction in the productivity of the public research system. By 1989 it was estimated that science capacity had been reduced by 20 percent in just four years (Shaw 1991).

Yet the cuts were not without some benefits. Along with the introduction of the "user-pays" principle, they led to some improvement in research management. Research managers became more conscious of costs, and accountability in the use of resources was improved. In the search for a "user," the extent, nature, and appropriability of the benefits of research were considered by managers who became more active in marketing their research. The adoption of the "user-pays" approach may also have dispelled, in part, the notion that agricultural research should be funded by government and that the results should be available gratis to all and may have begun to instill the perception of research as an investment.

Policy Reviews

A lengthy period of uncertainty and adjustment accompanied the funding cuts of the 1980s; the belief that science was in a crisis was widespread among the scientific community. This uncertainty was not alleviated by the numerous reviews of the public funding and organization of science that took place between 1986 and 1989. These included:

- "Key to Prosperity: Science and Technology," a report by Sir David Beattie (1986) that reviewed the role of government in science and technology;
- "Research and Development in New Zealand: A Public Policy Framework," a report by Alan Bollard and David Harper (1987) for the New Zealand Institute of Economic Research; and
- "Science and Technology Review: A New Deal," a report by the Science and Technology Advisory Committee (STAC 1988), which was set up following the Beattie Report.

The Beattie Report, which recommended a doubling of research spending within seven years and the introduction of the 150 percent write-off of research expenditures, became highly contentious. Rather than showing the way forward for public research, the report was viewed as having been captured by the science lobby in an attempt to turn back the clock and justify increased taxpayer funding. The report failed to reflect the new microeconomic reformist thinking driving changes in the approach to the role of government. The report was based on implicit assumptions that science and technology are desirable in themselves and, based on scant empirical evidence, that science and technology contribute to growth that benefits the whole economy.

The Bollard and Harper Report focused on the economic analysis of the appropriate role of the state in funding and providing R&D. It rejected most of the traditional market-failure arguments for government funding and provision and used transaction cost analysis to establish that different types of intervention were appropriate for different types of transactions.

Unlike the Beattie Report, the STAC document reflected the new microeconomic analysis, which was based on ideas of contestability and accountability, public ownership and provision, and the appropriability of research benefits. The proposals explicitly recognized that some of the benefits from research can be captured by the private sector: public-sector funding of research was therefore to be limited to those areas in which the private sector was unable to capture sufficient of the benefits to warrant investment in R&D. The comprehensive set of reforms recommended in the report were based on the following principles:

- there should be a clear separation of policy advice from the allocation of public funding and the provision of R&D;
- taxpayer funding should be allocated to "public-good research" and private groups should fund research where they capture the benefits;
- the government should purchase R&D services on a contestable basis; and
- public agencies should have the full range of commercial powers to act in a deregulated environment.

STAC recommended the establishment of a ministry to provide policy advice and develop national research priorities and an independent council to manage a contestable funding system in which public research funding was allocated among competing providers. Funding would be provided from current appropriations to government departments.

These recommendations received widespread support in the business and science communities. The government implemented them in 1989 with the establishment of the MRST responsible for policy advice and priority setting and the FRST responsible for managing a contestable fund to buy research from science providers in accordance with national priorities and government's research needs.

Contestable Funding

The establishment of a system of contestable funding through the FRST was a major policy development. FRST was to act as the government's agent in purchasing R&D services for research that generated widespread public benefits. Public-good research was considered to be research producing public benefits that, if not funded publicly, would not be undertaken or would be underprovided. Restricting public funding to such areas, while leaving firms and industries to fund research with private benefits, was designed to make research funding more equitable. Separating funding from the provision of research and allowing competition between research agencies for funds was intended to make the use of public funds more efficient.

Public funding would be allocated to the PGSF administered by FRST, which would purchase research outputs from government-owned research agencies, research associations, universities, and others on a competitive-grants basis. The PGSF was scheduled to increase incrementally from 20 percent of total government funding in 1990 to 50 percent in 1993, with funds pooled from the monies previously appropriated to DSIR, MAFTech, the Ministry of Forestry, and the Meteorological Service. In addition to the PGSF, part of government investment was channeled through government departments as operational funding to fund departmental policy research.

From the outset, the notion that government intervention should be limited to funding purely public-good research was compromised. It is difficult not to see the expansion of FRST's activities as the result of successful capture by industry and science lobbies. As soon as the establishment of FRST was announced, the definition of public-good science was a subject of considerable debate among the science community as providers sought to maintain access to the pool of funding. The definition eventually used in the legislation setting up the foundation was broader than the narrow economic view of public goods. Public-good research was defined as that which was likely to increase knowledge or understanding of the physical, biological, or social environment; likely to develop, maintain, or increase research skills or scientific expertise of particular importance to New Zealand; or of benefit to New Zealand but unlikely to be funded or adequately funded from nongovernment sources. This approach permitted a wide range of existing research activities aimed at increasing agricultural production to be defined as public goods.

The foundation has also managed a number of programs to promote innovative activity and business growth. The Technology for Business Growth scheme was intended to encourage industry R&D and improve international competitiveness through technological development. This and other programs effectively involve public subsidies to the private sector to stimulate private investment. These transfers are anomalous in an environment in which subsidies have been removed from other areas, particularly from agriculture, and reflect the ability of the science lobby to garner public subsidies not available to other industries.

Priorities for Research

The creation of the MRST was an attempt to separate the giving of policy advice to the minister from the agencies receiving public funding and providing R&D. The primary function of MRST was to establish national research priorities, thereby providing a framework within which FRST would allocate research funds.

In 1991 the government announced that the PGSF would be set at a minimum level of NZ$260 million for the next five years. It also established a Science and Technology Expert Panel to define research priorities over this period. The panel was largely drawn from providers and purchasers. In determining these priorities the panel took into account not only the ability of the private sector to capture benefits and the appropriateness of government funding, but also the socioeconomic importance and benefit to New Zealand of each output class as well as its R&D potential and capacity (Palmer 1993). The panel moved away from the principles of the STAC Report. Rather it focused on the use of science to promote sustainable growth and job creation by adding value to production and processing. To this end it recommended a change in emphasis from cost reduction and volume increases to improving the quality of primary products, to market-oriented processing, and to product development. The process involved wide consultation with both provider and user groups, which afforded them the opportunity for intensive rent seeking. The private sector, particularly in primary industries, had been concerned that the government would withdraw its support if private research investment were increased.

In response the government announced that it wished to maintain funding in areas where the private sector had funded heavily but would be more wary of investment where there had been little such investment. More important, the priorities were set in accordance with the government's politically dictated goals. The economy at this time was in a recession, with unemployment at record levels. The government's stated research strategies clearly reflected the political need for economic growth and increased employment, focusing as they did on adding value to primary production industries.

The overall funding for the primary production and processing sectors remained virtually unchanged, with some funding being shifted from production into processing (Science and Technology Expert Panel 1992). More funding was directed into "wealth-creating" outputs in sectors where New Zealand was seen to have an existing comparative advantage. Increased funding was directed to fisheries and forestry. These sectors had enjoyed significant output and export growth and past public support had been low. In the other outputs, funding was shifted from the physical and natural sciences into infrastructure.

Priority Setting and the Public-Good Principle

It is difficult to view many of the programs that have been funded as being in the public good, especially when the public is interpreted as the New Zealand taxpayer. For example, a program on the moth-proofing of wool enjoyed public funding of NZ$216,000 in 1994/95 alone, yet the primary beneficiaries are New Zealand producers and foreign consumers of wool carpet (Jacobsen 1995b). Nevertheless, the program proposals put forward to the FRST do not require any specific economic analysis of either the magnitude or the distribution of the expected benefits of research. The information against which the projects are assessed is primarily a peer review of scientific merit (as measured by the attributes of the suppliers of research) rather than the needs of the public. The implication is that the taxpayer is underrepresented. The potential for funding to be supply driven suggests that there is an inherent tendency for the capture of the process of funding by suppliers.

A striking feature of the funding pattern is the dominance of the allocation to CRIs, which emphasizes the continued hegemony of the government's science agencies. They are few in number and relatively large, and their transaction costs in lobbying FRST are relatively low, while the public users of research results are small and diffuse and frequently unaware of the potential benefits they are forgoing. However, there is a natural common interest between science providers and the agricultural sectors that they have traditionally served. Each has incentives to emphasize the importance of the research in those industries to achieving the government's goals, whatever they may currently be: environmental sustainability, value-added, or even competitive advantage.

In the absence of evaluations to assess the returns to public investment, the portfolio tends to be driven largely by the interests of the domestic science industry and its traditional client groups. FRST has not relied on ex ante economic analysis in screening proposals. It therefore has no means of ensuring that the portfolio will maximize the social return. Because a large proportion of research proposals will contribute both public and private benefits, there is no way for FRST to determine these or whether the private returns are sufficiently large to ensure that the research would be funded by the private sector in the absence of public investment. Nor is there any explicit mechanism to determine how each research program will contribute to the government's stated goals. However, AgResearch (the CRI for pastoral agriculture) has recently provided additional information to FRST on the expected benefits of projects to bolster its applications (Jacobsen 1995a). The government's 1996 budget also announced that ex post economic evaluation of public investment in R&D would be undertaken.

Contestability is limited to competition among domestic science providers. Rather than vigorously seeking tenders on a worldwide basis to ensure true competition, science of the highest quality, and the greatest return for the public investment, FRST favors domestic applicants. This exclusion of foreign competition is equivalent to an infinite level of tariff protection for science at a time when other industries are increasingly exposed to the international market. The argument that a domestic science capability should be maintained is enshrined in the government's statement of science goals issued in 1994. Yet in an environment of increasing international trade, it is not clear why a domestic capability should necessarily be maintained. Those services could be purchased internationally from the lowest-cost suppliers without compromising the quality or relevance of the outputs. There are surely areas of research that are highly specific to New Zealand, in which there is value in having a domestic capability; but it is in precisely those areas that domestic suppliers would have a comparative advantage in any event. They would flourish without the need for protection.

Although a contestable funding system does provide advantages in terms of increased competition among providers and greater transparency, the system adopted by FRST has a number of disadvantages. It can be time consuming and expensive for applicants, even though FRST has moved much of its funding to a multiyear basis. The system can bias contracts toward particular providers, and, most important, it can be deflected by ad hoc political pressures unrelated to the efficiency of the investment.

A Second Priority-Setting Exercise

A second priority-setting exercise was launched in 1994 for the five-year period from 1996 to 2001 by the Science Priorities Review Panel. The economic environment facing the government was substantially different from that which had prevailed earlier. The fiscal outlook was more favorable, the economy was growing, and unemployment was falling. The government therefore announced an increase in the PGSF and set an overall target for public investment in science of 0.8 percent of GDP in 2010. The Science Priorities Review Panel used a scoring method to assist in the overall priority-setting process and considered six factors determined by the government (see Table 8.3). It used a Delphi technique to elicit scores and weights from a consultative group of about 100, with two science providers and two users for each output group and the remainder made up of scientists with no particular affiliation and contributors representing other interests. The strategic importance of each output was scored in terms of its contribution to the government's goals, which had been expanded in the economic, environmental, and social areas (MRST 1994).

Overall the nominal amount of the PGSF was to increase by 28 percent over the five-year period. The amount allocated to animal industries increased very slightly in nominal terms because funding had been relatively high in the past. On the other hand, the funding for dairy research was scheduled to grow by 40 percent over the period (Science Priorities Review Panel 1995). The reasons given were that public funding had previously been low relative to other outputs and that funding should be increased to enhance the sector's strategic importance, despite the high levels of appropriability.

Inevitably there are drawbacks inherent in any scoring system. Most important, the results of a scoring system cannot be used to determine trade-offs between investment in different areas of research. Thus it is simply not possible to decide whether an additional dollar spent on dairy research would generate a greater return than if it were spent on research on animal industries. With no notion of a research production function, there is no way of knowing whether additional investment where past funding was low will generate more or better results or vice versa. Nor is there any means of determining whether the amount invested in research is optimal—why should 0.8 percent of GDP be the "right" amount? Yet clearly the actual amount voted by the government is politically determined and subject to economic fluctuations.

Neither of the priority-setting exercises was undertaken from a strong theoretical or methodological base. Both faced difficulties in translating the goals of government into research objectives. At the same time these goals were generally inappropriate for agricultural research, in particular, because

TABLE 8.3 Factors considered by the Science Priorities Review Panel in priority setting, 1996/97 to 2000/01

Factor	Explanation
1. Strategic importance	The potential contribution of the output to achieving the government's economic, environmental, and social goals
2. Potential of science	The likelihood that research will achieve results
3. Potential of users to capture benefits	The extent to which users of research in the output area will capture the benefit and their timeliness in doing so
4. Research capacity	The quantity and quality of resources available in New Zealand to support current and future research
5. Research intensity	The extent to which the sector is dependent on investing in research for its success
6. Appropriateness of PGSF funding[a]	The extent to which research should be funded from the PGSF compared to other sources

SOURCE: Science Priorities Review Panel (1995).
[a]PGSF, Public-Good Science Fund.

research is generally a blunt and ineffective tool for objectives other than economic efficiency (Alston and Pardey 1996). Inappropriate and incompatible objectives, together with overlapping criteria, generally combine to produce meaningless results from scoring systems (Alston, Norton, and Pardey 1995:Chapter 7). There can be little assurance that the suggested allocations have increased the social rate of return from the taxpayers' investment.

The government goals have clearly expanded beyond the original intention to fund research only where the private sector underinvests. They no longer have an economic rationale but instead appear to be principally politically driven. Witness the changes with each priority-setting round. Whether research is a public good is not used as a filter; rather it is secondary to the "strategic importance of the industry." There is no analysis of why science, rather than some other, more targeted policies, would achieve the government's goals, such as the support of Maori development aspirations. There has been scant examination of the reasons why the private sector would not fund research that would enhance its competitiveness, and no policies have been developed that would allow private or general underinvestment in research to be addressed.

Priority setting for research has become an explicitly political process. The government now has the ability to intervene politically in setting the research agenda in ways that were not possible under the bulk appropriations structures. So too is there additional scope for rent seeking by other interested parties, such as scientists, industry leaders, and environmentalists, at the goal-setting stage. Scientists have every incentive to stress the importance of science in generating benefits for the country. Managers of R&D funds collected from growers have every incentive to steer funding away from certain areas, thus making them appear to require government support.

The attempts by New Zealand to develop a comprehensive priority-setting system for all public research investment outside the research organizations themselves represents an important advance. One can be critical of the shortcomings, but it is a vast step forward from the days when historical precedence was the single most important determinant and advice on allocation came to the government from its own executing agencies.

Research Funding Patterns after 1990

The extent of agricultural research carried out by government, the private sector, and universities since 1990 is shown in Table 8.4. This information was obtained by means of surveys and shows that the business sector, which includes research associations funded by levies on producers, carried out a significant proportion of overall agricultural research. Overall half of the R&D in the business sector was carried out by just 3 percent of the private enterprises that perform research. Research associations alone carried out 23 percent of all research in the private sector. Research done in the private sector continues to be concentrated in processing. Of the NZ$75.4 million spent intramurally by the business sector in 1991, NZ$29.3 million was in dairy-processing research; NZ$10.4 in meat processing; and NZ$14.6 million in fiber, textiles, and skin processing. However, research carried out in the government sector is focused on production, with small amounts devoted to processing. It should be noted that Table 8.4 is not comparable with Tables 8.1 and 8.2, which show agricultural research expenditures by the source of funding. Table 8.4 shows the extent of research carried out within each sector, regardless of the source of the funding.

Crown Research Institutes

The first phase of the reform process focused on the separation of policy advice, funding, and provision of science. The second phase concentrated on the organizational structure of the government's research agencies. The STAC Report had recommended their reorganization into commercial entities. The traditional bureaucratic organization of science hampered efficiency in research and extension. A new, commercially oriented structure was seen as providing science more responsive to the needs of client groups and the priorities of the government.

CRIs were set up under the Crown Research Institutes Research Act 1992, with each institute covering a sector or group of natural resources (see Table 8.5). They were formed from the reorganization of government research, prin-

TABLE 8.4 Agricultural research activity, 1990–1994

Year Ending	Government	Business	Universities	Total: Nominal	Total: Real
	(New Zealand dollars)				(1984 New Zealand dollars)
1990	119,482,000	68,102,000	na	187,584,000	102,079,655
1991	120,970,000	75,420,000	16,290,000	212,680,000	110,725,155
1992	119,662,000	78,754,000	16,290,000	214,706,000	110,865,573
1993	120,340,000	84,010,000	29,200,000	233,550,000	119,496,201
1994	125,000,000	96,112,000	22,062,000	243,174,000	122,802,870

SOURCE: *New Zealand Research and Experimental Development Statistics* (Ministry of Research, Science, and Technology, various issues).

NOTE: na indicates not available.

cipally the DSIR, MAFTech, the Forest Research Institute, and the Meteorological Service (Ministerial Science Task Group 1991). These agencies had accounted for more than 90 percent of the PGSF in 1990/91. Agricultural research formerly carried out by MAF and DSIR is now undertaken by the Pastoral Agricultural Research Institute, the Horticultural and Food Research Institute, the Institute for Crop and Food Research Limited, and Landcare Research.

The CRIs operate under their own statute, modeled on the State Owned Enterprises Act 1986, which in turn is drawn from the Companies Act. Each institute has a board appointed by the cabinet and is responsible to the shareholding ministers. In effect the statutory provisions reflect the intent that the CRIs operate as independent financial entities, with the Crown as the sole shareholder. Capital injections were envisaged to come from new investments by the owners, commercial debt financing, and retained earnings. The act provides for the shareholding ministers to direct the CRIs to pay a certain rate of financial dividend to the Crown as owner. This is negotiated annually with each institute.

Clearly the implication of any given profit or dividend rate depends on the capital valuation. When the CRIs were established, a capital value was struck. However, much of the asset base was in land subject to potential claims by Maori as they seek to settle outstanding grievances concerning Crown appropriation of land belonging to the indigenous people under the Treaty of Waitangi.[1] As a result, the managers of CRIs were in many cases not free to alter the

1. New Zealand was colonized by Britain under the Treaty of Waitangi, which was signed in 1840 between the chiefs of the indigenous people (the Maori) and the British Crown. Although the treaty guaranteed the Maori ownership of their lands, fisheries, and forests, these resources were gradually lost. Over the last decade, successive governments have begun to address and redress Maori grievances. For more details see Orange (1987, 1990).

TABLE 8.5 Crown research institutes

Primary sector
New Zealand Pastoral Agricultural Research Institute Limited
The Horticultural and Food Research Institute of New Zealand Limited
New Zealand Institute for Crop and Food Research Limited
New Zealand Forest Research Institute Limited
Industrial sector
Institute of Industrial Research Development Limited
Service sector
Institute of Environmental Health and Forensic Science Limited
New Zealand Institute for Social Research and Development
Resource sector
Landcare Research (NZ) Limited
Institute of Geological and Nuclear Sciences Limited
National Institute of Water and Atmospheric Research Limited

asset structures in ways that a private corporation could. The whole issue of the dividend rate that the CRIs should pay to the Crown is currently under review.

The Crown has the dual roles of funder and purchaser of research. The rationale given for Crown ownership, rather than privatization, was the need to maintain strategic research capabilities and assets in New Zealand in efficient and effective institutional forms; the need to guarantee the production of the outputs required by the government to a specified quality, relevance, timeliness, and price; and a statement that reliance on a purchasing mechanism alone would not permit the government to achieve its desired outcomes (Palmer 1993).

The CRIs are set up under a company structure with the minister of CRIs and the minister of finance as the shareholders. They are governed by boards appointed by the ministers. Board members are not appointed on the basis of sectoral interests. CRIs are funded through research contracts from both the private sector and FRST, together with nonspecific output funding amounting to 10 percent of their PGSF allocation from government. The nonspecific output funding can be deployed flexibly and can be used to finance noncontracted research. CRIs have full commercial powers, and the government does not cover their liabilities. However, the performance of the CRIs is monitored by the Crown Company Monitoring Advisory Unit, which also monitors the performance of the state-owned enterprises in other sectors.

As public agencies, the CRIs seek to emulate the private sector, and efforts have been made to ensure that they compete "fairly" with private providers. However, the fact that they can use the nonspecific output funding as they choose gives them an unfair advantage relative to other private-sector research providers. This very structure raises questions about why, if they are truly commercial, they require public ownership. If CRIs were truly commercial, they would flourish in their own right with no need for protection.

Before the reforms of 1984, MAF provided extension services, information, and advice free to farmers. The reforms saw the withdrawal of some services, which were filled by the entry of private-sector consultancies, and the eventual reorganization of MAF's services into Agriculture NZ, which offers extension, facilitation, and consultation services to the primary sector on a commercial basis (Lovett 1994). Farmers are now expected to pay for technology transfer but are apparently unwilling to do so. Consultants report spending less than 20 percent of their time on extension work. However, the CRIs are developing their own extension services, and the DRI employs consultants involved in extension.

Universities

University research has traditionally been integrated with teaching. Two of the seven universities have close links with agriculture. Massey and Lincoln Universities offer degrees in agricultural science. Academic staff as individuals, small groups, or part of research teams have been responsible for developing, managing, and carrying out research projects. Much of the research conducted within universities is done through internal university grants financed through the Ministry of Education. Research contracts are an additional source of funding (New Zealand Vice-Chancellor's Committee 1991).

Universities were initially excluded from the PGSF and were able to apply for funds only in collaboration with other eligible organizations. In 1993/94 the universities together contributed NZ$10.7 million to the PGSF, to which only they as a group had access. These funds were to be used only for research unrelated to teaching. The universities will have full access to the PGSF after 1996/97. There is some concern over competition with the CRIs for funding because, while the CRIs must price their bids according to their true cost, universities are able to cross-subsidize their activities.

Private Financing of R&D

The principal characteristic of agricultural industries is that there are many small producers. Market-failure arguments for government intervention in R&D frequently contend that in these circumstances there will be free-riding and underinvestment and that the costs of collective action prevent voluntary funding.

Yet many agricultural industries fund their own research. Often the organization costs are low because industry associations exist for other purposes as well, such as marketing. Free-riding may be tolerated, as it is throughout the marketplace, where the return to the investor is sufficiently high to make the investment worthwhile. The level of funding represents a rational choice by the investors, who choose among a portfolio of opportunities in deciding to invest in research. If they considered that marginal funds could be more profitably invested elsewhere, then there is no underinvestment.

Statutory marketing authorities are one mechanism by which the private agricultural sector has provided funds for research. They operate under industry-specific legislation that grants them significant monopoly powers, including the power to acquire the product; to control the form, destination, and price of exports; to regulate shipping; and to fund research. To fund these activities, marketing authorities were also granted the power to levy producers. The levies on producers are relatively substantial. In 1991/92 the levy income of the Meat and Wool Boards represented an estimated 11 percent of farm profits from sheep and beef farms. More than one-half of all New Zealand's agricultural exports are currently influenced by statutory powers. The boards still use their statutory power to levy funds for research. The research funding is principally directed to their own research organizations, MIRINZ, WRONZ, and DRI. These organizations are now also free to seek funds elsewhere and have also won contracts from FRST.

The optimal form of government intervention to support private research in the agricultural sector is currently under debate in New Zealand. There is no explicit justification for the use of the power of the state to force producers to fund research, nor is there an explicit mandate from the producers. The inherent problem with the structure of the boards is that they neither provide incentives for maximizing the returns to producers nor generate sufficient information for producers to assess the worth of their investment. Although the board is politically accountable to producers, this alone is insufficient to ensure that research investment maximizes the returns to producers. The very nature of the boards accounts for their structural weakness. Therefore, financing R&D through the use of statutory powers does not necessarily result in the optimal level of investment in R&D, does not create incentives for efficient R&D management, coerces funding on an ongoing basis without a clear or finite mandate, and is not directly accountable to producers. Nevertheless, existing evidence suggests that investment is highly satisfactory for producers (Scobie and Jacobsen 1994). The total investment by producers over the period from 1962 to 1993 was estimated to generate net present value of NZ$58 million in 1993 terms, even when only the benefits from some 12 selected projects were considered.

There may be some justification for some government intervention where agricultural producers cannot voluntarily fund R&D. The core issues are whether statutory authorities provide the most efficient institutional arrangement or whether other options, such as the use of the Commodity Levies Act 1990, would produce better outcomes for investors.

The Commodity Levies Act 1990 was established to facilitate the voluntary funding through levies of various activities, including R&D, by industry groups. Under the act, a referendum of producers is necessary before a compulsory levy can be imposed on them all. The onus is on those who propose a levy to show that voluntary funding would be impossible or impracticable or that those who did not pay under a voluntary system would derive unearned benefits. The rate of levy is decided by producers, who implicitly make an assessment of the relative rates of return from their investment in R&D and other opportunities. The levy expires after six years and must be renewed with a new referendum of producers, when the rate of the levy can be changed.

The Commodity Levies Act is increasingly being used to fund R&D by industries that had not previously made such investments. For example, the arable industries do not have a statutory board through which to fund their R&D. The reduction in government funding of research in this area, however, has recently necessitated the formation of the Foundation for Arable Research under the Commodity Levies Act. The levy is set at a maximum of 1.5 percent of the farm-gate value and is used for both promotion and R&D.

Conclusion

The market-failure view of R&D formed the basis for science policy in New Zealand for most of this century. This view was predicated on elaborate theories of underinvestment by the private sector, supported by a perspective in which the government was benevolent and omniscient and capable of providing research that benefited the nation as a whole. However, a view richer in political elements than the notion of market failure is surely necessary to explain the allocation of public-good funding for R&D. Public-choice theory sees the state as the locus of a distributional struggle between interest groups, including both bureaucrats and politicians. In this latter view, the funding institutions can well become subject to capture by science providers, while politicians respond to the needs of sectoral interest groups. On the other hand, the government has a degree of autonomy that permits it to pursue goals independent of the distributional interests of competing claimants. By reducing transaction costs, it can contribute to a net gain in social efficiency, in this case by reaching an optimum in the funding of research.

The reforms to R&D policy were one aspect of the process of liberalization and structural change that transformed New Zealand after 1984. The basis of the reform process was a coherent theoretical framework that permitted policymakers to reconsider the appropriate role of the state in providing services, including R&D. Changes in the role of the state in science were predicated on the need to achieve greater equity and efficiency in publicly funded R&D. They were based on new economic theories that progressed beyond simplistic notions of market failure. It was acknowledged that a range of government actions might be justified where the private sector was unable to capture sufficient amounts of the benefits of research to make investment worthwhile.

After an initial period in which funding was reduced to public research agencies in the expectation that they would make up the shortfall from private sources, a restructuring of the science industry was finally announced in 1989. The reforms were in three parts. First was the separation of policy advice from the provision of R&D, with the formation of the MRST, which is responsible for policy advice to the government and the setting of research priorities. Second, the FRST was established to manage a fund to be allocated to R&D that would not otherwise be funded by the private sector. Finally, the existing government science agencies were restructured into CRIs along the lines of private firms but with ministerial shareholding. In addition, the Commodity Levies Act has provided a vehicle for private-sector producers to fund their own research, in which the role of government is limited to providing the enabling legislation.

This new institutional framework, which provides increased attention to contestability and accountability, gives grounds for cautious optimism. However, it would be premature to conclude that a large dividend is imminent. There are a number of reasons for this.

- By its very nature, R&D is a long-term process. It could very well take decades before changes to the way that funds are allocated show up as higher net social yields to public investment.
- The process has involved substantial changes to institutions; this in itself is a slow process and will undoubtedly involve some false starts and corrections along the way.
- Publicly funded research has continued to be funded largely through state agencies, the CRIs. Despite an emulation of the private sector, they remain creatures of government protected from true competition.
- The new system of funding is more flexible than the bulk appropriations of the past. However, there is little reason to conclude that the incentives for capture by science providers have changed in any fundamental way.
- Whether the system encourages a higher proportion of private funding is far from clear. Little has been done to diagnose the causes of any private and industry underinvestment or to prescribe appropriate cures.
- Entrenched interests have proved sufficiently strong to preclude the development of a comprehensive system of priority setting and resource allocation that allows trade-offs across all areas to maximize the payoff to the taxpayers' investment.
- There has been no development of a system that would result in the investment of an optimal amount of public research. Public funding remains subject to political vagaries.

The direction of public funding is subject to the government's political goals. Rather than funding research only in those areas in which the private sector would not otherwise invest, at times research has been directed to stimulate growth and employment. Research has been used as a tool to achieve objectives for which other policies might be more suitable. There is no mechanism in place to prevent this from happening again.

Overall the reforms of the system of public funding of research were soundly based on economic principles. Public-sector funding was seen as necessary where the gains could not be appropriated by the private sector and where other policies were not more appropriate for facilitating private investment. The very title of the "Public-Good Science Fund" reflects this view. It was considered that efficiency and equity would be enhanced by the funding of public-good research and that an accountable and contestable funding system would permit that investment to be directed to the research areas with the highest expected payoff.

These initiatives have fundamentally changed the conceptual framework in which science is carried out in New Zealand. Science funding is now widely seen as an investment, and those who benefit are considered to be those who should pay. Ideas about how science should be organized have also changed. Notions of accountability, competition, transparency, and efficiency are dominant. National priorities for research are being set in a comprehensive way that permits all investment opportunities to be viewed through the same lens and assessed according to how they meet the government's goals. Public funding is being allocated among competing research providers to promote cost-effective research. The Commodity Levies Act provides agricultural producers with a mechanism for funding their own research. These innovative changes place New Zealand public research in the forefront of policy reform and may provide a sound foundation for an efficient and equitable research system.

Like any process of reform, a period of maturation supported by monitoring and evaluation will be needed to assess the true social payoff from this major investment in institutional reform. Overall the change to contestable funding for public-good science represents a promising development that has yet to prove its worth. Whether potential efficiencies are realized depends on how the institutional framework determines how much should be invested, which programs are funded, and who undertakes the research. The goals of government have moved beyond enhancing efficiency and equity by funding research where there was private-sector underinvestment to include politically convenient but vague objectives such as "value added," "competitiveness," and "sustainability." Much of the research funded by the PGSF seems to be appropriable and thus capable of attracting private funding. In this context, agricultural production and processing research continues to enjoy substantial public investment. Although the priority-setting exercises represent a significant improvement over the past, they have not yet addressed the difficult issue of how trade-offs are made between competing research areas to maximize the return to public investment. The method of allocating funds among projects is also less than ideal, as there is no requirement of applicants to present an economic assessment of the expected net benefits from the research. Finally, the system has endogenized ample opportunity for rent-seeking behavior by all concerned, in particular those who benefit from government largesse in agricultural research (Scrimgeour and Pasour 1994).

9 A Synthesis

JULIAN M. ALSTON, PHILIP G. PARDEY, AND
VINCENT H. SMITH

This book is about the evolution of agricultural science policy in developed countries since the 1950s, especially during the past 25 years. Policy principles were presented to provide a framework for discussion, before turning to a review of facts and figures and the stories behind them. Trends in public and private funding and expenditures on agricultural R&D for OECD member countries, in a global context, suggest a general picture of rapid expansion in public agricultural R&D between 1950 and about 1975, with subsequent slowdowns in public funding and expenditures on publicly executed research thereafter. In contrast private funding and execution of agricultural R&D has expanded quite rapidly worldwide, and especially in some developed countries, so that the private sector has become a much more important player in agricultural R&D. These changes in funding and expenditure on agricultural R&D have implied substantial shifts in agricultural R&D policy for some national agricultural research systems. The trends in public and private expenditures on agricultural R&D tell only part of the story, however. The allocation of public funds has shifted substantially away from more traditional research agendas that focus on enhancing agricultural productivity toward new research agendas that concern the environment, natural resources, human nutrition, food safety, and other issues. Moreover, many countries have made important changes in funding mechanisms and the management and execution of publicly funded research.

The nature and pace of change have varied among countries. The five case studies presented in this volume for Australia, the Netherlands, New Zealand, the United Kingdom, and the United States provide more detailed insights into the changes in funding and institutional innovations in agricultural research policy. Each of these countries is of interest in its own right, and together they account for more than 40 percent of total agricultural R&D expenditures by developed economies. In particular, the United States has the largest publicly funded national agricultural research system among all developed countries, while the other countries are important leaders and innovators with respect to agricultural research, agricultural science policy, or both.

Policy Developments in the Five Selected Countries

Australia, the Netherlands, New Zealand, the United Kingdom, and the United States are similar in some significant respects: they all enjoy relatively high per capita incomes and can generally be regarded as having some significant comparative advantage in agriculture. However, these countries differ substantially in terms of their structure of government, the nature and importance of agriculture in the economy, the sizes of their science sectors, and their place in the world. In spite of these differences, the five countries experienced remarkably similar patterns of change in their policies for funding, organizing, and managing agricultural R&D from the mid-1970s to the mid-1990s, but with some significant differences in the details of the changes.

Total Funding

In all five countries, total funding stalled or fell in real terms during the 1980s. However, except in New Zealand (where plans for increases in agricultural research budgets were forestalled by governmentwide budget crises), funding began to grow again in real terms in the early 1990s, although at slower growth rates than in the 1960s and 1970s. These developments were driven by generally tighter total government budgets in each country and changing attitudes in society toward science and agriculture and the role of the public sector.

Broadening the Agenda

The declining relative political and economic importance of agriculture in the economy, combined with increased interest in newer issues, has affected the research agenda in each of the five countries. These newer issues include the environment (in particular, soil erosion, water and air quality, and wildlife habitat), natural resources (especially fisheries and forestry), food safety, human nutrition, and food-processing technologies (sometimes in the guise of value-adding research). Alston and Pardey (1996:317–318) discuss the notion of "value added" as it pertains to research priorities and point out that proponents of value-adding R&D seem to be arguing that secondary activity (such as adding value to primary products) is intrinsically better in some sense than farming. Some proponents of value-added production have proposed other forms of government support for it, not just R&D. Cashin (1988) persuasively criticized this value-added movement from a more general, comparative advantage perspective.

The shifts in resource allocation are difficult to quantify and may have been uneven among the different countries because the general importance of environmental and food-safety issues seems to vary substantially among the countries as does the influence of the food-processing sector. However, the pressure seems to be continuing to shift resources in the newer directions in spite of little hard evidence that the returns to the research investments in these areas will be as great as those from investments in agricultural productivity research.

More Emphasis on Public Funding of Basic Research

In the Netherlands, the United Kingdom, and the United States over the past decade public agricultural research funding has been redirected toward a heavier emphasis on basic research, and in the Netherlands and the United Kingdom, public funds for near-market research have been cut substantially. The economic arguments for shifting in this direction are clear and have been enunciated in public reviews in several of the countries. What is not so clear is the extent to which the reported shifts of resources reflect real changes in research programs rather than mere relabeling. In Australia, the pendulum may have actually swung the other way, with more of the public agricultural R&D budget being controlled by research and development corporations.

Industry Levies

In parallel with the slowdown in the growth of general public funding for agricultural research and a shift away from near-market research, institutions have been developed to enhance industry funding of more-applied research. In particular, Australia, the Netherlands, New Zealand, and the United Kingdom have expanded their use of mandatory commodity levies, or checkoffs, to support commodity-specific, near-market R&D. The United States has not yet made any similar changes. Interestingly, only the Netherlands levies a general tax on all agricultural output to partially fund public agricultural research. This has been a longstanding policy.

Private-Public Roles

A related development, especially in Australia, the Netherlands, and the United Kingdom, has been an increase in the representation of industry groups on key public agricultural research oversight committees. This development has provided benefits, to the extent that scientists have become more focused on industry concerns in their research, but it may have also imposed costs in the form of some degree of industry capture of publicly funded research agendas. Some countries have made recent attempts to develop joint public-private research ventures in agricultural research, a practice that has a long history in the area of defense-related research. However, some have also moved toward implementing full-cost pricing for research performed by public research facilities on behalf of private-sector organizations, applying the principle that where research provides private goods rather than public goods, the private beneficiaries of the research should pay the full cost of its provision.

Competitive Processes

The Netherlands and the United Kingdom have sought to infuse more competition into the supply side of the market for research by partially or fully privatizing previously purely public agricultural research entities. These countries have shown a distinct shift away from the use of block grants toward competitive grants and an increased emphasis on using evaluations of research performance to determine future funding levels. These policies are more in line with procedures used in Australia. A more modest shift in this direction has occurred in the United States, where formula funding arrangements still play a significant although declining role. In all five countries, competitive processes also have been extended to allow a broader pool of potential applicants for agricultural research funds (including purely for-profit private research organizations as well as government agencies, universities, and not-for-profit research institutions).

Organization

Another important issue concerns the optimal structure of publicly funded agricultural research institutes. Changes in the economics of research imply changes in the optimal research organization. Particularly in the Netherlands and the United Kingdom, over the past 15 years public agricultural research facilities have been reorganized and rationalized through a process of mergers and privatization to exploit economies of scale, size, and scope. The pace of change has been much slower in the United States.

Causes of Change in Agricultural Research Policies

The changes in agricultural research policies over the past 20 years have been broadly similar among the five countries, largely because the changes resulted from similar developments in general political economic philosophies, budgetary pressures, the relative economic and political importance of the farm sector, the nature of science and scientific inquiry, and other elements of the policymaking environment. The five case studies all show that these changes in public agricultural R&D policy have been formulated in the context of governmentwide policies and, increasingly, science and technology R&D policies, which also have been heavily affected by changes in general government policies and political economic philosophies. Similar changes in political economic philosophies and government policies in the five countries have led to similar shifts in agricultural R&D policies.

In all five countries, relatively market-oriented governments came to power in the late 1970s and early to mid-1980s. These governments reduced the growth rates of government spending on R&D in general and on science and agricultural R&D in particular. These governments also tended to adopt a more "economic rationalist" approach to other elements of R&D policy by increasing competition among researchers; increasing the accountability of research institutes and university academic departments tied to future block-grant funding; increasing the responsiveness of scientists to client needs by shifting more of the responsibility for near-market research to commodity-specific industry councils; increasing industry representation on the boards of funding agencies and funding committees; and, in the Netherlands and the United Kingdom, privatizing some research institutes with more-applied research agendas, such as seed development and horticultural institutes.

Links between shifts in governmentwide policy, general science policy, and agricultural R&D policy are stronger in some countries than others, and seem to depend on differences in the structure of government. Australia and the United States both have federal-state systems of government in which states have substantial autonomous rights to implement spending programs and, in the United States, to raise taxes. In both of these countries, public agricultural research policy, at least with respect to funding, seems to be less influenced by shifts in general national government policy. In contrast, the Netherlands, New Zealand, and the United Kingdom have much more centralized systems of government. In those countries, especially New Zealand and the United Kingdom, the link between general government R&D policy and agricultural R&D policy seems to be stronger. The fact that government decisionmaking is more centralized has meant that the pace of policy change for agricultural R&D has been more rapid in these three countries than in Australia and the United States.

Other forces have also played important roles in generating change in agricultural research policies. In richer countries, the role of farming generally has declined in relative importance in the economy as a whole, as well as in the production and marketing chain that extends from natural resources to the ultimate consumer of the products of the food and fiber industry. Thus the scope for agricultural R&D that was once devoted primarily to the farming sector has been expanded to include a broader range of prefarming and postharvest technologies. In addition, science itself has changed. The opportunities for agricultural research are different as a result of general developments such as modern biotechnology.

Another element of change concerns what people, and therefore policymakers, care about. In richer countries, as per capita incomes have risen over time, demand has grown for goods with larger income elasticities—such as environmental amenities, food safety, variety, and convenience. These developments have all had implications for the economic and scientific opportunity for different types of agricultural R&D and the changing demands for production of different types of R&D in the public sector. As a result, the environment, food quality, and the economic and social vitality of rural communities have become more important issues in the agricultural research agendas of developed countries. Pressures from groups interested in these issues have led to considerable change and demands for further change in the balance of agricultural research conducted and funded by the public sector. It should be noted, however, that whether such reallocations are justified on the basis of relative social returns is less clear.

Conclusion

Agricultural research policy does not develop in a vacuum. In particular, where government spending and taxation authority is highly centralized (as in the Netherlands, New Zealand, and the United Kingdom), shifts in political economic philosophies and general science policy tend to result in similar changes in agricultural research policy, although agricultural policy is also influenced by sector-specific events. Where the authority to tax and spend is more decentralized, as in Australia and the United States, the links between changes in political economic philosophies and general science policy and agricultural research policy seem to be less pronounced.

The fundamental forces for change in agricultural R&D policy in one country are likely to be shared with other, similar countries. This means that it is probably useful for any one country to pay attention to its neighbors, learn from the institutional experiments conducted in other countries, imitate their successes, and avoid a repetition of the mistakes made abroad as well as at home. However, we have emphasized that countries also differ in important ways in their fundamental political institutions, infrastructure, and the organization of the agricultural sector. Thus it would be folly to assume that all countries should adopt similar agricultural R&D funding and institutional policies.

Unfortunately, assessing whether research policies work requires a long time (because their effects take place with long lags) and a great deal of information. Although it is possible to learn much from the rich tapestry of international patterns of research institutions and investments, this book represents just one step in collecting and assessing the information needed. Currently we can only make fairly general, although useful, international comparisons of productivity patterns and their links with research policies, institutions, and investments. The ultimate measure—net social benefits—is beyond our immediate reach and likely to remain so for some time.

Most of the institutional changes in agricultural R&D policies documented here have been relatively recent, whereas the effects of research on measures of economic output are visible only after lags of many years. Clearly it is too soon to make any definitive calls on whether these changes have been good or bad for society as a whole. Whether the reductions in the rate of growth of public agricultural R&D funding have been appropriate is also something of a moot point. Most studies have estimated social rates of return to public agricultural R&D as very high, between 30 and 100 percent per year. Some recent work has raised questions about those estimates, although no major study has indicated that rates of return are unacceptably low (see Chapters 2 and 3). Thus the only reasonable response to the question of whether much has been gained or lost by slowing down the growth rate of public agricultural R&D funding and reorganizing the management and structure of public agricultural research is that we will have to wait and see. Policy choices and change will nevertheless take place in the meantime. We hope that this book will help to inform and thereby shape those policies.

References

ABARE (Australian Bureau of Agricultural and Resource Economics). 1995a. *Commodity statistical bulletin 1994.* Canberra, A.C.T.: Australian Government Printing Service.

———. 1995b. Rural research and development: The role for government. A response to the Industry Commission's draft report on research and development. Canberra, A.C.T. Mimeo.

ABRC (Advisory Board to the Research Councils). 1982. *A study of commissioned research.* London: Her Majesty's Stationery Office.

———. 1986. *Report of the Working Party on the private sector of scientific research.* London: Her Majesty's Stationery Office.

ABS (Australian Bureau of Statistics). 1993a. *1992–93 research and experimental development: Business enterprises in Australia.* Catalogue no. 8112.0. Canberra, A.C.T.: Australian Government Printing Service.

———. 1993b. *1992–93 research and experimental development: All-sector summary, Australia.* Catalogue no. 8104.0. Canberra, A.C.T.: Australian Government Printing Service.

ADAS (Agricultural Development Advisory Service). 1994. *Annual report.* Oxford: Kidlington.

AFRC (Agricultural and Food Research Council). 1987–1995. *Annual report.* London.

———. 1988. *Report to the Working Group: AFRC needs for scientific and management information.* London.

Agricultural Research Council. Various years. *Annual report.* London.

Alston, J. M., J. R. Anderson, and P. G. Pardey. 1995. Perceived productivity, foregone future farm fruitfulness, and rural resource research rationalization. In *Agricultural competitiveness: Market forces and policy choice. Proceedings of the twenty-second International Conference of Agricultural Economists,* ed. G. H. Peters and D. D. Hedley. Aldershot, U.K.: Dartmouth.

Alston, J. M., J. Chalfant, and P. G. Pardey. 1993. *Structural adjustment in OECD agriculture: Government policies and technical change.* Working Paper WP93-3. St. Paul, Minn., U.S.A.: Center for International Food and Agricultural Policy, University of Minnesota.

Alston, J. M., B. J. Craig, and P. G. Pardey. 1998. *Dynamics in the creation and depreciation of knowledge, and the returns to research.* EPTD Discussion Paper 35. Washington, D.C.: International Food Policy Research Institute.

Alston, J. M., M. S. Harris, J. D. Mullen, and P. G. Pardey. 1995. Paying for productivity: Financing agricultural investment in Australia. Paper prepared for the U.S. Congress Office of Technology Assessment. Davis, Calif., U.S.A.: Department of Agricultural Economics, University of California at Davis. Mimeo.

Alston, J. M., M. C. Marra, P. G. Pardey, and T. J. Wyatt. 1998. *Research returns redux: A meta-analysis of the returns to agricultural R&D.* EPTD Discussion Paper 38. Washington, D.C.: International Food Policy Research Institute.

Alston, J. M., and J. D. Mullen. 1992. Economic effects of research into traded goods: The case of Australian wool. *Journal of Agricultural Economics* 43 (1992): 268–278.

Alston, J. M., G. W. Norton, and P. G. Pardey. 1995. *Science under scarcity: Principles and practice for agricultural research evaluation and priority setting.* Ithaca, N.Y., U.S.A.: Cornell University Press.

Alston, J. M., and P. G. Pardey. 1994. Distortions in prices and agricultural research investments. In *Agricultural technology: Policy issues for the international community,* ed. J. R. Anderson. Wallingford, U.K.: CAB International.

———. 1995a. Revitalizing R&D. In *Agricultural policy: Rethinking the role of government,* ed. D. A. Sumner. Washington, D.C.: American Enterprise Institute for Public Policy.

———. 1995b. Agricultural R&D in the public interest. Paper prepared for the U.S. Congress Office of Technology Assessment. Davis, Calif., U.S.A.: University of California at Davis. Mimeo.

———. 1996. *Making science pay: The economics of agricultural R&D policy.* Washington, D.C.: American Enterprise Institute.

Alston, J. M., P. G. Pardey, and J. Roseboom. 1998. Financing agricultural research: International investment patterns and policy perspectives. *World Development* 26 (6): 1057–1071.

Alston, J. M., and G. M. Scobie. 1983. Distribution of research gains in multistage production systems: Comment. *American Journal of Agricultural Economics* 65 (2): 353–356.

Anderson, J. R., and G. Gryseels. 1991. International agricultural research. In *Agricultural research policy: International quantitative perspectives,* ed. P. G. Pardey, J. Roseboom, and J. R. Anderson. Cambridge, U.K.: Cambridge University Press.

Anonymous. 1950. Het Landbouwkundig Onderzoek, de Landbouwvoorlichting, en het Landbouwonderwijs in Nederland, 1950. *Landbouwkundig Tijdschrift* 62: 865–902.

Arnon, I. 1989. *Agricultural research and technology transfer.* New York: Elsevier.

Baker, I., A. Baklien, and A. S. Watson. 1990. *A review of agricultural research in Victoria.* Melbourne: Department of Agriculture, Victoria.

Baum, W. C. 1986. *Partners against hunger: Consultative Group on International Agricultural Research.* Washington, D.C.: World Bank.

BBSRC (Biotechnology and Biological Sciences Research Council). 1994. *Exploitation of AFRC research: Agricultural and Food Research Council annual report.* Swindon, U.K.

———. 1995. *Food Directorate strategy document.* Swindon, U.K.

Beattie, D. 1986. *Key to prosperity: Science and technology.* Report of the Ministerial Working Party. Wellington, N.Z.: New Zealand Government Printer.

Bollard, A., D. Harper, and M. Theron. 1987. *Research and development in New Zealand: A public policy framework.* Research Monograph 39. Wellington, N.Z.: New Zealand Institute of Economic Research.

Bollard, A., and D. Mayes. 1993. Lessons for Europe from New Zealand's liberalisation experience. *National Institute Economic Review* No. 143 (February): 81–97.

Bos, J. T. M. 1989. *Privatization and reorganization of the Dutch Agricultural Extension Service.* The Hague: Ministry of Agriculture and Fisheries.

Boyce, J. K., and R. E. Evenson. 1975. *National and international agricultural research and extension programs.* New York: Agricultural Development Council.

Brennan, J. P., and P. N. Fox. 1995. *Impact of CIMMYT wheats in Australia: Evidence of international research spillovers.* Economic Research Report 1/95. Wagga Wagga: New South Wales Agriculture.

Broekema, C. 1938. Organisatie van het Landbouwkundig Onderzoek. *Landbouwkundig Tijdschrift* 50: 173–191.

Bundesamt für Statistik. 1982. *Forschung und Entwicklung an Den Schweizerischen Hochschulen.* Beitrage zur Schweizerischen Statistik, Heft 98. Bern.

———. 1983. *Forschung und Entwicklung des Bundes 1978–1981.* Beitrage zur Schweizerischen Statistik, Heft 104. Bern.

Bundesminister für Forschung und Technologie. 1979. *Bundesbericht Forschung VI. Reihe: Berichte und Dokumentationen,* Band 4. Bonn.

Bureau Bartels B.V. and Bakkenist Management Consultants. 1996. *Evaluatie van de Wet Bevordering Speur- en Ontwikkelingswerk.* Utrecht/Assen/Voorburg/Diemen.

Burell, A., and J. Medland. 1990. *Artifacts: A handbook of U.K. and EEC agricultural and food statistics.* London: Harvester Wheatsheaf.

Buri, M., R. Suarez de Miguel, and B. Walder. 1988. Forschung und Entwicklung 1986 in der Schweiz. Bericht des BFS für den Basisbericht. Bundesamt für Statistik, Bern. Mimeo.

Busch, L., W. B. Lacy, J. Burkhardt, and L. R. Lacy. 1991. *Plants, power, and profit: Social, economic, and ethical consequences of the new biotechnologies.* Malden, Mass., U.S.A: Blackwell.

Business Bureau of Agriculture, Forestry, and Fishery Technical Conference. 1994. *Abstract of research and development in agriculture, forestry, and fishery.* Data collected by Shenggen Fan. Tokyo.

Butler, L. J., and B. W. Marion. 1985. *The impacts of patent protection on the U.S. seed industry and private breeding.* North Central Regional Research Publication 304, North Central Project 117, Monograph 16. Madison, Wis., U.S.A.: Research Division, College of Agricultural and Life Sciences, University of Wisconsin.

Byerlee, D., and G. Traxler. 1995. National and international wheat improvement research in the post–Green Revolution period: Evolution and impacts. *American Journal of Agricultural Economics* 77 (2): 268–278.

Campden Food and Drink Research Association. 1994. *Annual report.* Chipping Campden, U.K.

Cashin, P. A. 1988. Is there any value in "high-value" commodities? *Australian Economic Papers* 7 (June): 21–32.

CBS (Centraal Bureau voor de Statistiek). Various years. *Speur- en Ontwikkelingswerk in Nederland 1971–93.* Voorburg, the Netherlands.

Central Statistical Office of Finland. Various years. *Research activity.* Official Statistics of Finland. XXXVIII:1–7, biannually 1971–1983. Helsinki.

Charles, D. 1994. *Role of government in funding research and development.* Consultancy Report to the Rural Research and Development Corporations. Melbourne: Allen Consulting Group Pty. Ltd.

Christian, J., P. G. Pardey, and J. M. Alston. 1995. Rural R&D investments and institutions in the United States. Davis, Calif., U.S.A.: Department of Agricultural Economics, University of California at Davis. Mimeo.

Chudleigh, P. D., K. A. Bond, and J. C. McColl. 1993. *Evaluation of the impact of SRDC.* Report to the Sugar Research and Development Corporation. Canberra, A.C.T.: Sugar Research and Development Corporation.

Cockburn, I., and Z. Griliches. 1987. *Industry effects and appropriability measure in the stock market's valuation of R&D and patents.* Working Paper 2465. Cambridge, Mass., U.S.A.: National Bureau of Economic Research.

———. 1988. Industry effects and appropriability measure in the stock market's valuation of R&D and patents. *American Economic Review* 78 (May): 419–423.

Commonwealth of Australia. 1965. *Report of the Committee of Economic Enquiry.* Canberra, A.C.T.: Australian Government Printing Service.

Congress of the United States, Office of Technology Assessment (OTA). 1995. Challenges for U.S. Agricultural Research Policy. Washington D.C.: U.S. Government Printing Office, September.

Constantine, J. H., J. M. Alston, and V. H. Smith. 1994. Economic impacts of California's one variety cotton law. *Journal of Political Economy* 102 (October): 66–89.

Cooke, G. W., ed. 1991. *Agricultural research 1931–81: A history of the Agricultural Research Council and a review of developments in agricultural science during the last fifty years.* London: Agricultural Research Council.

Cooperative Research Centres Program—Australia. 1996. *Compendium, 1996.* Canberra, A.C.T.: Department of Industry, Science and Tourism.

———. 1997. *Compendium, 1996.* ⟨http://www.dist.gov.au/crc/compend/newctres.html#forestry⟩ July 28, 1997.

Crafts, N. 1996. Post-neoclassical endogenous growth theory: What are its policy implications? Working Paper. London: London School of Economics. Mimeo.

Craig, B. J., and P. G. Pardey. 1996a. Productivity measurement in the presence of quality change. *American Journal of Agricultural Economics* 78 (5): 1349–1354.

———. 1996b. Inputs, outputs and productivity developments in U.S. agriculture. Paper presented at the conference *Global agricultural science policy for the twenty-first century,* August 1996, Melbourne, Australia.

Craig, B. J., P. G. Pardey, and J. Roseboom. 1997. International agricultural productivity patterns: Accounting for input quality, infrastructure and research. *American Journal of Agricultural Economics* 79 (November): 1064–1079.

Cremers, M. W. J., and J. Roseboom. 1997. *Agricultural research in government agencies in Latin America: A preliminary assessment of investment trends.* ISNAR Discussion Paper 97-7. The Hague: International Service for National Agricultural Research.

Croon, I. 1986. Personal communication. Department of Economics and Statistics, Swedish University of Agricultural Sciences. Uppsala, Sweden.

———. 1988. Resursinsatser i Offentlig Jordbruksforskning i Sverige under Perioden 1945–85. Draft. Swedish University of Agricultural Science, Uppsala, Sweden. Mimeo.

Cunningham, C. M., and R. H. Nicholson. 1991. Central government organisation and policy making for U.K. science and technology since 1982. In *Science and technology in the United Kingdom,* ed. Sir. R. Nicholson, C. M. Cunningham, and P. Gummett. London: Longman.

Dalrymple, D. G. 1990. The excess burden of taxation and public agricultural research. In *Methods of diagnosing research system constraints and assessing the impact of agricultural research.* Vol. II, *Assessing the impact of agricultural research,* ed. R. G. Echeverría. The Hague: International Service for National Agricultural Research.

Davey, M. E. 1994. *Research and funding: Fiscal year 1995.* CRS Issue Brief. Washington, D.C.: Congressional Research Service, Library of Congress.

Davis, J., P. A. Oram, and J. G. Ryan. 1987. *Assessment of agricultural research priorities: An international perspective.* ACIAR Monograph 4. Canberra, A.C.T.: Australian Centre for International Agricultural Research.

DIST (Department of Industry, Science, and Technology), Cooperative Research Centres Secretariat. 1995. *Changing research culture Australia: 1995* (Myers Report). Canberra, A.C.T.: Australian Government Publishing Service.

———. 1999. ⟨www.dist.gov.au/crc/html/centres.html⟩ April.

DLO (Agricultural Research Department). n.d. *Research institutes for advanced agriculture.* Wageningen, the Netherlands.

———. 1994. *Guide to DLO-NL 1994.* Wageningen, the Netherlands.

Donaldson, G. F. 1964. The financing of agricultural research by levies on farm produce. *Review of Marketing and Agricultural Economics* 32 (4): 3–36.

DPIE (Department of Primary Industries and Energy). n.d. *Review of rural research.* Report of the Task Force on Review of Rural Research (including Fisheries Research). Canberra, A.C.T.

Duncan, I., and A. Bollard. 1992. *Corporatisation and privatisation: Lessons from New Zealand.* Auckland: Oxford University Press.

Economist, The. 1997. The knowledge factory: A survey of universities. *The Economist* (October 4).

Edgerton, D. 1988. The relationship between military and civil technologies: A historical perspective. In *The relationship between defence and civil technologies,* ed. P. J. Gummer and J. Reppy. Dordrecht, the Netherlands: Kluwer.

Edwards, G. W., and J. W. Freebairn. 1981. *Measuring the country's gains from research: Theory and application to rural research and extension.* Report to the Commonwealth Council for Rural Research and Extension. Canberra, A.C.T.: Australian Government Printing Service.

Enzing, C. M. 1989. Het Meerjarenplan Landbouwkundig Onderzoek 1987–91. In *Verkennen in Nederland, Deel 2: Vijf Case-studies naar Verkennende Activiteiten van Verschillende Typen Organisaties in het Nederlandse Wetenschaps en Technologiesysteem.* Apeldoorn, the Netherlands: TNO-Studiecentrum voor Technologie en Beleid.

Enzing, C. M., and R. E. H. M. Smits. 1989. *Research foresight in the Netherlands: An analysis.* Apeldoorn, the Netherlands: TNO Center for Technology and Policy Studies.

EUROSTAT. Various years. *Government financing of research and development.* Luxemburg: Office for the Official Publications of the European Communities.

Executive Office of the President. 1994. *Budget of the United States government: Historical tables fiscal year 1995.* Washington, D.C.: U.S. Government Printing Office.

Fan, S., and P. G. Pardey. 1992. *Agricultural research in China: Its institutional development and impact.* The Hague: International Service for National Agricultural Research.

FAO (Food and Agriculture Organization of the United Nations). 1986. *Intercountry comparisons of agricultural production aggregates.* Economic and Social Development Paper 61. Rome.

———. 1995. *AGROSTAT diskettes.* Rome.

———. 1997. *AGROSTAT diskettes.* Rome.

Finkel, E. 1997. Australia enters deep water in devising management plan. *Science* 277 (5): 1428.

Forsknings og Teknologiministeriet. 1993. *Forskning og Udviklingsarbejde i den Offetlige Sektor 1991.* Copenhagen.

Forskningssekretariatet. 1986. *Forskningsstatistik 1982.* Copenhagen.

Fox, G. C. 1985. Is the United States really underinvesting in agricultural research? *American Journal of Agricultural Economics* 67 (November): 806–812.

Fuglie, K., N. Ballenger, K. Day, C. Klotz, M. Ollinger, J. Reilly, U. Vasavada, and J. Yee, with contributions from J. Fisher and S. Payson. 1996. *Agricultural research and development: Public and private investments under alternative markets and institutions.* USDA-ERS-NRED Agricultural Economic Report 735. Washington, D.C.: United States Department of Agriculture.

Fullerton, D. 1991. Reconciling recent estimates of the marginal welfare cost of taxation. *American Economic Review* 81: 302–308.

Galante, E., and C. Sala. 1988. Personal communication. Instituto Biosintesi Vegetali, Consiglio Nazionale delle Ricerche. Milan, Italy.

Gleeson, T., and A. Lascelles. 1992. *Review of the Research and Development Corporation model.* Unpublished report prepared for the Primary Industries and Energy Council. Canberra, A.C.T.

Grantham, G. 1984. The shifting locus of agricultural innovation in nineteenth-century Europe: The case of the agricultural experiment stations. In *Technique, spirit, and form in the making of modern economies: Essays in honor of William N. Parker,* ed. G. Saxonhouse and G. Wright. Research in Economic History Supplement 3. Greenwich, Conn., U.S.A.: JAI Press.

Grauls, L. 1987. Head study-statistics-inventory. December. Personal communication. Brussels.

GRDC (Grains Research and Development Corporation). 1992. *Gains for grain.* Vol. 1, *Overview.* Report on an evaluation of returns from research and development investments in the Australian grains industry. Occasional Paper Series 1. Canberra, A.C.T.

———. 1993. *Gains for grain.* Vol. 2, *The case studies.* Report on an evaluation of returns from research and development investments in the Australian grains industry. Occasional Paper Series 2. Canberra, A.C.T.

———. 1996. *Annual report 1996.* Canberra, A.C.T.

Gregory, R. G. 1993. The Australian innovation system. In *National innovation systems: A comparative analysis,* ed. R. R. Nelson. New York: Oxford University Press.

Griffiths, A., and S. Wall, eds. 1993. *Applied economics,* 5th ed. London: Longman.

Grigg, D. 1989. *English architecture: An historical perspective.* Oxford: Blackwell.

Groenewegen, P. 1990. *Public finance in Australia: Theory and practice,* 3d ed. Sydney: Prentice Hall.

Gummett, P. 1980. *Scientists in Whitehall.* Manchester, U.K.: Manchester University Press.

———. 1991. History, development and organisation of U.K. science and technology up to 1982. In *Science and technology in the United Kingdom,* ed. Sir R. Nicholson, C. M. Cunningham, and P. Gummett. London: Longman.

Harris, M., and A. G. Lloyd. 1991. The returns to agricultural research and the underinvestment hypothesis: A survey. *Australian Economic Review* (third quarter): 16–27.

Harris, S., J. G. Crawford, F. H. Gruen, and N. D. Honan. 1974. *Rural policy in Australia: Report to the Prime Minister by a working group.* Canberra, A.C.T.: Australian Government Printing Service.

Hayami, Y., and V. W. Ruttan. 1985. *Agricultural development: An international perspective.* Baltimore, Md., U.S.A.: Johns Hopkins University Press.

Herruzo, A. C., M. C. Fernandez, and R. G. Echeverría. 1993. El sistema español de ciencia y tecnología agrarias. *Investigación Agraria: Economía* 8 (3): 465–483.

Hissink, D. J. 1916. *De reorganisatie van het Proefstationwezen in Nederland.* Kniphorst, Wageningen. (Overdruk uit De Indische Mercuur van 7, 14, en 28 January 1916).

HMSO (Her Majesty's Stationery Office). 1990. *Annual review of government funded research and development.* London: Office of Science and Technology.

———. 1992. *Annual review of government funded research and development.* London: Office of Science and Technology.

———. 1993. *Realising our potential: A strategy for science, engineering and technology.* London: Her Majesty's Stationery Office.

———. 1995a. *Forward look at government-funded science and technology.* London: Office of Science and Technology.

———. 1995b. *Technology foresight: Progress through partnership.* Vol. 11. London: Office of Science and Technology.

Holderness, B. 1985. *British agriculture since 1945.* Manchester, U.K.: Manchester University Press.

House of Lords. 1988. *Agricultural and food research.* Vol. 1. Report of the Select Committee on Science and Technology. London: Her Majesty's Stationery Office.

Huffman, W. E., and R. E. Evenson. 1993. *Science for agriculture: A long-term perspective.* Ames, Iowa, U.S.A.: Iowa State University.

———. 1994. *Agricultural research and education: An economic perspective.* Ames, Iowa, U.S.A.: Department of Economics, Iowa State University.

Huffman, W. E., and R. E. Just. 1994. Funding, structure and management of agricultural research in the United States. *American Journal of Agricultural Economics* 76 (November): 744–759.

IAC (Industries Assistance Commission). 1976. *Financing rural research.* Canberra, A.C.T.: Australian Government Printing Service.

IKC (Informatie en Kennis Centrum Landbouw). 1995. *IKC Jaarplan 1995.* Ede, the Netherlands.

IMF (International Monetary Fund). 1995. *Government finance statistics yearbook.* Volume XIX. Washington, D.C.

Industry Commission. 1994a. *Research and development.* Vol. 1, *The report.* Canberra, A.C.T.: Australian Government Printing Service. Mimeo.

———. 1994b. *Research and development.* Vol. 3, *Quantitative appendices.* Draft report. Canberra, A.C.T.: Australian Government Publishing Service. Mimeo.

———. 1995. *Research and development.* Industry Commission Report. Canberra, A.C.T.: Australian Government Publishing Service.

INE (Instituto Nacional de Estadística). n.d. *Estadística sobre las actividades en investigación científica y desarrollo tecnológico: Años 1971–1972.* Madrid: Ministerio de Planificación del Desarrollo, INE.

———. 1978. *Estadística sobre las actividades en investigación científica y desarrollo tecnológico: Años 1973–1974.* Madrid: Ministerio de Economía, INE.

———. 1986. *Estadística sobre las actividades en investigación científica y desarrollo tecnológico: Años 1982 y 1983.* Madrid.

INRA (Institut National de la Recherche Agronomique). 1986a. *Analyse rétrospective des moyens de l'INRA de 1970–1985.* Paris.

———. 1986b. *INRA 1986.* Paris.

———. 1988. *INRA rapport d'activités 87.* Paris.

———. 1991. *Regard sur l'INRA 1991.* Paris.

———. 1994. *Regard sur l'INRA de 1992–1994.* Paris.

Jacobsen, V. 1995a. *The returns to investment by WRONZ for the New Zealand wool industry: An analysis of additional projects.* Hamilton, N.Z.: SER Consulting Economists.

———. 1995b. *Economic analysis of selected research projects.* Hamilton, N.Z.: SER Consulting Economists.

Jardine, D. V. A. 1986. Funding and conduct of research and development, theory and its application to New Zealand agriculture: A transaction costs approach. M.Sc. thesis, University of Waikato, Hamilton, N.Z.

Jarrett, F. G. 1990. Rural research organisations and policies. In *Agriculture in the Australian economy,* 3d ed., ed. D. B. Williams. Melbourne: Sydney University Press in association with Oxford University Press.

Johnston, W. E., and G. A. G. Frengley. 1991. The deregulation of New Zealand agriculture: market intervention (1964–84) and free market readjustment (1984–90). *Western Journal Of Agricultural Economics* 16 (1): 132–143.

Junta Nacional de Investigacao Cientifica e Tecnologica. n.d. *Recursos em ciencia e tecnologia: Inventario de 1971.* Lisbon.

———. n.d. *Recursos inventario em ciencia e tecnologia 1972.* Lisbon.

———. n.d. *Investigacao e desenvolvimento Portugal 1976.* Lisbon.

———. 1981. *Recursos em ciencia e tecnologia Portugal 1978.* Lisbon.

———. 1986. *Investigacao e desenvolvimento experimental: Inquerito ao potencial cientifico e tecnologico nacional actualizacao a 84.12.31: Sector estado (dados provisorios).* Lisbon.

Just, R. E., and W. E. Huffman. 1992. Economic principles and incentives: Structure, management and funding of agricultural research in the United States. *American Journal of Agricultural Economics* 74 (December): 1102–1108.

Kennedy, R. M., S. L. Pollack, and L. Lynch, eds. 1990. *Title XVI: Research in provisions of the Food, Agriculture, Conservation and Trade Act of 1990.* Agricultural Information Bulletin No. 624. Washington, D.C.: Agriculture and Trade Analysis Division, ERS, United States Department of Agriculture.

Kenwood, A. G. 1995. *Australian economic institutions since federation: An introduction.* Melbourne: Oxford University Press.

Kerin, J., and P. Cook. 1989. *Research innovation and competitiveness: Policies for reshaping Australia's primary industries and energy portfolio.* Department of Primary Industries and Energy. Canberra, A.C.T.: Australian Government Printing Service.

Kerr, N. A. 1987. *The legacy: A centennial history of the State Agricultural Experiment Stations, 1887–1987.* Columbia, Mo., U.S.A.: Missouri Agricultural Experiment Station.

Khanna, J., W. E. Huffman, and T. Sandler. 1994. Agricultural research expenditures in the United States: A public goods perspective. *Review of Economics and Statistics* 76 (2): 267–277.

Khatri, Y. J. 1994. Technical change and the returns to research for U.K. agriculture, 1953–90. Ph.D. diss, University of Reading, Reading, U.K.

Khatri, Y. J., and C. Thirtle. 1996. Supply and demand functions for U.K. agriculture, 1953–90: Biases of technical change and the returns to public R&D. *Journal of Agricultural Economics* 47 (September): 338–354.

Klotz, C., K. Fuglie, and C. Pray. 1995. *Private-sector agricultural research expenditures in the United States, 1960–92.* ERS Staff Paper No. AGES9525. Washington D.C.: Economic Research Service, October.

Knopke, P., L. Strappazzon, and J. D. Mullen. 1995. Productivity growth on Australian broadacre farms. ABARE Conference Paper 95.31 at the New England Conference on Efficiency and Productivity, November 1995, Armidale, Australia.

Kurath, R. 1994. Landwirtschaftliche Forschung im Umbruch. *Agrarforschung* 1 (1): 4–7.

Landbouwhogeschool. Various years. *Jaarverslag 1975–84.* Wageningen, the Netherlands.

———. Various years. *Wetenschappelijk Verslag 1979–82.* Wageningen, the Netherlands.

———. 1977. *Ontwikkelingsplan Landbouwhogeschool 1980–83.* Wageningen, the Netherlands.

Landbouwhogeschool Vaste Commissie voor de Wetenschapsbeoefening. 1974. *Op Weg naar een Onderzoekbeleid van de Landbouwhogeschool.* Wageningen, the Netherlands: Landbouwhogeschool.

Leatherhead Food Research Agency. 1993. *Annual report.* Leatherhead, U.K.

Lee, H., J. M. Alston, H. F. Carman, and W. Sutton. 1996. *Mandated marketing programs for California commodities.* Giannini Foundation Information Series 96-1. Davis, Calif., U.S.A.: Giannini Foundation of Agricultural Economics.

Leskien, D., and M. Flitner. 1997. *Intellectual property rights and plant genetic resources: Options for a sui generis system.* Issues in Genetic Resources 6. Rome: International Plant Genetic Resources Institute.

Lindner, R. K. 1993. Privatizing the production of knowledge: Promise and pitfalls of agricultural research and extension. *Australian Journal of Agricultural Economics* 37 (3): 205–225.

Lloyd, A. G. 1989. Value added for primary products. *The Australian Quarterly* 61 (1): 50–58.

Lloyd, A. G., M. S. Harris, and D. E. Tribe. 1990. *Australian agricultural research: Some policy issues.* Melbourne: Crawford Fund for International Agricultural Research.

Lovett, S. E. 1994. *Evaluating reform in the New Zealand science, research and development system: New deal or dud hand?* Discussion Paper 42. Canberra, A.C.T.: Public Policy Program, Australian National University.

LUW (Landbouwuniversiteit Wageningen). Various years a. *Jaarverslag 1985–93.* Wageningen, the Netherlands.

———. Various years b. *Wetenschappelijk Jaarverslag 1985–93.* Wageningen, the Netherlands.

———. 1987. *Wetenschappelijk Jaarverslag 1986.* Wageningen, the Netherlands.

———. 1994. *Jaarverslag 1993—Onderwijs en Onderzoek.* Wageningen, the Netherlands.

Macaulty Land Use Research Institute. 1993. *Annual report.* Aberdeen, U.K.

McLean, I. W. 1982. The demand for agricultural research in Australia: 1970–1914. *Australian Economic Papers* 21 (December): 294–308.

MAFF (Ministry of Agriculture, Fisheries and Food). Various years. *Agriculture in the United Kingdom.* London: Her Majesty's Stationery Office.

———. Various years. *Annual review of agriculture.* London: Her Majesty's Stationery Office.

———. Various years. *Departmental net income calculation: Annual review.* London: Statistical Division I.

———. 1961. Economic trends, 1961. In *Productivity measurement in agriculture.* London: Central Statistical Office.

———. 1967. *A century of agricultural statistics: Great Britain, 1866–1966.* London: Her Majesty's Stationery Office.

———. 1969. Economic trends, 1969. In *Productivity measurement in agriculture.* London: Central Statistical Office.

———. 1972/73–1983/84. *Report on research and development.* London: Her Majesty's Stationery Office.

———. 1984a. *Departmental net income calculation: Sources and methods.* London: Statistical Division I.

———. 1984b. *Report of a study of ACDS by its director general, Professor Ronald L. Bell.* London: Ministry of Agriculture, Fisheries and Food.

———. 1984c. *Report on research and development, 1983–84.* London: Her Majesty's Stationery Office.

———. 1985. *Review of ADAS.* London.

———. 1986a. *Reference information on agricultural economic research within the United Kingdom.* London: Her Majesty's Stationery Office.

———. 1986b. *Report on research and development.* London.

———. 1989. *Farm incomes in the United Kingdom.* London: Her Majesty's Stationery Office.

———. 1994. *Research strategy.* London.

Maltha, D. J. 1976. *Honderd Jaar Landbouwkundig Onderzoek in Nederland 1876–1976.* Wageningen, the Netherlands: PUDOC.

Marketing International. 1984. *Commercial biotechnology: An international analysis.* Amsterdam: Elsevier.

Mathews, R. L., and W. R. C. Jay. 1972. *Federal finance: Intergovernmental financial relations in Australia since federation.* Melbourne: Thomas Nelson.

Mauldon, R. G. 1990. Price policy. In *Agriculture in the Australian economy,* ed. D. B. Williams. Sydney: Sydney University Press in association with Oxford University Press.

Meer, C. L. J. van der, H. Rutten, and N. A. Dijkveld Stol. 1991. *Technologiebeleid in de Landbouw: Effecten in het Verleden en Beleidsoverwegingen voor de Toekomst.* The Hague: Wetenschappelijke Raad voor het Regeringsbeleid.

Miller, R. J., and C. L. Harris. 1994. *Trends in agricultural research: Thoughts for discussion.* Washington, D.C.: CSRS, United States Department of Agriculture.

Ministerial Science Task Group. 1991. *Crown Research Institutes: Research companies for New Zealand.* Wellington, N.Z.

Ministerie van Landbouw en Visserij. 1977. *Meerjarenvisie 1977–81 voor het Landbouwkundig en het Visserij Onderzoek.* The Hague.

———. 1985. *Aard en Omvang van het Landbouwkundig Onderzoek.* Notitie van de Minister aan de Tweede Kamer der Staten-Generaal. The Hague.

———. 1987. *Landbouwkundig Onderzoek in Perspectief: Ontwikkelingsplan voor de Instituten en Proefstations van het Ministerie van Landbouw en Visserij, 1978–90.* Ontwikkelingsplan voor het Departementale Landbouwkundig Onderzoek van de Minister aan de Tweede Kamer der Staten-Generaal. The Hague.

Ministerie van Landbouw en Visserij—DLO (Directie Landbouwkundig Onderzoek). 1972. *Bijdrage tot een Meerjarenplan voor het Landbouwkundig Onderzoek 1972–76.* The Hague.

———. 1985. *Aard en Omvang van het Landbouwkundig Onderzoek.* Notitie van de Minister aan de Tweede Kamer der Staten-Generaal. The Hague.

Ministerie van Landbouw, Natuurbeheer, en Visserij. Various years. *Rijksbegroting Ministerie van Landbouw, Natuurbeheer, en Visserij 1970–95.* The Hague.

———. 1992. *Beleidsplan Wetenschap en Technologie 1991–94.* The Hague.

———. 1993. *LNV-Kennisbeleid: Eenheid in Verscheidenheid.* The Hague.

———. 1996. *LVN-Kennisbeleid tot 1999.* The Hague.

Ministerie van Landbouw, Natuurbeheer en Visserij—Directie Wetenschap en Kennis-Overdracht. 1995a. De Financiering van Proefstations en Regionale Onderzoekscentra. The Hague. Mimeo.

———. 1995b. *Ontwerp Kennisbeleidsplan 1996–98.* The Hague.

Ministry of Agriculture and Fisheries. 1988. *The agricultural extension system in the Netherlands.* The Hague.

Ministry of Agriculture, Nature Management and Fisheries. 1991. *Agricultural structure memorandum: Policy of agriculture in the Netherlands in the 1990s.* The Hague.

———. 1992. *Science and technology policy plan 1991–94 summary.* The Hague.

Ministry of Research, Science, and Technology. Various years. *New Zealand research and experimental development statistics.* Wellington, New Zealand.

Moore, E. G. 1967. *The Agricultural Research Service.* New York: Praeger.

MRC (Meat Research Corporation). 1996. *Annual report 1995–1996.* Canberra, A.C.T.

MRST (Ministry of Research, Science and Technology). Various years. *New Zealand research and experimental development statistics.* Wellington, N.Z.

———. 1994. *Science and technology: The way forward, 1996–2000.* Wellington, N.Z.

Mullen, J. D. 1995. The returns from public investment in research and development. Invited paper presented to the AIAS Research Forum, 3–4 April 1995, Warwick Farm. New South Wales, Australia.

Mullen, J. D., and T. L. Cox. 1994a. Alternative approaches to measuring productivity growth: An application to Australian broadacre agriculture. Invited paper presented to the Symposia on Productivity at the second International Conference on Asia-Pacific Modeling, August 1994, Sydney.

———. 1994b. The contribution of research to productivity growth in Australian broadacre agriculture. Contributed paper presented to the second International Conference on Asia-Pacific Modeling, August 1994, Sydney.

———. 1995. The returns from research in Australian broadacre agriculture. *Australian Journal of Agricultural Economics* 39 (August): 105–128.

Mullen, J. D., K. Lee, and S. Wrigley. 1996a. *Agricultural production research expenditures in Australia: 1953–1994.* Agricultural Economics Bulletin 14. Orange, Australia: NSW Agriculture, Industry Economics Unit.

———. 1996b. Financing agricultural research in Australia. Invited paper presented at REGAE Workshop on Economic Evaluation of Research in Australia and New Zealand at AARES Conference, February 1996, Melbourne University, Melbourne.

National Agricultural Chemicals Association. 1990. *Industry profile survey.* Washington, D.C.

National Research Advisory Council. Various years. *Report of the National Research Advisory Council for the year ended 31 March 1986.* Presented to the House of Representatives Pursuant to Section 11 of the National Research Advisory Council Act 1963. Wellington, N.Z.

National Research Council, Board on Agriculture. 1989. *Investing in research: A proposal to strengthen the agricultural, food and environmental system.* Washington, D.C: National Academy Press.

———. 1994. *Investing in the National Research Initiative: An Update of the Competitive Grants Program of the U.S. Department of Agriculture.* Washington D.C.: National Academy Press.

———. 1995. *Colleges of Agriculture at the Land Grant University: A Profile.* Washington D.C.: National Academy Press.

———. 1996. *Colleges of Agriculture at the Land Grant University: Public Service and Public Policy.* Washington D.C.: National Academy Press.

National Science Board. 1996. *Science and engineering indicators, 1996.* Washington, D.C.: U.S. Government Printing Office.

National Science Foundation. 1995. *Federal R&D funding by budget function: Fiscal years 1993–95.* Washington, D.C.: NSF Science and Technology Information

———. 1996. *Science and Engineering Indicators 1996.* Washington D.C.: U.S. Government Printing Office.

Nelson, R. R. 1982. *Government and technical progress.* London: Pergamon Press.

———. 1997. Why the Bush Report has hindered an effective civilian technology policy. In *Science for the 21st century: The Bush Report revisited,* ed. C. E. Barfield. Washington, D.C.: American Enterprise Institute.

Nelson, R. R., ed. 1993. *National innovation systems: A comparative analysis.* New York: Oxford University Press.

New Zealand Vice-Chancellors Committee. 1991. *Research priorities for New Zealand: A university perspective.* Discussion Paper. Wellington, N.Z.

Nicholson, R., C. M. Cunningham, and P. Gummett, eds. 1991. *Science and technology in the U.K.* London: Longman.

Norton, G. W., P. G. Pardey, and J. M. Alston. 1992. Economic issues in agricultural research priority setting and evaluation. *American Journal of Agricultural Economics* 74 (December): 1089–1095.

NRLO (Netherlands National Council for Agricultural Research). 1981. *Meerjarenvisie Landbouwkundig Onderzoek 1982–86.* The Hague.

———. 1988a. *NRLO-Gids 1988.* The Hague.

———. 1988b. *What is and who form the National Council for Agricultural Research (NRLO)?* The Hague.

———. 1990a. *Meerjarenvisie Landbouwkundig Onderzoek 1991–94.* The Hague.

———. 1990b. *Netherlands agricultural research plan 1991–94: Summary of essentials.* The Hague.

———. 1993. *Facts and figures 1993: Highlights of Dutch agriculture, nature management and fisheries.* The Hague.

———. 1995. *Landbouwkundig Onderzoek op Weg naar de 21e Eeuw—Meerjarenvisie Landbouwkundig Onderzoek 1995–1998.* The Hague.

NWO (Nederlandse Organisatie voor Wetenschappelijk Onderzoek). 1994. *NWO Jaarboek 1993.* The Hague.

———. 1995. *NWO Jaarboek 1994.* The Hague.

OECD (Organization for Economic Cooperation and Development). 1972. *Reviews of national science policy: Iceland.* Paris.

———. 1974. *Survey of the resources devoted to R&D by OECD member countries, international statistical year 1971.* Vol. 5, *Total tables, statistical tables, and notes.* Paris.

———. 1979. *International survey of the resources devoted to R&D by OECD member countries, international statistical year 1975.* International Volume, *Statistical tables and notes.* Paris.

———. 1981. *Science and technology indicators: Basic statistical series.* Vol. A, *The objectives of government R&D funding 1969–1981.* Paris.
———. 1983. *Science and technology indicators: Basic statistical series.* Vol. A, *The objectives of government R&D funding 1974–1985.* Paris.
———. 1991. *Basic science and technology statistics.* Paris.
———. 1992. *Agricultural policies, marketing and trade: Monitoring and outlook.* Paris.
———. 1994. *The measurement of scientific and technical activities: Frascati manual, 1993.* Paris.
———. 1995a. *Tables of producer and consumer subsidy equivalents, 1979–94.* Paris.
———. 1995b. *Technological change and structural adjustment on OECD agriculture.* Paris.
———. 1995c. *Basic science and technology statistics, edition 1995.* Paris.
———. 1996. Basic science and technology statistics 1981–1995. Paris. Computer disk.
OECD, Directorate for Science, Technology, and Industry. 1985. Public funding of R&D by socioeconomic objectives in Austria. Paris. Mimeo.
OECD, STIID Data Bank. 1987a. Public funding of R&D by socio-economic objective. Paris. Mimeo.
———. 1987b. R&D expenditure in the higher education and private non-profit sectors. Paris. Mimeo.
Office of Management and Budget. 1992. *Budget of the United States government: Fiscal year 1992.* Washington, D.C.: U.S. Government Printing Office.
———. 1993. *Budget of the United States government: Fiscal year 1993.* Washington, D.C.: U.S. Government Printing Office.
Offut, S. 1993. The National Research Initiative: Competing at the margin. In *Agricultural research: Strategic challenges and options,* ed. R. D. Weaver. Bethesda, Md., U.S.A.: Agricultural Research Institute.
Ohashi, T. 1987. Personal communication. Director, International Research Division, Agriculture, Forestry and Fisheries Research Council. Tokyo.
Orange, C. 1987. *The Treaty of Waitangi.* Wellington, N.Z.: Allen and Unwin.
———. 1990. *The story of a treaty.* Wellington, N.Z.: Allen and Unwin/Port of Nicholson.
OST (Office of Science and Technology). 1990. *Annual review of government funded research and development.* London: Her Majesty's Stationery Office.
———. 1992. *Annual review of government funded research and development.* London: Her Majesty's Stationery Office.
———. 1995a. *Forward look at government-funded science, engineering and technology.* London: Her Majesty's Stationery Office.
———. 1995b. *Technology foresight: Progress through partnership.* London: Her Majesty's Stationery Office.
Palmer, C. 1993. *The reform of the public science system in New Zealand.* No. 33. Wellington, N.Z.: Ministry of Research, Science and Technology.
Pardey, P. G., J. M. Alston, J. Christian, and S. Fan. 1996. *Hidden harvest: U.S. benefits from international aid.* Food Policy Report. Washington, D.C.: International Food Policy Research Institute.

Pardey, P. G., B. J. Craig, and M. Hallaway. 1989. *U.S. agricultural research deflators, 1890–1985.* Research Policy 18 (5): 289–296.

Pardey, P. G., and J. Roseboom. 1989. *ISNAR agricultural research indicator series.* Cambridge, U.K.: Cambridge University Press.

Pardey, P. G., J. Roseboom, and J. R. Anderson, eds. 1991. *Agricultural research policy: International quantitative perspectives.* Cambridge, U.K.: Cambridge University Press.

Pardey, P. G., J. Roseboom, and B. J. Craig. 1992. A yardstick for international comparisons: An application to national agricultural research expenditures. *Economic Development and Cultural Change* 40 (2): 333–349.

Pardey, P. G., J. Roseboom, and S. Fan. 1998. Trends in financing Asian and Australian agricultural research. In *Financing agricultural research: A sourcebook,* ed. S. R. Tabor, W. Janssen, and H. Bruneau. The Hague: International Service for National Agricultural Research.

Peper, B. 1996. *Duurzame Kennis, Duurzame Landbouw: Een Advies ann de Minister van LNV over de Kennisinfrastructuur van de Landbouw in 2010.* The Hague: Ministerie van Landbouw, Natuurbeheer, en Visserij.

Perrin, R. K. 1994. Intellectual property rights in agricultural development. In *Agricultural technology: Policy issues for the international community,* ed. J. R. Anderson. Wallingford, U.K.: CAB International.

Peterson, W. L. 1967. Return to poultry research in the United States. *Journal of Farm Economics* 49 (August): 656–670.

Piggott, R. R. 1990. Agricultural marketing. In *Agriculture in the Australian economy,* 3d ed., ed. D. B. Williams. Melbourne: Sydney University Press in association with Oxford University Press.

Post, J. H., and H. Rutten. 1991. Technologiebeleid voor de Nederlandse Landbouw. In *Landbouwpolitiek tussen Diagnose en Therapie: Liber Amicorum for Prof. J. de Hoogh,* ed. H. J. Silvis and L. H. G. Slangen. Wageningen, the Netherlands: Landbouwuniversiteit Wageningen.

Pray, C. E., and M. Knudson. 1994. Impact of intellectual property rights on genetic diversity: The case of U.S. wheat. *Contemporary Economic Policy* XII (January): 102–112.

Pray, C. E., M. Knudson, and L. Masse. 1994. Impact of intellectual property rights on U.S. plant breeding. In *Evaluating agricultural research and productivity in an era of resource scarcity,* ed. W. B. Sundquist. Staff Paper P94-2. St. Paul, Minn., U.S.A.: University of Minnesota, Department of Agricultural and Applied Economics.

Priorities Board for Research and Development in Agriculture and Food. 1990. *Third report to the agriculture ministers and the chairman of the Agricultural and Food Research Council.* London: Ministry of Agriculture, Fisheries and Food.

Psacharopoulos, G. 1993. *Returns to investment in education: A global update.* Policy Research Working Papers WPS 1067. Washington, D.C.: World Bank.

———. 1994. Returns to investment in education: A global update. *World Development* 22 (September): 1325–1343.

Ralph, W. 1994. Meeting expectations: An analysis of the activities of the R&D Corporations. *Agricultural Science* 7 (September/October): 33–39.

Rasmussen, W. D., and G. L. Baker. 1972. *The Department of Agriculture.* New York: Praeger.

Rattigan, G. A. 1986. *Industry assistance: The inside story.* Melbourne: Melbourne University Press.

Read, N. 1989. The "near market" concept supplied to U.K. agricultural research. *Science and Public Policy* 16 (4): 233–238.

RIRDC (Rural Industries Research and Development Corporation). 1993. *Annual report, 1992–1993.* Canberra, A.C.T.

———. 1996. *Annual report, 1995–1996.* Canberra, A.C.T.

Roe, T. L., and P. G. Pardey. 1991. Economic policy and investment in rural public goods: A political economy perspective. In *Agricultural research policy: International quantitative perspectives,* ed. P. G. Pardey, J. Roseboom, and J. R. Anderson. Cambridge, U.K.: Cambridge University Press.

Roseboom, J., and H. Rutten. 1995. Financing agricultural R&D in the Netherlands: The changing role of government. Paper prepared for the U.S. Congress Office of Technology Assessment. Davis, Calif., U.S.A.: Department of Agricultural Economics, University of California at Davis. Mimeo.

———. 1998. The transformation of the Dutch agricultural research system: An unfinished agenda. *World Development* 26 (6): 1113–1126.

Rost, E. 1988. Personal communication. Bundesministerium für Forschung und Technologie. Bonn.

———. 1993. Personal communication. Bundesministerium für Forschung und Technologie. Bonn.

Rothwell, R. 1978. *Technical change and competitiveness in agricultural engineering: The performance of the U.K. industry.* Occasional Paper Series 9. Sussex, Brighton, U.K.: University of Sussex.

Ruttan, V. W. 1982. *Agricultural research policy.* Minneapolis, Minn., U.S.A.: University of Minnesota Press.

Rutten, H. 1992. *Productivity growth of Dutch agriculture, 1949–89.* Mededeling no. 470. The Hague: Landbouw-Economisch Instituut.

Sala, C. 1993. Personal communication. Instituto Biosintesi Vegetali, Consiglio Nazionale delle Ricerche. Milan, Italy.

Salmon, S. C., and A. A. Hanson. 1964. *The principles and practice of agricultural research.* London: Leonard Hill.

Sandrey, R. 1990. The regulatory environment. In *Farming without subsidies,* ed. R. Sandrey and R. Reynolds. Wellington, N.Z.: Ministry of Agriculture and Fisheries.

School of Agriculture, Aberdeen. 1989. *Annual report.* Aberdeen.

Schweikhardt, D. B., and J. F. Whims. 1993. Trends and issues in agricultural research funding at the State Agricultural Experiment Stations. In *U.S. agricultural research: Strategic challenges and options.* Bethesda, Md., U.S.A.: Agricultural Research Institute.

Science and Technology Expert Panel. 1992. *Long-term priorities for the Public-Good Science Fund.* Wellington, N.Z.: Minister for Science, Research, and Technology.

Science Priorities Review Panel. 1995. *Establishing priorities for the Public-Good Science Fund: Discussion document.* Wellington, N.Z.: Ministry of Research, Science and Technology.

Scobie, G. M., and W. M. Eveleens. 1987. *The return to investment in agricultural research in New Zealand: 1926/27–1983/84.* MAFCorp Policy Services Research Report No. 1/87. Hamilton, N.Z.: Ruarura Agriculture Centre, Economics Research Station, Ministry of Agriculture and Fisheries.

Scobie, G. M., and V. Jacobsen. 1994. *The returns to investment by WRONZ in R&D for the New Zealand wool industry.* Hamilton, N.Z.: SER Consulting Economists, 1994.

Scottish Agricultural Science Agency. 1994. *Annual report.* London: Her Majesty's Stationery Office.

Scottish Crop Research Institute. 1993. *Annual report.* London: Her Majesty's Stationery Office.

———. 1994. *Annual report.* London: Her Majesty's Stationery Office.

Scrimgeour, F. G., and E. C. Pasour Jr. 1994. The public choice revolution and New Zealand farm policy. *Review of Marketing and Research Economics* 62 (2): 273–283.

Sexton, R. J. 1990. Imperfect competition in agricultural markets and the role of cooperatives: A spatial analysis. *American Journal of Agricultural Economics* 72 (August): 709–720.

Shattock, M. 1991. Higher education and the research councils. In *Science and Technology in the United Kingdom,* ed. Sir R. Nicholson, C. M. Cunningham, and P. Gummett. London: Longman.

Shaw, R. 1991. *Purchasing science.* No. 6. Wellington, N.Z.: Ministry of Research, Science and Technology.

Sieper, E. 1982. *Rationalising rustic regulation.* St. Leonards, Australia: Centre for Independent Studies.

Smith, D. F. 1992. Joint government-producer funding of agricultural R&D: The Australian scheme and its lessons. The Hague: International Service for National Agricultural Research. Mimeo.

Snelling, D. E. 1976. The establishment and growth of an agricultural research system in Great Britain to 1937. Manchester, U.K.: University of Manchester.

Spedding, C. 1984. Agricultural research policy. In *U.K. science policy: A critical review of policies for publicly funded research,* ed. M. Goldsmith. London: Longman.

Spriggs, J. D. 1990. Transparency versus protectionism: The Australian way. Seminar Paper 90-03. Adeleide, Australia: Centre for International Economic Studies, University of Adeleide.

STAC (Science and Technology Advisory Committee). 1988. *Science and technology review: A new deal.* Wellington, N.Z.

State Government of Victoria. 1994. Submission to the Industry Commission inquiry into research and development in Australia. Melbourne, Australia: Department of Agriculture. Mimeo.

Sundsbo, S., and M. Villa. 1987. Personal communication. Agricultural Research Council of Norway. Oslo.

Thirtle, C. 1986. *Problems in the definition and management of technical change and productivity growth in the U.K. agricultural sector.* Manchester Working Papers in Agricultural Economics WP86/03. Manchester, U.K.: Manchester University.

———. 1989. *Agricultural research, development and extension expenditure in the United Kingdom, 1947–93.* EPARD Working Paper 14. Reading, U.K.: University of Reading.

———. 1996. *Producer funding of R&D: Productivity and returns to R&D in British sugar, 1954–93.* Working Paper 1996/1. Reading, U.K.: Department of Agricultural Economics, University of Reading.

Thirtle, C., H. S. Beck, M. Upton, and W. S. Wise. 1991. Agriculture and food. In *Science and technology in the United Kingdom,* ed. Sir R. Nicholson, C. M. Cunningham, and P. Gummett. London: Longman.

Thirtle, C., and P. Bottomley. 1989. The rate of return to public sector agricultural R&D in the U.K., 1965–80. *Applied Economics* 21 (8): 1063–1086.

———. 1992. Total factor productivity in U.K. agriculture, 1967–90. *Journal of Agricultural Economics* 43 (September): 381–400.

Thirtle, C., J. Bureau, and R. Townsend. 1994. Accounting for efficiency differences in European agriculture: Cointergration, multilateral productivity indices and R&D spillovers. In *Agricultural competitiveness: Market forces and policy choice. Proceedings of the twenty-second International Conference of Agricultural Economists,* ed. G. H. Peters and D. D. Hedley. Aldershot, U.K.: Dartmouth.

Thirtle, C., J. Piesse, and V. H. Smith. 1995. Agricultural R&D and technology in the United Kingdom. Paper prepared for the U.S. Congress Office of Technology Assessment, Davis, Calif., U.S.A.: Department of Agricultural Economics, University of California at Davis.

TNO (Netherlands Organization for Applied Scientific Research). 1991. *TNO in profile.* Delft, the Netherlands.

True, A. C., and D. J. Crosby. 1902. *Agricultural experiment stations in foreign countries.* USDA Office of Experiment Stations Bulletin 112. Washington, D.C.: U.S. Government Printing Office.

U.S. Bureau of the Census. 1998. ⟨www.census.gov/population/estimates/nation/popclockest.txt⟩ "Historical National Population Estimates." Release date April 2.

U.S. Congress. 1993. *Conference report to accompany H.R. 5847, Agriculture, Rural Development, Food and Drug Administration and Related Agencies Appropriation Bill 1993.* Washington, D.C.: U.S. Government Printing Office.

United Kingdom Department of Agriculture and Forestry, Scotland. 1989. *Strategy for agricultural research and development.* Edinburgh, Scotland.

USDA (United States Department of Agriculture). 1994. *USDA news October–December 1994.* No. 53. Washington, D.C.: USDA Office of Communications.

USDA, CRIS (Current Research Information System). 1998. *Inventory of agricultural research, fiscal year 1997.* Washington, D.C.: U.S. Government Printing Office.

USDA, CSRS (Cooperative State Research Service). 1993. *Dynamics of the Research Investment: Issues and Trends in the Agricultural Research System.* Washington, D.C.: U.S. Government Printing Office, July.

USDA, CSREES (Cooperative State Research, Education, and Extension Service). 1998. *National Research Initiative Competitive Grants Program. Annual Report Fiscal Year 1997.* ⟨www.reeusda.gov/nri⟩ January.

USDA, CSREES, Competitive Research Grants and Awards Management. 1999. Personal communication, March.

USDA, ERS (Economic Research Service). 1995. *Agricultural research.* AREI Updates 5 (revised). Washington, D.C.

USDA, ERS. 1998. ⟨www.econ.ag.gov/briefing/fbe/fi/fi.htm⟩ Farm cash receipts data. Updated September 24, 1998.

USDA, National Agricultural Statistics Service. 1997. ⟨http://usda.mannlib.cornell.edu/reports/nassr/other/zfl-bb/farms_and_land_in_farms_07.30.97⟩ Number of farms, land in farms, and average farm size. Revised September 30.

Vaage, R., and E. Bjorgum. 1988. Personal communication. Norwegian Fisheries Research Council. Trondheim.

Valdés, A. 1993. Mix and sequencing of economywide and agricultural reforms: Chile and New Zealand. *Agricultural Economics* 8 (4): 295–311.

Venner, R. J. 1997. An economic analysis of the U.S. Plant Variety Protection Act: The case of wheat. Ph.D. diss., Department of Agricultural and Resource Economics, University of California at Davis, Davis, Calif., U.S.A.

Verkaik, A. P. 1972. *Organisatiestructuur Landbouwkundig Onderzoek en Achtergronden van Haar Totstandkoming.* Research Management Studies 3. The Hague: Nationale Raad voor Landbouwkundig Onderzoek.

Weaver, R. D., ed. 1993. *U.S. agricultural research: Strategic challenges and options.* Bethesda, Md., U.S.A.: Agricultural Research Institute.

Webster, A. J. 1989. Privatisation of public sector research: The case of a plant breeding institute. *Science and Public Policy* 16 (4): 224–232.

Williams, D. B., ed. 1990. *Agriculture in the Australian economy,* 3d ed. Melbourne: Sydney University Press in association with Oxford University Press.

Williams, R., and G. Evans. 1988. *Commonwealth policy for rural research, past and present.* Bureau of Rural Resources. Canberra, A.C.T.: Australian Government Printing Service.

Winnifrith, J. 1962. *The Ministry of Agriculture, Fisheries and Food.* London: Allen & Unwin.

Woods, A., J. P. Taylor, D. C. Harley, D. N. Housdon, and A. N. Lance. 1988. *The reform of the Common Agricultural Policy: New opportunities for wildlife and the environment.* Conservation Topic Paper 24. London: Royal Society for the Protection of Birds.

World Bank. 1992. *World tables,* 1991–92 ed. Washington, D.C. Computer diskette.

———. 1995. *World tables,* 1995 ed. Washington, D.C.

———. 1997. *World tables,* 1997 ed. Washington, D.C. Computer diskette.

Wright, B. D. 1983. The economics of invention incentives: Patents, prizes and research contracts. *American Economic Review* 73 (September): 691–707.

Wright, B. D., and D. Zilberman. 1993. Agricultural research structures in a changing world. In *U.S. agricultural research: Strategic challenges and options,* ed. R. D. Weaver. Bethesda, Md., U.S.A.: Agricultural Research Institute.

Zulauf, C. R., and L. G. Tweeten. 1993. Reordering the mission of agricultural research at land-grant universities. *Choices* (2/4): 31–33.

Contributors

Julian M. Alston is a professor in the Department of Agricultural and Resource Economics, University of California–Davis, and associate director of the University of California Agricultural Issues Center. His main research interest is the economics of agriculture and policies that affect it.

Jason E. Christian is principal economist, California Independent System Operator; he was previously a research associate in the Department of Agricultural and Resource Economics, University of California–Davis.

Barbara J. Craig is an associate professor in the Economics Department, Oberlin College. Her research interests include econometrics, international economics, and agricultural productivity.

Michael S. Harris is a research fellow at the Melbourne Institute of Applied Economic and Social Research, University of Melbourne. His research interests include the economics of agricultural and industrial research and natural resource accounting.

Veronica Jacobsen is a senior manager with Arthur Andersen in Wellington, New Zealand. Dr. Jacobsen works on law, economics, and public policy.

John D. Mullen is economics coordinator with the New South Wales Department of Agriculture. His research relates to the economics of agricultural research and development, agricultural productivity, and natural resource issues.

Philip G. Pardey is a research fellow at the International Food Policy Research Institute and an associate professor in the Department of Applied Economics, University of Minnesota. He leads a research program on agricultural science and technology policy in less-developed countries.

Michael Phillips is director of the U.S. National Reseach Council's Board on Agriculture and Natural Resources. He was formerly director and senior associate at the Food and Renewable Resources Program at the U.S. Congress Office of Technology Assessment. Dr. Phillips is the author of numerous studies of U.S. food and agricultural policy.

Jenifer Piesse is a senior lecturer at the Clore Management Centre, Birkbeck College, University of London. Her research focuses mainly on productivity and income distribution.

Johannes Roseboom is a research officer at the International Service for National Agricultural Research, the Hague. His research program concerns the economics of agricultural research and development policy with an emphasis on less-developed countries.

Hans Rutten is a senior advisor at the Netherlands' National Council for Agricultural Research. His main area of interest is strategic management of agricultural research and innovation processes.

Grant M. Scobie is director general of the Centro Internacional Agricultura Tropical in Cali, Colombia, which conducts research on tropical agriculture.

Vincent H. Smith is a professor in the Department of Economics and Agricultural Economics, Montana State University. His main research interests include the economics of voluntary giving, agricultural policy, and the economics of research and development, technology, and policy.

Colin Thirtle is a reader in the Department of Agricultural and Food Economics, the University of Reading, U.K., and extraordinary professor, Department of Agricultural Economics, Extension, and Rural Development, University of Pretoria, South Africa. His main research concerns are the economics of agricultural research and development and agricultural productivity.

Index

Page numbers for entries occurring in figures are followed by an *f;* those for entries occurring in notes, by an *n;* and those for entries occurring in tables, by a *t.*

LIBRARY OF CONGRESS CATALOGING-IN-PUBLICATION DATA

Paying for agricultural productivity / edited by Julian M. Alston, Philip G. Pardey, Vincent H. Smith.
p. cm.
"Published for the International Food Policy Research Institute."
Includes bibliographical references and index.
ISBN 0-8018-6185-3 (alk. paper) ISBN 0-8018-6278-7 (pbk.: alk. paper)
1. Agriculture—Research—Finance—Case studies. I. Alston, Julian M. II. Pardey, Philip G. III. Smith, Vincent H. IV. International Food Policy Research Institute.
HD1410.5.P39 1999
338.1—dc21 99-17315
CIP